$19.50

ROOF FRAMING

by
Marshall Gross

Craftsman Book Company
6058 Corte del Cedro,
Carlsbad, CA 92008

Acknowledgements

The author would like to thank the following for their assistance in preparing the text.

Mr. Philip L. Starrett, advertising and sales promotion manager of the L. S. Starrett Company, 121 Crescent Street, Athol, Massachusetts 01331

Mr. Douglas A. Lashmit, copyright counsel, Texas Instruments Incorporated, P.O. Box 225474, Dallas, Texas 75265

Dedicated to

My father Abba, for all his help.

Library of Congress Cataloging in Publication Data

Gross, Marshall.
Roof framing.

Includes index.
1. Roofs. 2. Framing (Building) I. Title.
TH2393.G76 1984 690'.15 84-12123
ISBN 0-910460-40-X

Photos by Rob Colombo

Contents

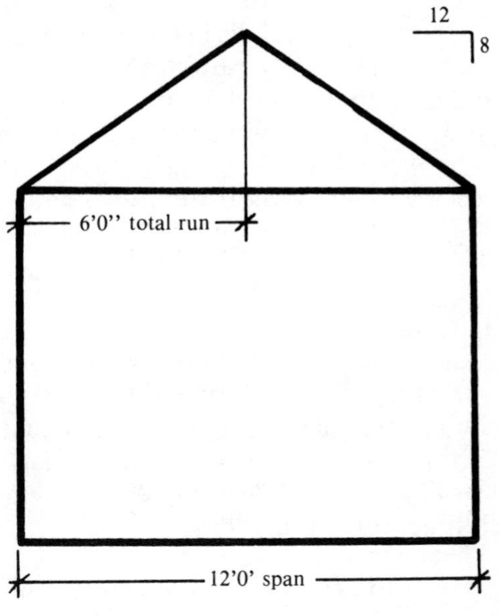

1

Introduction to a Simple Roof

Roof framing is the Ph.D. of carpentry. Most carpenters would agree that it requires more knowledge and skill than any other framing task. Many experienced carpenters, even master carpenters who have put a roof on many homes, don't claim to be expert roof framers. There are too many roof styles and there's too much mathematics for most carpenters to feel like they can handle any roof job that's likely to come along.

And if I had to select a single framing job on which carpenters waste the most time and material, it would be roof framing without a doubt.

Having admitted right at the start that roof framing (or roof cutting, as I'll call it) isn't as easy as framing a partition or floor, I'm going to set out to prove that any diligent carpenter with the intelligence to read and understand the pages in this book can become an expert roof cutter. Even if you've never driven a straight nail in your life, this book can make you a skilled roof cutter. It isn't hard if you have a knowledgeable and patient teacher. And I intend to be exactly that.

I learned roof cutting from a master carpenter by the name of Florin Alder. He perfected his skills in Germany over 50 years ago. I was lucky. There are few really expert roof cutters working in the construction industry today. And I know of no book or other source for most of the information presented in this manual. But I expect that this book will keep the fine art of roof cutting available to any carpenter or apprentice roof cutter who wants to master the trade.

From Simple to Complex

Don't get discouraged if something in this book seems too complicated at first. My goal is to make you a master roof cutter capable of framing irregular, octagon and unequal pitch roofs. Knowledge like this doesn't come overnight. A lawyer or doctor spends years learning and perfecting his skills. A craftsman needs nearly as much time to learn his trade.

Give yourself time to get comfortable with the procedures and recommendations in this book. Build the models I describe. Work through the problems until your answers match my answers at the back of the book. Master each type of roof as that kind of roof job comes along. When you can frame any roof discussed in this book, you should have no trouble making a good living as a master roof cutter.

The First Few Chapters

If you've worked as a roof cutter or carpenter, you already know much of what's in the first few chapters. But the apprentice programs I'm familiar with don't do an adequate job of explaining many of the important points that you'll find in Chapters 1 through 6. You may want to review these chapters even if you feel reasonably certain that you can handle gable and hip roofs. These chapters include information that will help even experienced roof cutters.

In Chapter 2 I'm going to suggest that you use one of the most powerful tools a roof cutter can own . . . an inexpensive hand-held calculator. It will free you from dependence on rafter length tables, increase your accuracy, and provide correct rafter lengths for all of the irregular roofs that no rafter table could possibly cover. Modern hand-held calculators make the tables on a framing square a poor second choice for modern craftsmen.

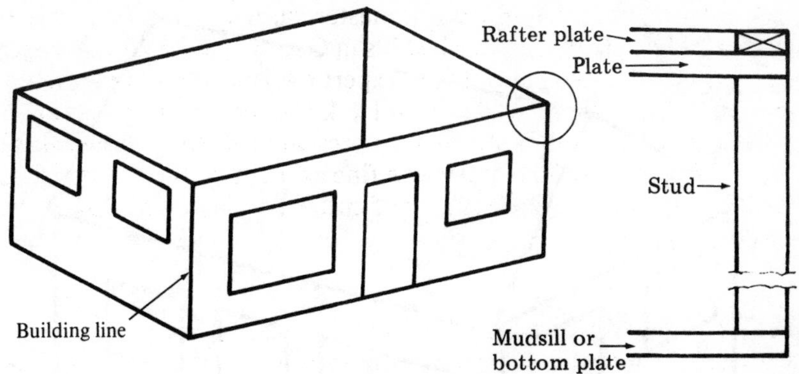

The completed wall
Figure 1-1

Your Calculator

I use a Texas Instruments calculator, the TI-35, and have based my examples on it. But many others are available, at a very reasonable cost, at most drug and discount stores. If you buy a calculator for roof cutting, be sure it has keys that will calculate square root, square, sine, cosine, tangent and that will store and recall figures in memory.

Before we begin, note that there's a Reference Section near the end of this manual. Appendix B in the Reference Section may be especially helpful if you didn't take trigonometry in high school or need a quick brush-up on terms used to describe sides of a right triangle.

Now, let's start at the beginning and take it one easy step at a time.

In the Beginning

Figure 1-1 shows a building with the wall framing completed. The stage is set for the roof cutter to begin his work.

At the top of the wall studs are two horizontal members called plates. The first horizontal member above the studs is simply called *the plate*. The plate above that is called the *rafter plate* because this is the resting place for the rafters. The outside edge of the rafter plate is the reference plane for all roof cutting work. It's the line from which many important roof dimensions are measured. We'll call this the *building line*.

The completed wall and roof
Figure 1-2

A Simple Roof

Figure 1-2 shows a simple roof added to the framing in Figure 1-1. The roof shown would finish the roof cutter's work on this building. Notice that the roof has only two slopes. This is called a *gable roof*. By the end of the next chapter you'll know how to cut this simple roof.

Now we're going to look at this building from the direction of the arrow in Figure 1-2.

Span and Total Run

From the direction of the arrow in Figure 1-2 we can see two right triangles formed by the roof. These are *right* triangles because each has one right (90 degree) angle.

Look at Figure 1-3. Notice that both triangles are identical in every aspect. Whatever we calculate for one triangle will apply to the other.

The width of the building is called the *span*. For calculation purposes, we'll divide the span in half (as in Figure 1-3) to get the base of one right triangle. We'll call half the span distance the *total run*. This is an important dimension to the roof cutter. See Figure 1-4.

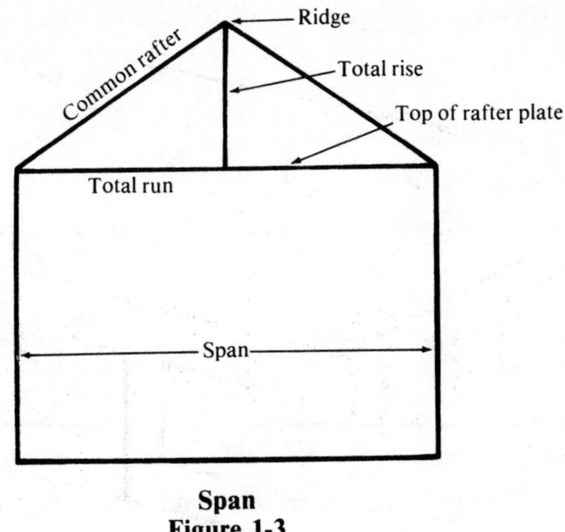

Span
Figure 1-3

Total Rise
This is the vertical height of the roof measured at the midpoint between opposite rafter plates (Figure 1-3). The word *total* tells us that this is the overall dimension to the highest point. This highest point is called the *ridge*.

Total run is expressed only in feet (as in 14.75'), while total rise is usually expressed in feet and inches (as in 4'3½").

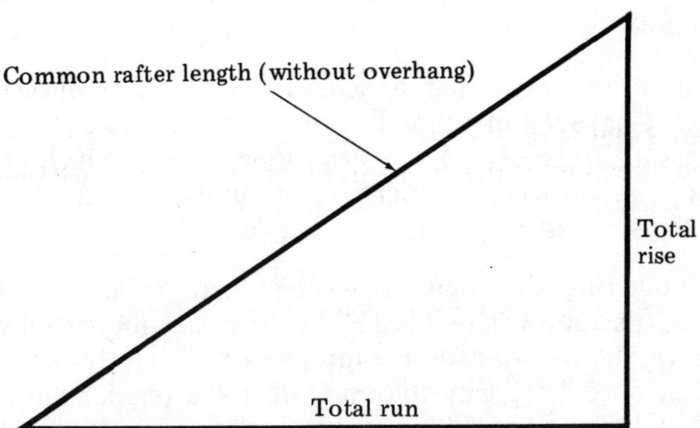

Total run and total rise in a right triangle
Figure 1-4

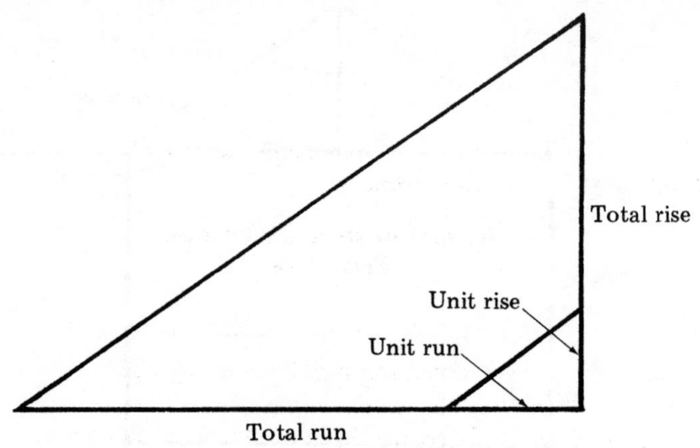

Relationship of "unit" to "total"
Figure 1-5

Here are the terms we've used so far:

Total Run: half the span of the building (expressed in feet).

Total Rise: the apparent height of the roof ridge above the rafter plate.

Unit Run and Unit Rise

Unit run and *unit rise* are also key terms used in roof cutting. They're smaller segments, or building blocks, of the roof triangle. See Figure 1-5.

The unit rise is expressed in inches from 1" to 24" of rise. When we say, "I have a 4 in 12 pitch roof," it means that the roof surface rises 4" for every 12" moved along the line which identifies total run. In carpenter's language, the unit rise and unit run indicate the slope of the roof.

Since our English system of linear measure is based on 12", or one foot, it's appropriate that 12" be the basic unit in roof cutting. Therefore, the *unit run* for a common rafter is always 12", or one foot. Later we'll see why the unit run for a regular hip rafter is 16.97" and the unit run of a regular octagon hip rafter is 12.988". These numbers are not arbitrary. They're fixed mathematical relationships based on the 12" unit run of the common rafter.

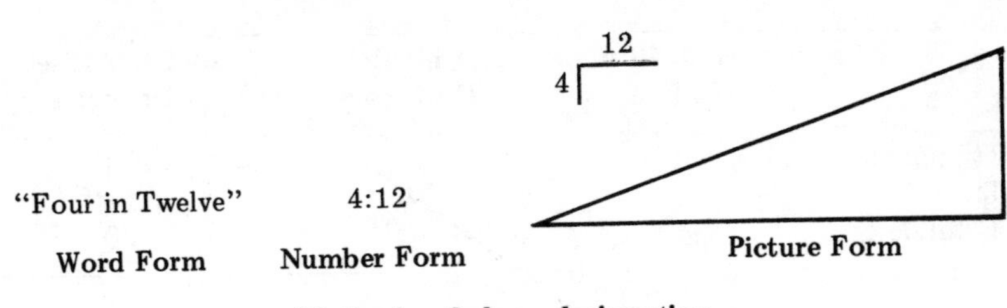

Methods of slope designation
Figure 1-6

The *unit rise* can be anything the designer of the building wants. The unit rise expresses the steepness of the roof's slope as related to the 12" unit run. There are three common ways to note the particular slope: in words, such as "four in twelve," in numbers, expressed as a ratio such as "4:12," and a symbol, showing a horizontal line with 12 above the line and a vertical line with 4 beside that line. See Figure 1-6.

Calculating Total Rise

Figure 1-7 shows a 4 in 12 roof. The total rise increases 4" every time an additional foot (12") is added to the total run.

All of the lines (a) through (e) in Figure 1-7 represent a 4 in 12 rafter, and each line makes a successively larger triangle. If the total run for a particular roof is known and the unit rise is given on the blueprint, it's easy to find the height of the total rise. Simply multiply the unit rise by the number of feet in the total run. Figure 1-8 shows examples.

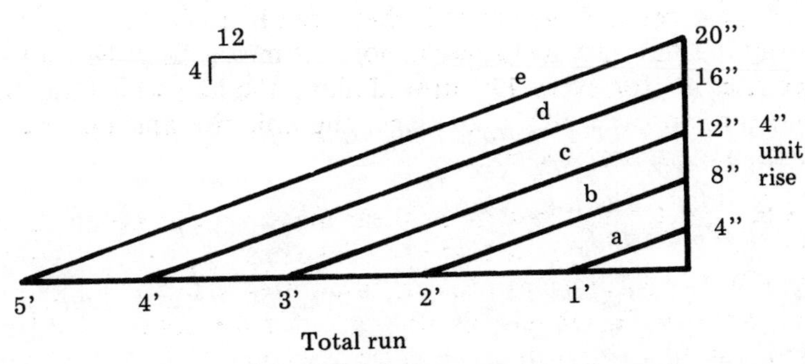

Unit rise times total run
Figure 1-7

	Total Run (In Feet)	x	Unit Rise (In Inches)	=	Total Rise (In Inches)
Triangle a	1’	x	4”	=	4”
Triangle b	2’	x	4”	=	8”
Triangle e	5’	x	4”	=	20”

Calculating total rise
Figure 1-8

Test your understanding of the information presented so far by working the following example:

A 26’ wide building is to have a 6 in 12 gable roof. Find the a) *unit run,* b) *unit rise,* c) *total run,* and d) *total rise.*

Here’s how to do it:

a) The unit run is the basic run of 12”, which is always used for common rafters. Since this is to be a gable roof, there will only be common rafters in this building.

b) The unit rise is given as 6”.

c) The total run is another name for half the span or width of a building. Since the span is given as 26’, the total run must be 13’.

d) To find total rise, multiply unit rise by the total run: 6” times 13 equals 78”, or 6’6”.

Problems

Here are two more problems. The answers are in the back of the book.

1) A 22’-wide building has a gable roof that rises 8” for every 12” of run.

a) What’s the common rafter total run?
b) How high is the peak?
c) What’s the unit rise?
d) What’s the unit run?

2) A regular gable roof with a 4 in 12 pitch has a span of 17’.

a) What’s the total run?
b) What’s the unit run?
c) What’s the unit rise?
d) What’s the total rise?

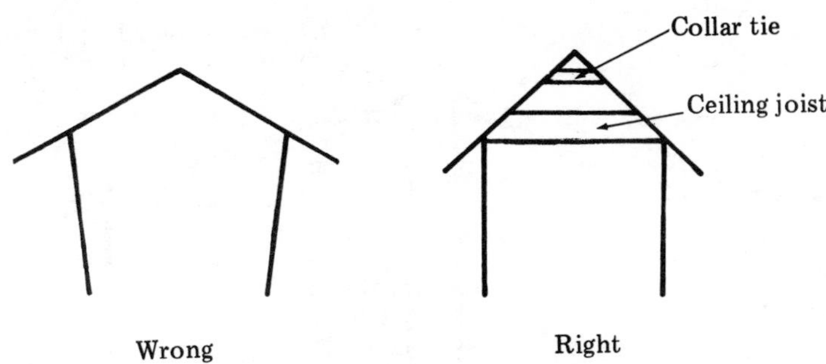

Necessity of joists and collar ties
Figure 1-9

Two Types of Roof Framing

Type 1 - Conventional: If a roof rests solely on opposite rafter plates, the weight of the roof will tend to push the upper part of the walls outward. See Figure 1-9. To keep this from happening, *ceiling joists* are installed across the building span. They're nailed into the rafter plate on each side of the building, and into each rafter. Usually there's a ceiling joist for every set of common rafters. These ceiling joists also provide a support for the first story ceiling and the second story flooring.

Collar ties can also help hold the walls together. See the right-hand illustration in Figure 1-9. Collar ties are generally made from 1 x 6 material and connect every third set of rafters at a point one-third of the distance down the rafter from the ridge.

Type 2 - Post and Beam: In this type of construction, a post is built into the framing of the wall at either end of the house. See Figure 1-10. These posts support a heavy beam which is the ridge board for the roof. The beam supports the upper end of the common rafters and the roof load. Since the rafters will be exposed to view from the room below, you'll probably want to use rough lumber to create the rustic look that's popular in exposed beam ceilings.

The posts hold up the beam and the beam holds up the roof. That eliminates the need for ceiling joists. If the span is very large, an occasional joist or metal rod will be added for strength.

Design Considerations

Selecting the roof pitch isn't purely a matter of design preference. Roof pitch determines what type of roof covering can be used, the size of rafters required, the snow and wind load the roof can carry

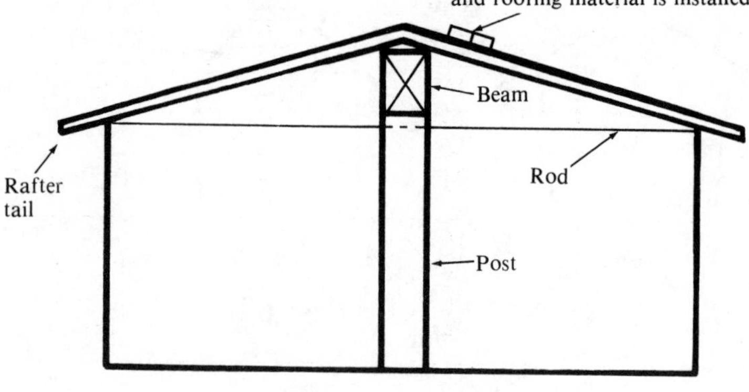

Post in wall supporting beam
Figure 1-10

with safety, and more. Here are some of the terms that influence the choice of roof pitch:

Dead Load

Dead load refers to the weight of the building roof: roof framing members, the roofing material and any equipment permanently mounted on the roof.

Live Load

Live loads are weights placed on the roof after construction is complete: people, ice and snow, and the pressure of strong winds.

A steep (high pitch) roof holds less snow. But the wind stress is greater on a high pitch roof than on a flatter (low pitch) roof.

Allowable Span

Allowable span means the greatest horizontal distance permitted between two bearing points. This is the distance of total run. Allowable span varies with the type of lumber, rafter spacing, and rafter dimension. Table 1-11 shows allowable spans for Douglas Fir lumber.

This table is only an example. The span permitted depends on the snow, wind and rain loads expected in your area. The building code enforced where you work will establish the allowable span.

Douglas fir grade	Douglas fir size	Spacing center to center	Slope 3 in 12 or less	Slope 4 in 12 or more
S. Str.	2 x 4	16"	---	10'- 1"
S. Str.	2 x 4	24"	---	8'- 6"
No. 2	2 x 4	16"	---	10'- 1"
No. 2	2 x 4	24"	---	8'- 6"
S. Str.	2 x 6	16"	14'- 2"	14'- 4"
S. Str.	2 x 6	24"	12'- 8"	11'- 8"
No. 2	2 x 6	16"	13'- 3"	12'- 4"
No. 2	2 x 6	24"	10'-10"	10'- 0"
No. 1	2 x 8	16"	18'- 9"	17'-10"
No. 1	2 x 8	24"	15'- 9"	14'- 7"
No. 2	2 x 8	16"	17'- 6"	16'- 3"
No. 2	2 x 8	24"	14'- 4"	13'- 3"
No. 1	2 x 10	16"	23'-11"	22'-10"
No. 1	2 x 10	24"	20'- 1"	18'- 7"
No. 2	2 x 10	16"	22'- 4"	20'- 8"
No. 2	2 x 10	24"	18'- 3"	16'-11"

		F-b
S. Str.	= (Select Structural)	1,800
No. 1	= (Middle Grade)	1,500
No. 2	= (Most generally used construction grade)	1,250

Allowable span
Table 1-11

Problems

Use Table 1-11 to answer the following questions. Correct answers are in the back of the book. Remember, use the smallest member at the widest on-center spacing and of the lowest lumber grade that's acceptable for the span of your building.

3) A 30' span building is to be built with No. 2 and better Doug Fir. What's the maximum rafter spacing for 2 x 8 lumber if the roof is 4 in 12?

4) What grade of material must be ordered for 2 x 6 rafters on an 8 in 12 pitch roof with a total run of 10'6"?

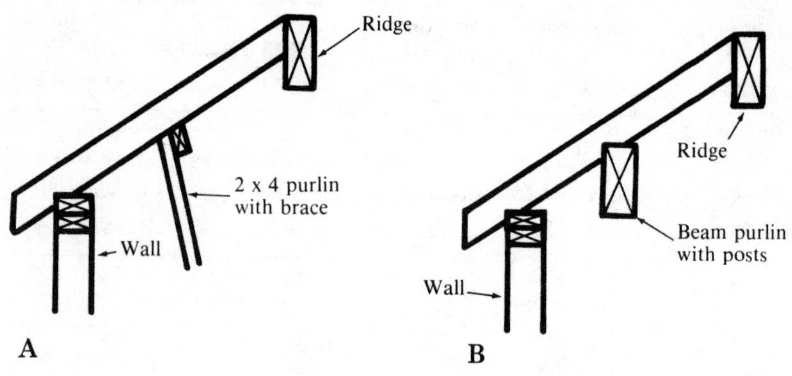

Types of purlins
Figure 1-12

Purlins

If it's difficult to stay within the allowable span, consider installing a support called a *purlin* part way between the ridge and the building line. See Figure 1-12. This support divides the allowable span so you can use smaller rafter material.

The purlin could be a long 2 x 4 nailed to the underside of the rafters and then braced to a bearing wall. It could also be a beam with each end set on posts. If you use a beam, be sure to make a seat cut on the rafter at the purlin beam.

Fascia Board

The *fascia board* is a horizontal board that's nailed against the lower end of the rafters or rafter tails. Fascia boards are joined with a miter cut of 45 degrees at all corners.

Barge Board

On gable ends, the fascia board turns the corner and runs up along the roof edge to the ridge. The section of fascia running up the gable end is called the *barge board*.

Pitch

You'll hear carpenters say, "I'm building a 4 in 12 pitch roof," meaning that the roof rises 4 inches in every 12 inches of total run. This accurately describes what the carpenter is doing and won't create any confusion.

But to a mathematician, pitch is the ratio between total rise and the total span expressed as a fraction.

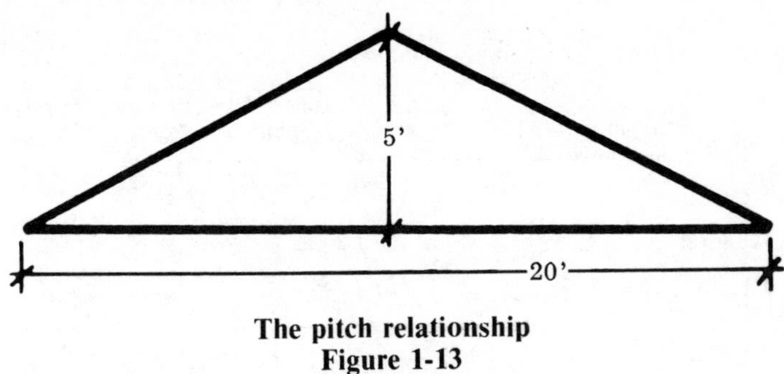

The pitch relationship
Figure 1-13

The roof in Figure 1-13 might be said to have a 1/4 pitch. But to the carpenter on the job, this is bound to create confusion.

As already explained, you have to know the unit rise and the total run before beginning work. This information will usually be on the plans. But sometimes you'll see only a fraction like "1/4 pitch." If so, simply multiply the fraction by 24 to find the unit rise in 12 inches of run.

For our example, 1/4 times 24 equals 6.

A 1/4 pitch roof is the same as a 6 in 12 pitch roof. The 6 and 12 accurately describe the angle of the roof and are the numbers used on the framing square to cut this particular roof.

Pitch expressed as a fraction comes from looking at a gable roof as one large triangle rather than two identical right triangles with their 90-degree angles directly below the ridge line.

Unit rise is based on a right triangle with a unit run of 12". If two right triangles are put together, you get one isosceles triangle with a base of 24". That's why 24 is used to convert from pitch expressed as a fraction to pitch expressed as inches of rise in 12 inches of run.

Common Rafter Roof Angle Chart

Figure 1-14 shows angles for common rafters. It gives the unit rise in inches, pitch expressed as a fraction, degrees in decimal form, and the secant of the angle. The secant is the relation between horizontal distance and vertical distance for any common rafter. This roof angle chart is for common rafters only.

Rise	Pitch	Degrees	Secant
24	1	63.43°	2.2360
23	23/24	62.45°	2.1618
22	11/12	61.39°	2.0883
21	7/8	60.26°	2.0156
20	5/6	59.04°	1.9436
19	19/24	57.72°	1.8727
18	3/4	56.31°	1.8028
17	17/24	54.78°	1.7340
16	2/3	53.13°	1.6667
15	5/8	51.34°	1.6008
14	7/12	49.40°	1.5366
13	13/24	47.29°	1.4743
12	1/2	45.00°	1.4142
11	11/24	42.51°	1.3566
10	5/12	39.80°	1.3017
9	3/8	36.87°	1.2500
8	1/3	33.69°	1.2018
7	7/24	30.26°	1.1577
6	1/4	26.56°	1.1180
5	5/24	22.62°	1.0833
4	1/6	18.43°	1.0541
3	1/8	14.04°	1.0308
2	1/12	9.46°	1.0138
1	1/24	4.76°	1.0035

Common rafter roof angle chart
Figure 1-14

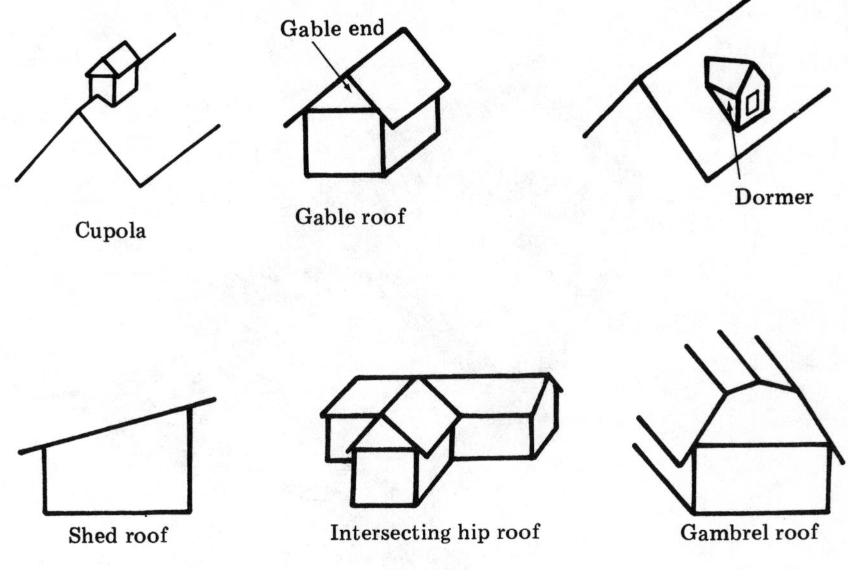

Styles of roofs
Figure 1-15

Figures 1-15 and 1-16 give you a handy guide to roof styles and roof framing members. Look at them so you will be familiar with the terms as they come up in later chapters.

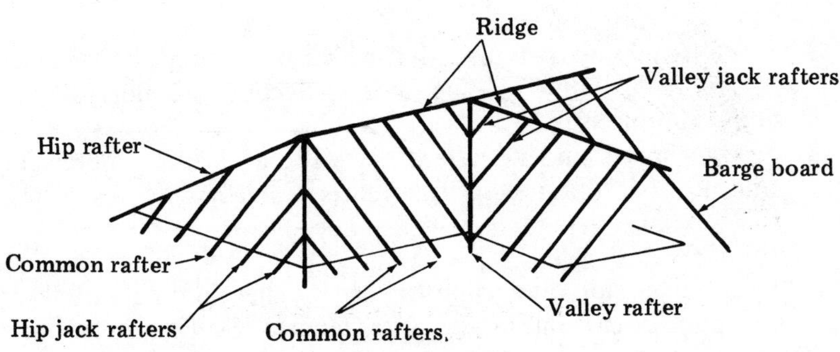

The name and location of roof framing members
Figure 1-16

2

Cutting Common Rafters

In this chapter we'll explain how to calculate common rafter lengths. This is the most basic concept in roof cutting. You'll need a framing square (sometimes called a ''steel square'') to do the work outlined here. I'm also going to suggest that you build the model that's explained in this chapter, especially if you haven't framed a roof before.

We'll start by assuming that we've been hired to frame a 12' x 20' garage as shown in Figure 2-1. Because of snow load conditions, the pitch is to be 8 in 12. This pitch will also give a little storage space under the rafters without greatly increasing the amount of framing or roofing material needed.

Span

The shorter side dimension (width) of the building will always be the span for calculating the rafter lengths. So 12' will be our span.

Total Run

As a roof cutter, you know that the total run is 1/2 the span, or 6', as in Figure 2-2.

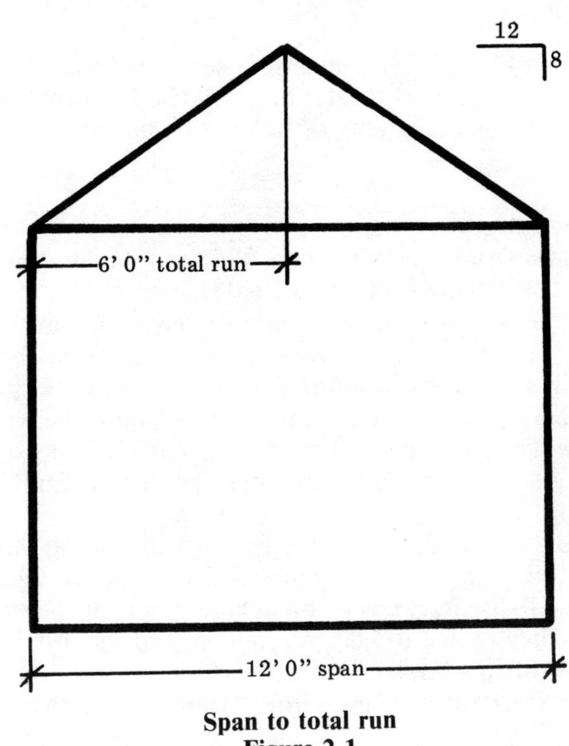

Span to total run
Figure 2-1

12
8
Rafter
4' 0"
6' 0"

$$\left[\frac{12'}{2} \quad = \quad 6' \quad = \quad \text{Total Run}\right]$$

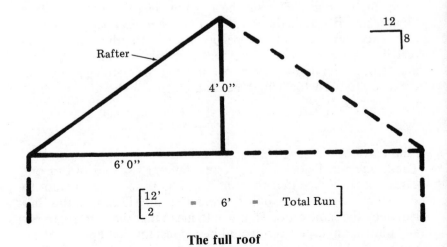

The full roof
Figure 2-2

Total Rise

This is an 8 in 12 roof. That means it rises 8'' for every 12'' of total run. Since there are 6' of total run, it will rise 8'' a total of 6 times. Eight inches times six equals 48 inches of total rise.

The Rafter Length

On this simple roof the total run is 6'0'' and the total rise is 4'0''. Now, what's the length of the rafter?

Here's one way to find out. On a driveway, lay out two chalk lines that meet at a right angle. Measure out from the meeting point of the two lines 6' on one line and 4' on the other. Make a mark at these points. Then measure the distance between the 6' and 4' points. The tape will read 7'2½'' if you laid out the 90 degree angle very carefully and marked the 4' and 6' points precisely.

You could figure all rafter lengths this way. But as you've probably guessed already, there are easier ways to figure rafter lengths. We'll discuss a few of them here. Some of the ways aren't practical. They're just to help you understand the problems in the second half of this book.

Let's get back to our garage roof. Here are the first two questions a roof cutter will ask: What's the pitch? What's the total run? (These are the main points of Chapter 1.)

Here we're working on an 8 in 12 pitch roof that has a total run of 6'0''. With these two facts, the roof cutter can find the mathematical length of each common rafter.

The mathematical length is the apparent length of the rafter. See Figure 2-3. It's sometimes called the *theoretical* length. If the rafter were only as thick as a pencil line, this length would be correct. But since all framing members are thicker than a line, you must (for common rafters) remove some of the material by cutting the rafter to the actual length needed. Otherwise the roof won't fit together properly.

Finding the Unit Length

The triangle in Figure 2-4 expresses the unit rise and run of our garage roof. Notice that the unit length is 14 7/16 inches. I want to show that expanding the unit rise and unit run will expand this unit length by the same ratio. That would mean that the unit length for any unit rise and unit run would be a constant that applies on all roofs.

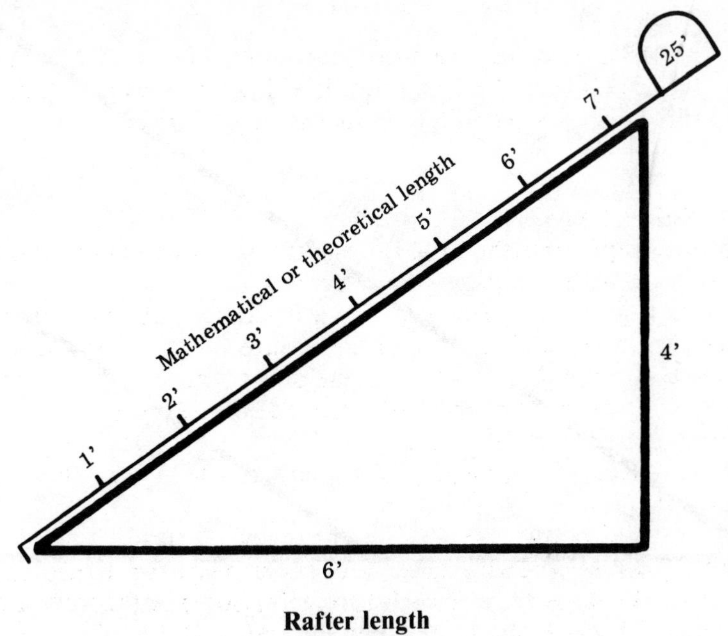

Rafter length
Figure 2-3

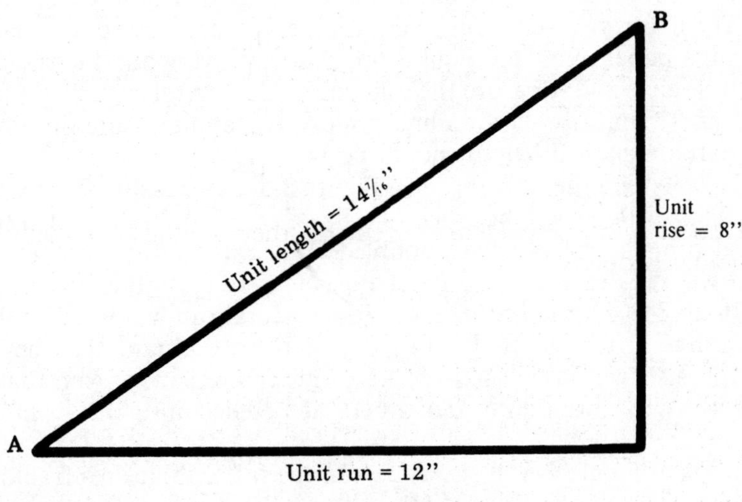

A unit triangle
Figure 2-4

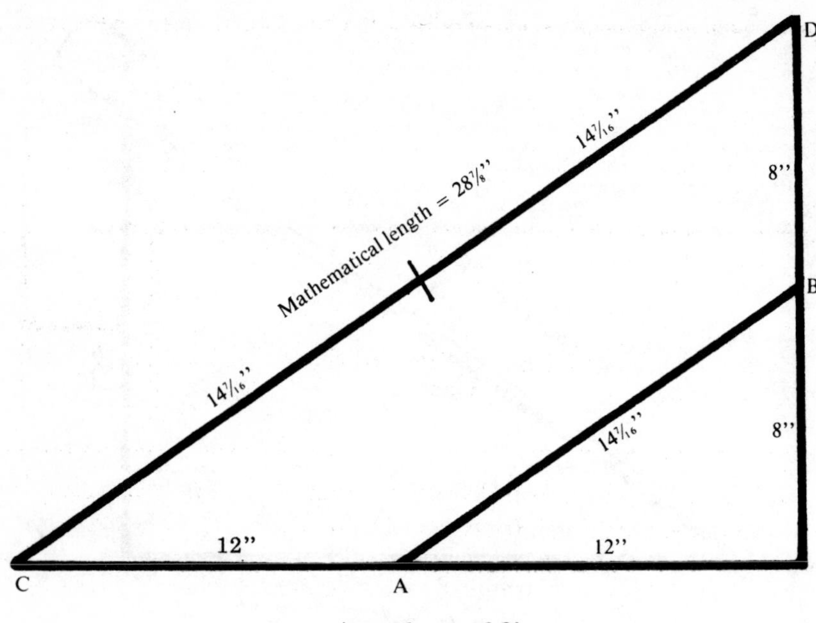

A total run of 2'
Figure 2-5

Figure 2-5 is an expansion of the triangle in Figure 2-4. The unit rise and unit run remain the same. But the lengths have been doubled. Notice that the mathematical length has also doubled.

Try it for yourself. Cut open a large grocery bag so you have enough paper to lay out Figure 2-5. Start by laying out the triangle as in Figure 2-4. Measure the unit run of 12". At a right angle lay out an 8" unit rise. Draw a line from A to B and measure this line. The mathematical length should be 14⁷⁄₁₆".

Now extend the drawing as in Figure 2-5 and measure from C to D. This distance will be twice 14⁷⁄₁₆", or 28⅞". As we doubled the total run, the rafter length doubled. You can see that changes in run will create proportionate changes in rafter length.

In an 8 in 12 pitch roof, every foot of total run will increase the common rafter length by 14⁷⁄₁₆". That's a constant that never changes. It would be useful to keep this figure handy. And that's exactly why this figure and others are etched into the framing square.

Look at Figure 2-6. It shows part of the rafter table on a framing square. The number under 8" of unit rise is filled in because we know that common rafters in an 8 in 12 roof are 14⁷⁄₁₆" long for each foot of total run.

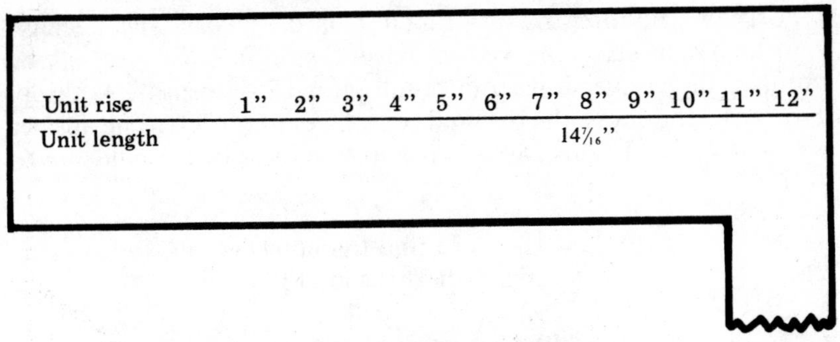

Unit rise
1" 2" 3" 4" 5" 6" 7" 8" 9" 10" 11" 12"
Unit length
14 7/16"

Common rafter unit lengths
Figure 2-6

If we made a lot of drawings using different pitches and measured each unit length, we could fill in the rest of Figure 2-6. That would give us a handy table for almost any roof we came across. The makers of framing squares have done exactly that.

Unit Length on the Framing Square

Hold a framing square in front of you as shown in Figure 2-7. Notice that the rafter scales have already been worked out. Numbers on the top edge are understood as inches of unit rise. The first line below this is called "Length common rafters per foot run." The words "common rafter run" indicate a run of 12". On this line, under the 8" rise, the unit length of 14.42 is given. This means fourteen and forty-two hundredths of an inch, or about 14 7/16".

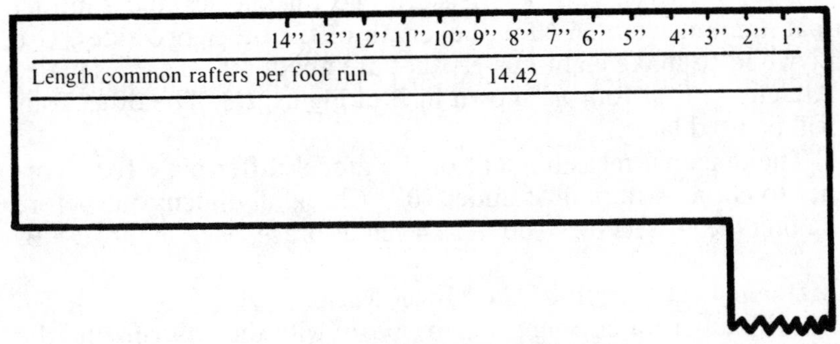

The common rafter scale on the framing square
Figure 2-7

Forty-two hundredths of an inch is hard to work with because your tape measure is marked in sixteenths of an inch. Here's how to make the conversion from hundredths to sixteenths. Multiply 0.42 by 16 and move the decimal two places to the left. The answer is in sixteenths. In this case it's 6.7 sixteenths, which rounds up to 7/16.

```
   .42    to find the number
 x 16     of 16ths in .42"
  252
  42
 6.72  =  7/16ths
```

The Model Rafter Plate

Suppose we decided to make a scale model of the garage roof just to check the technique we intend to use on the actual project. On a very complex roof, you might want to build a scale model before beginning actual construction. Of course, this garage roof isn't a complex job, but building a model will be instructive at this point, especially if you've never framed a roof before.

You'll remember that the garage we're working on measures 12' by 20' and will have an 8 in 12 gable roof. If we built a scale model of this project at one fourth its actual size, the model would be 3' by 5'.

If you want to practice on a model roof, start by cutting construction grade 2 x 6 to form the rafter plates, as shown in Figure 2-8. Accuracy is important. Work to the nearest sixteenth of an inch so that the model fits together quite perfectly.

Cut two pieces of 2 x 6 exactly 5' long. Nominal 2-inch lumber is either 1½" or 1⅝" thick. Measure two thicknesses and subtract that amount from 36" for the length of the other two sides. Use plywood to make eight gussets that are about 10" on the square. Attach the first four as shown in B in Figure 2-8. The other four will be used later.

The diagonal measurement of the model rafter plate from corner to corner will be just under 70". Check all dimensions before nailing the gussets on. Add a 2 x 4 center brace, as in Figure 2-8 B.

Mathematical Length of the Model Rafter

As with any roof cutting problem, begin with the unit run and the unit rise. Then we can compute the mathematical length of the rafter.

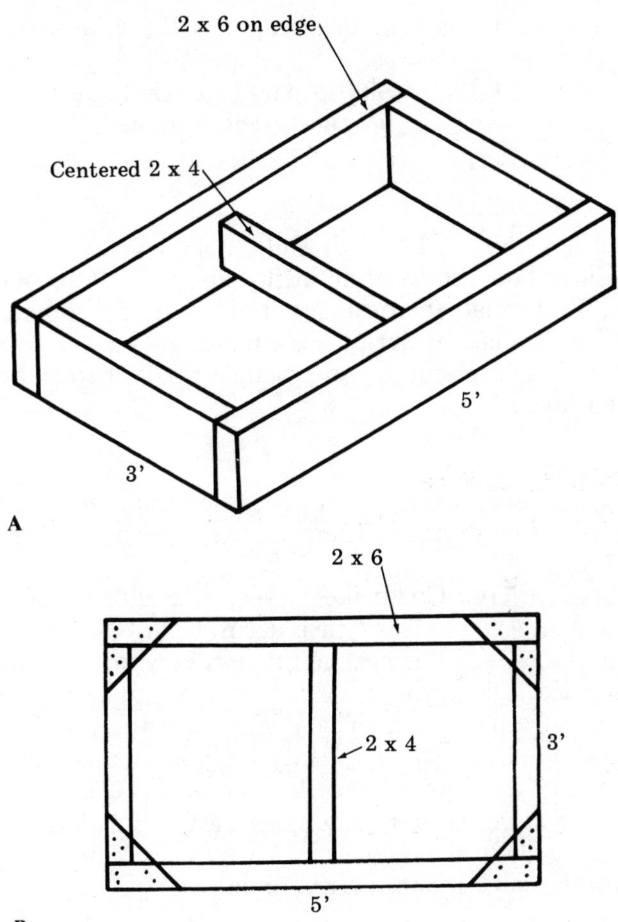

The building rafter plate
Figure 2-8

The model roof will have a total run of 1½ feet (3 feet divided by 2 which equals 1.5'). You know that the unit rise is 8''. Looking at the rafter tables on the framing square under 8, we find 14.42. But 14.42 is for only one foot of total run. We have:

1 foot	=	14.42"
½ foot	=	7.21"
		21.63"

21.63" will be the mathematical length of the rafter.

```
    .63"     Multiply by 16 to find out
 x  16       how many 16ths are in .63"
 ------
   378
    63
 ------
 10.08   =   10.08/16ths
```

Sixty-three hundredths of an inch is about 10/16ths of an inch, or 5/8". So the mathematical length of this model common rafter is 21⅝". We'll make it out of 2 x 4 material.

Now, let's talk about a few practical points before beginning our rafter layout.

Stair Gauges

A stair gauge is a metal clip which can be attached to a framing square with a thumb screw. There are two types on the market. See Figure 2-9.

The Starrett Tool Co. makes a very fine stair gauge which is their tool number 111. This type is accurate and can be set directly to the mark desired. The hexagon type is very inaccurate and not recommended. The metal of the corners tends to keep the framing square away from the work. That changes the angle slightly. If you were to lay out a 13 step stair stringer with hexagonal stair gauges, the whole stairway might be off by as much as 2".

If you use hexagon stair gauges, set them a little larger than the number needed. In the second part of this book or when cutting stairs, you'll really appreciate the finger-type gauges.

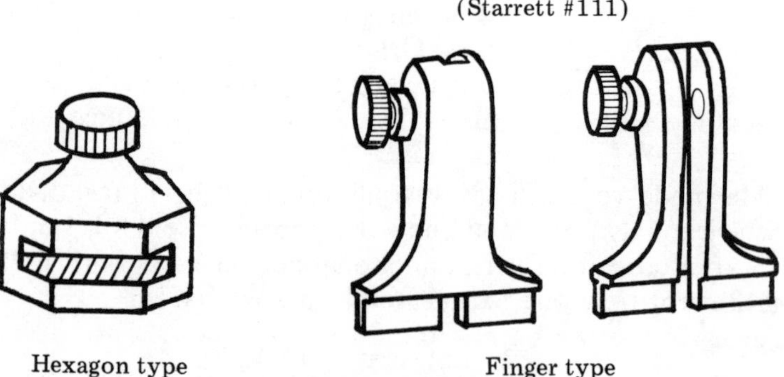

Stair gauges
Figure 2-9

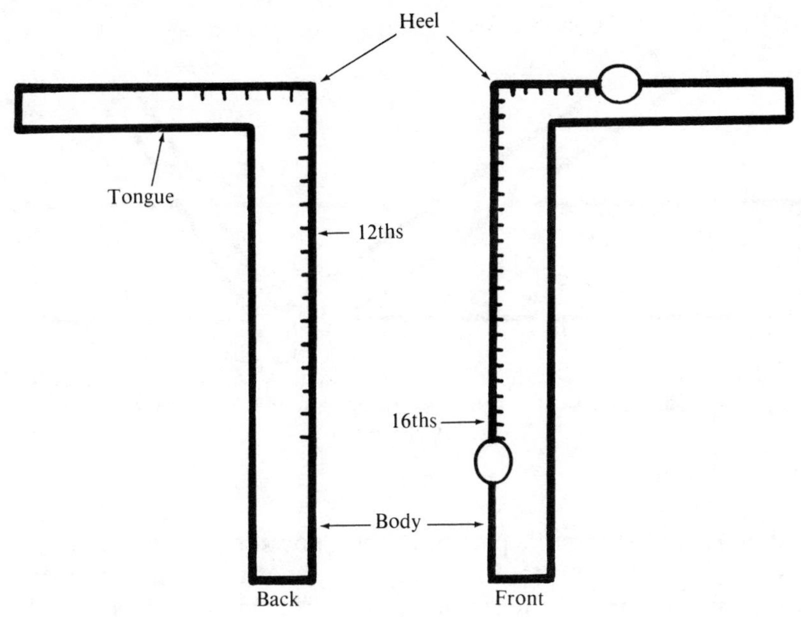

The framing square
Figure 2-10

Setting the Stair Gauges

The long fat blade of the framing square is called the *body*. See Figure 2-10. It's 2' long and 2'' wide. On the back of the body, on the outside edge, is a twelfths scale. Don't use this scale when you intend to measure in sixteenths; you'll get the wrong setting.

The short thin blade of the framing square is called the *tongue*. It's 16'' long and 1½'' wide. The outside corner is called the *heel*. On the front side of the tongue and body, on the outside edge, is a sixteenths scale where the stair gauges are set. Notice the stair gauges set in Figure 2-10. These gauges make marking off the length of rafters faster and more accurate.

Usually, you'll hold the square in the back position with the twelfths scale up, as in the left-hand drawing in Figure 2-10 and in Figure 2-11. Be careful not to use this scale for roof cutting. Turn the square over to the sixteenths scale. Move the stair gauges so that the unit rise is measured off on the tongue and the unit run on the body. Then tighten the gauges securely in place.

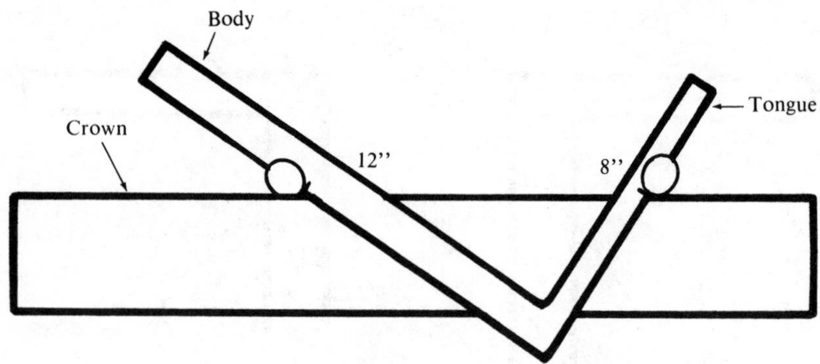

The back position of the square for marking rafters
Figure 2-11

Figure 2-12 shows the part of the framing square that's used for making the various cuts. For a very steep roof with a unit rise over 14'', use the body of the framing square for the rise and set the 12 inch unit run on the tongue.

Crowning

Pick up any long board and hold it flat rather than on edge. Sight down each side. You'll probably notice a bow, or crown, in the board near the middle. See Figure 2-13. When framing any roof, always turn the crowned edge up. This improves the strength of the roof.

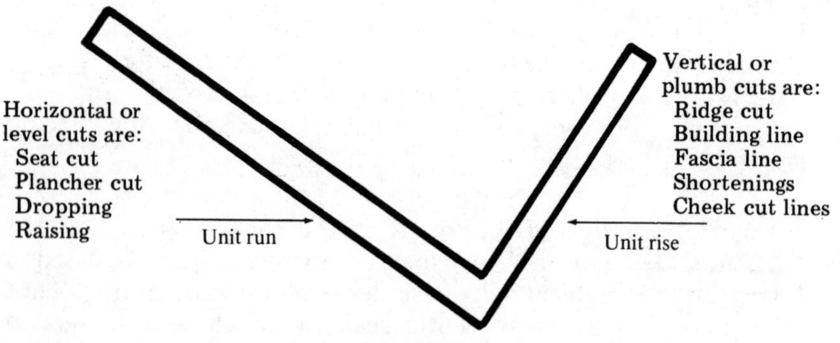

The square and its cuts
Figure 2-12

Crowning and how to crown
Figure 2-13

Drawing the Ridge Line

Lay the 2 x 4 flat with the crown away from you, as in Figure 2-14. A 2 x 4 as short as you're working with for the model won't have a crown that you can notice. But be sure that the crown is *away* from you when cutting full-size rafters.

Place the framing square so that the unit rise is directly on the corner of the wood. If the end is split, trim it back first. Now draw the ridge line as indicated in Figures 2-14 and 2-17.

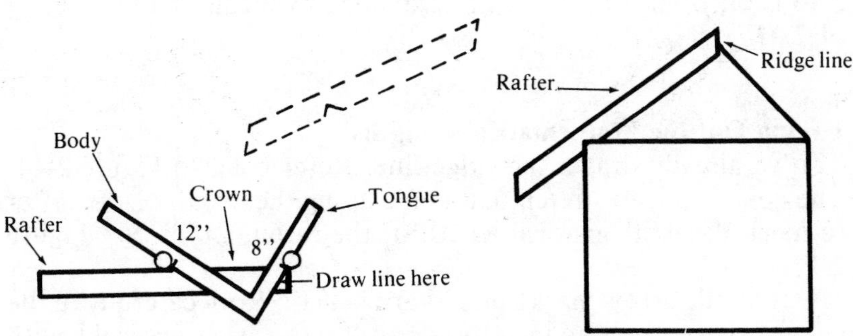

From laying out to placing the rafter
Figure 2-14

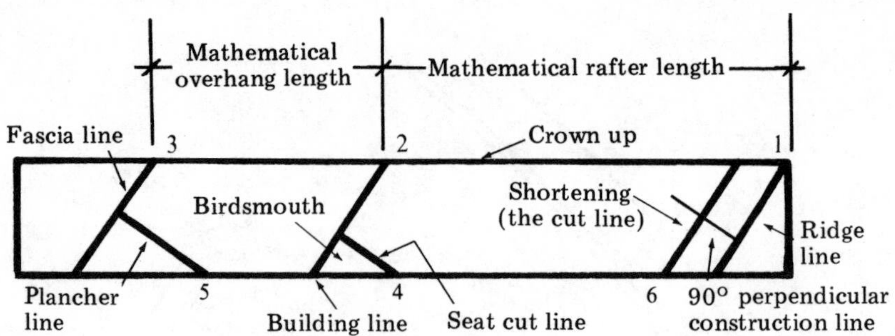

The layout lines for a common rafter
Figure 2-15

Layout for Common Rafters

Names for the lines used in laying out a common rafter are given in Figure 2-15. It shows what the rafter will look like when you finish the layout. At this point you would be ready to begin cutting. But before we do, let's learn a little more so we can finish the layout.

Projection or Overhang

The *projection* is the horizontal distance that the eave extends beyond the house. See Figure 2-16. The word *overhang* refers to the rafter tail length and will be more than the projection in anything but a flat roof.

Only the roof cutter is concerned with the overhang distance. He has to calculate it the same way he calculates the mathematical length of the rafter. The projection distance is given on the plans.

For an 8 in 12 roof, the unit length of the common rafter for every foot of run is always 14.42". Since the projection in Figure 2-16 is only half of a foot, 14.42 divided by 2 equals an overhang of 7.21", or 7³⁄₁₆".

Laying Out the Mathematical Lengths

You've already drawn the ridge line. Refer back to Figure 2-14. The next step is to stretch out a tape along the length of the rafter to mark the mathematical length of the rafter, 21⅝". See Figure 2-17.

Use a little arrow mark and a sharp pencil. Work carefully to the nearest sixteenth of an inch. Next, add to the mathematical length of the rafter (21⅝") the length of the overhang (7³⁄₁₆"). This equals 28⅞". Mark this point on the rafter also.

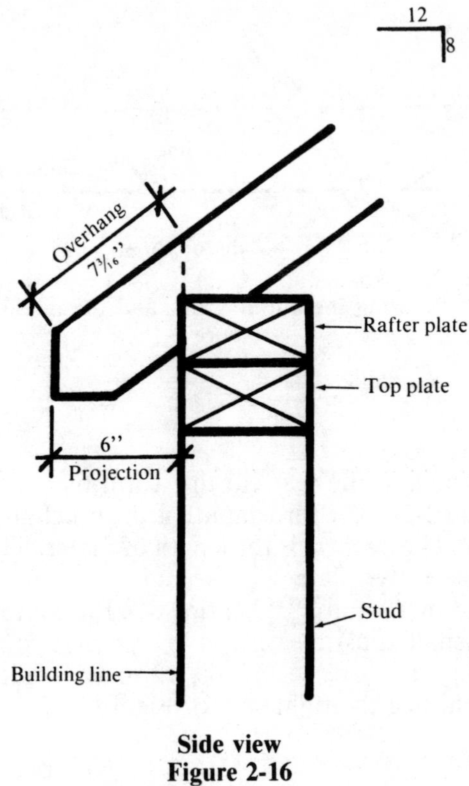

Side view
Figure 2-16

Hold the framing square as shown in Figure 2-18 and draw the building line and the fascia line on the rafter.

Notice that we marked the ridge line first at the right end. Then the mathematical lengths are measured along the top edge of the material. Only these two measurements are made along the material this way. Don't measure anything else along this top edge. See Figure 2-28.

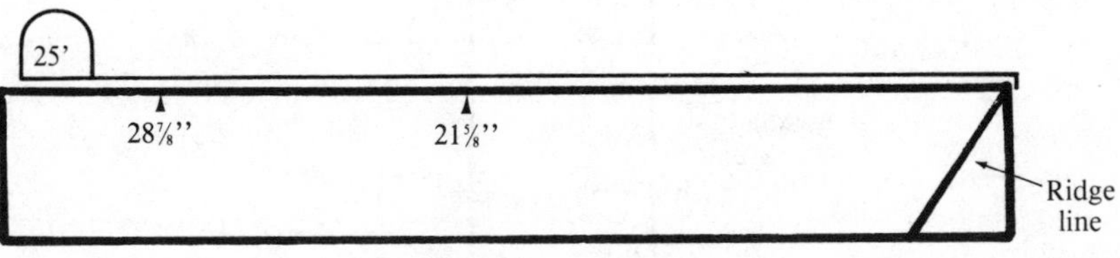

Mathematical lengths
Figure 2-17

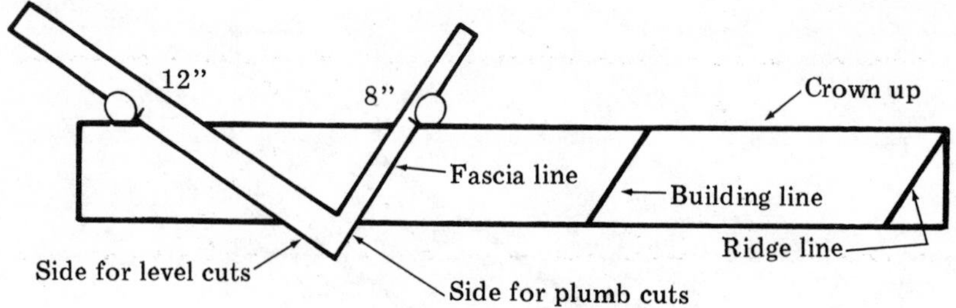

Drawing the building line and the fascia line
Figure 2-18

Bearing and Seat Cut

The building line and the seat cut line form the *birdsmouth*. See Figures 2-19 and 2-20. The birdsmouth is the notch that fits against the rafter plate. The *bearing* is the length of material in the seat cut that rests on the rafter plate.

Figure 2-19 shows a full 3½" bearing . . . the width of the rafter plate. That's usually possible only in low-pitched roofs. Making a full 3½" bearing in a steep roof like 12 in 12 would weaken the rafter too much. See the right side of Figure 2-20.

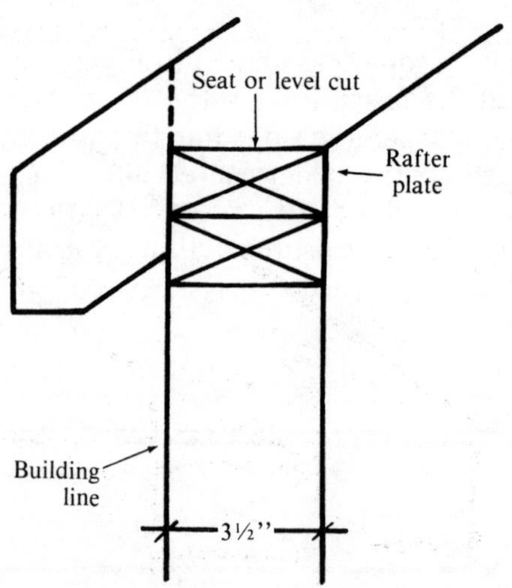

The birdsmouth with full bearing
Figure 2-19

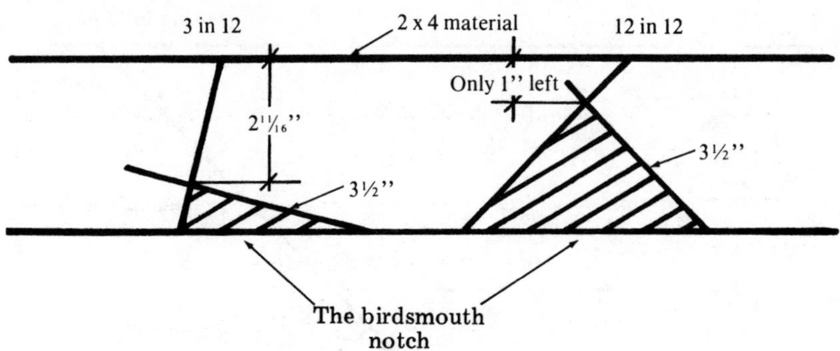

The difference full bearing makes in higher pitched roofs
Figure 2-20

Both birdsmouths in Figure 2-20 have a 3½" bearing. But notice that the 12 in 12 rafter has only 1" of material left and is in danger of breaking off. The 3 in 12 rafter has plenty of material left above the building line.

Since the amount of bearing varies widely with the pitch, we need some way other than the width of the rafter plate to determine the proper length of the bearing.

Height Above Plate (*HAP*)

Notice the *HAP* in Figures 2-21 and 2-22. Here's a good rule for most roof cutting. If two-thirds of the material is retained and only

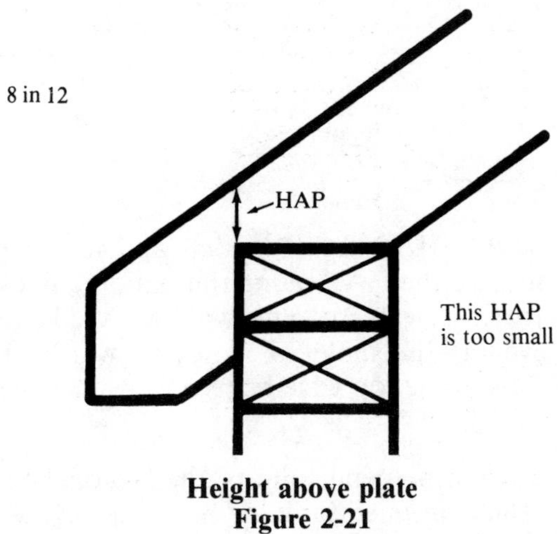

Height above plate
Figure 2-21

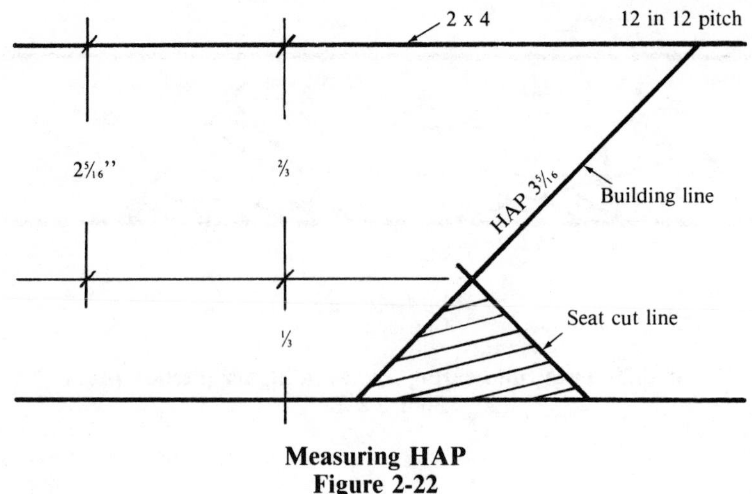

Measuring HAP
Figure 2-22

one-third is cut away, strength will be maintained along with adequate bearing. Two-thirds of a 3½" board is equal to 2⁵⁄₁₆". But the distance of the HAP will vary on steeper pitches if the two-thirds rule remains. See Table 2-23.

Pitch of	HAP on 2 x 4	HAP on 2 x 6
4 in 12	2 ⁷⁄₁₆	3⅞
5 in 12	2 ½	4
8 in 12	2¹³⁄₁₆	4⁷⁄₁₆
12 in 12	3 ⁵⁄₁₆	5³⁄₁₆

HAP using the two-thirds rule
Table 2-23

Select a reasonable HAP for the pitch and the material that you're using. Measure the HAP along the building line, from the top edge of the rafter downward. This will keep the HAP uniform on all rafters even if the material varies in width. Therefore, measuring down from the top edge helps ensure that the roof line isn't wavy.

HAP isn't measured perpendicular to the top of the rafter. It's measured along the continued building line. See Figure 2-22.

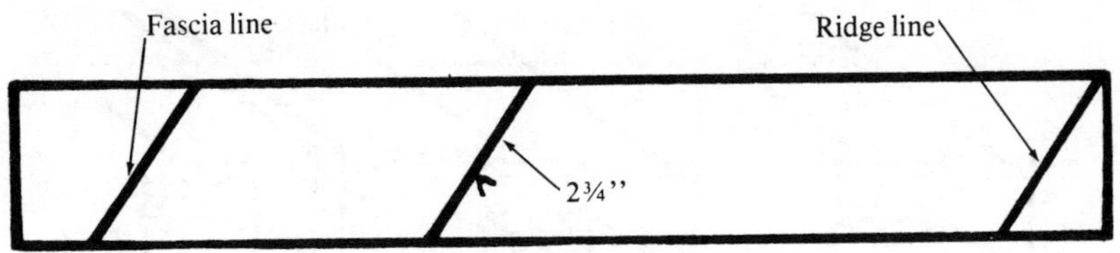

Laying out the HAP
Figure 2-24

Laying Out the HAP

On all regular roofs you'll have only one HAP for all rafters. On your model, all HAPs will be 2¾'' measured only from the top down and along the building line. Measure and make the mark as in Figure 2-24.

Drawing the Seat Cut

Follow Figure 2-25 to position the square for drawing the seat cut line from the HAP measurement to the bottom of the rafter. Now the birdsmouth is complete.

The Need for a Plancher Cut

Sometimes the rafter tail is left square, as in A of Figure 2-26. On other jobs you'll cut the tail, as in B in the figure. This is called a *plumb cut.*

It's common to have a gutter or fascia board hung on the rafter tails. The fascia board has to be wider than the rafter material so

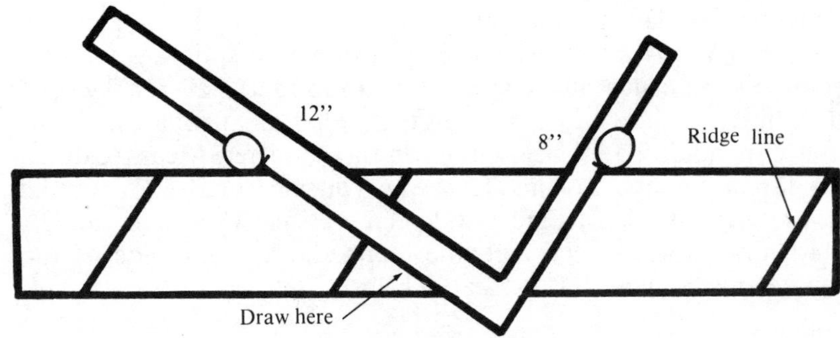

Drawing the seat cut
Figure 2-25

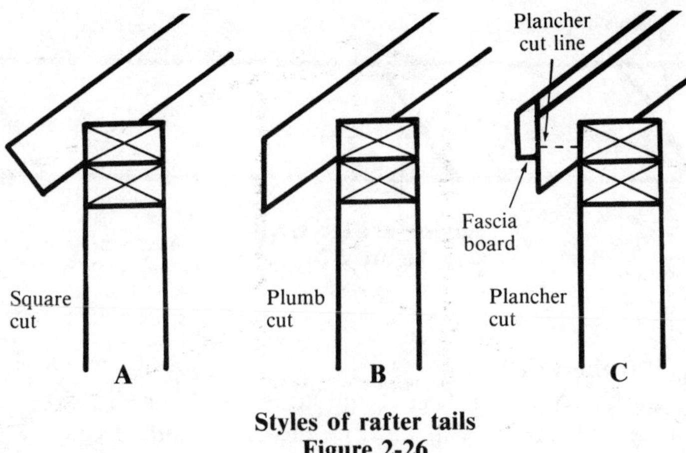

Styles of rafter tails
Figure 2-26

the fascia hangs below the lower edge of the rafter tails. This gives a nice finished look.

If the fascia material is the same width as the rafter material, the plumb cut will be longer than the fascia is wide. That's because the angled plumb cut is longer than the rafter is wide. The lower point of the rafter will extend below the fascia board, as in C of Figure 2-26. It's best if the fascia board extends about 1" below the rafter. Ensure that it is by making a *plancher cut*, shown as the dotted line in Figure 2-26 C.

Our model uses 2 x 4 rafters. A 2 x 6 fascia would be too wide for our model. Instead, we'll use a 2 x 4 fascia and make a plancher cut.

Drawing the Plancher Line

Drawing the plancher line on the rafter tail is just like measuring for the HAP. A 2 x 4 fascia is 3½" wide and it should extend 1" below the plancher cut. You must also subtract 1/2" for the roof sheathing, as shown in Figure 2-26 C. That means that the plancher cut should be 2" below the top of the rafter. Measure down the fascia cut line 2" the same way you measured 2¾" for the HAP. Refer back to Figure 2-24 if necessary. Through this 2" mark draw the plancher cut line along the level cut side of the square, just as you did the seat cut line.

Why the Common Rafter Shortens

Almost every roof you frame will have a ridge board. Because there is a ridge board, every rafter has to be shortened to allow for

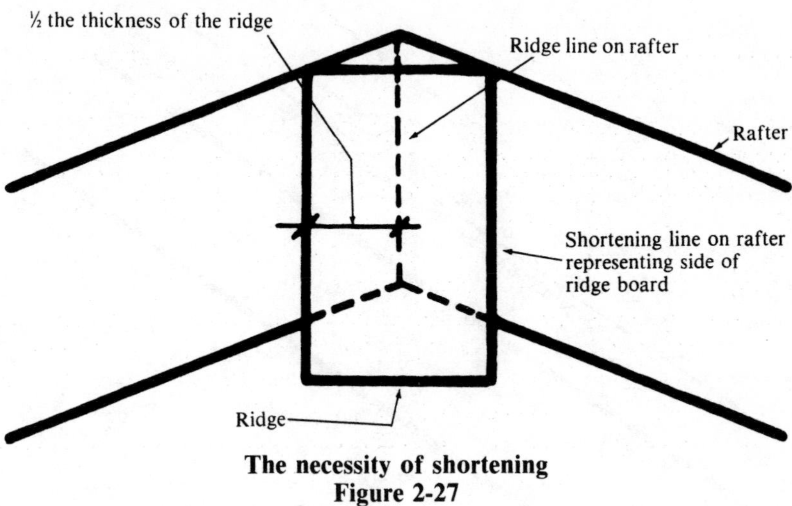

The necessity of shortening
Figure 2-27

one-half of the thickness of that ridge. See Figure 2-27. All common rafters are shortened one-half the thickness of the ridge they nail to. There is only one exception to this, which is discussed in Chapter 3.

Examples of Shortening

For a 2 x 6 main ridge, the ridge material measures 1½" thick. Shorten the calculated length of each common rafter by 3/4". For a 1 x 6 secondary ridge, the ridge material measures 3/4" thick. Shorten the calculated length of each common rafter by 3/8". Note that the shortening measurement is made at a right angle to the ridge line, not along the top edge of the rafter. See Figure 2-27.

Do's and Don't's

Figure 2-28 reviews some of the important do's and don't's of rafter layout. Look at it and make sure you understand where to lay out the various lines before we go on to laying out the shortening and making the cuts.

Laying Out the Shortening

Lay out a line perpendicular to the ridge line and measure back 3/4" on this line, as shown in Figure 2-29. Through this point draw the shortening line.

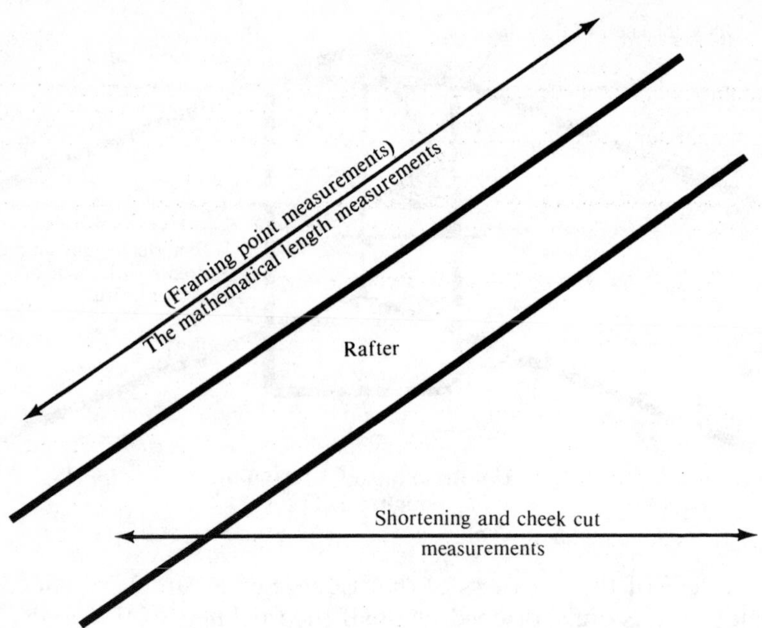

Don't mix mathematical length measurements for the building line and the fascia line with shortening and cheek-cut measurements.

Don't try to short-cut rafter layout by subtracting a shortening distance from the calculated rafter length. Lay out the cuts in their proper order. The shortening is laid out a different way.

Do lay out the ridge line, building line, and fascia line along the top edge of the rafter. Next, parallel to their respective lines, lay out the seat cut line and the plancher cut line. Finally, lay out the shortening cuts (and other cuts we'll cover later).

Chart of measurements
Figure 2-28

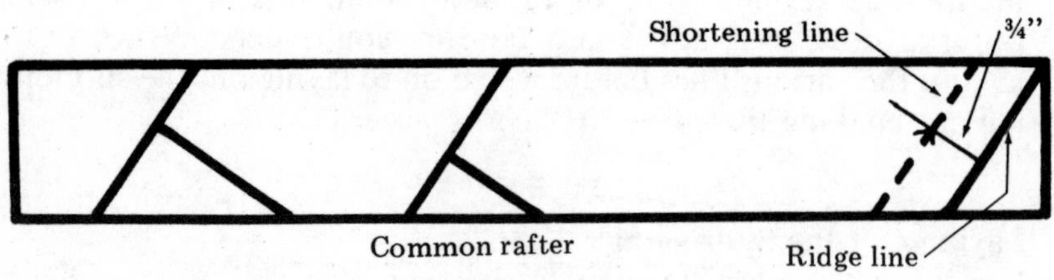

Locating half the thickness of the ridge
Figure 2-29

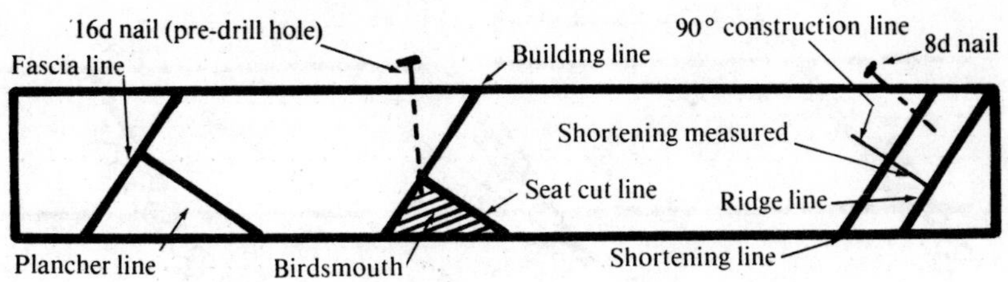

All the lines for a common rafter layout
Figure 2-30

Label each of the lines as indicated in Figure 2-30. Along the shortening line write "3/4 shortening" so you'll remember where to cut and that the shortening has already been marked.

Using a Circular Saw

A 7¼" circular saw is probably the best tool for cutting rafters. But, like any powerful tool, it can be dangerous. Be conscious of safety. Many roof cutters work a full career without a serious accident. Plan on doing the same yourself.

Here's how to make a safe cut. Prop the rafter off the floor with another 2 x 4 as shown in Figure 2-31. Put your foot on the rafter to hold it securely. Then make your cut.

Most saw blades remove 1/8" of material when they cut. Roof cutters work to a 1/16th inch accuracy. We'll work to 1/16" on this model. Be careful about where you place the blade. Figure 2-32 shows that we cut on the outside of the line, leaving the exact length marked. Because the model is small you must be most accurate.

Cutting the Common Rafter

When cutting common rafters, the blade of the circular saw is set

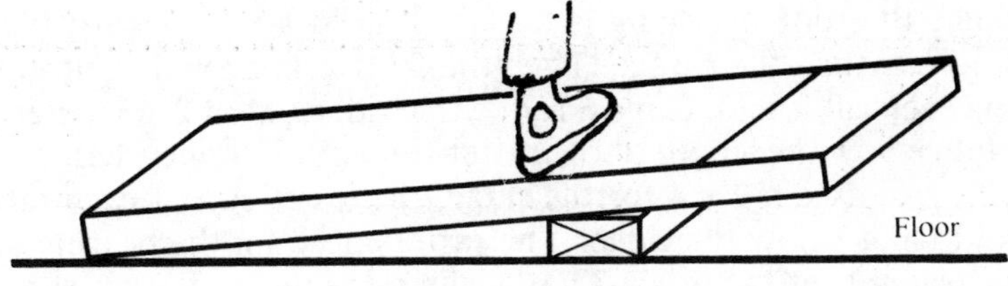

Cutting position
Figure 2-31

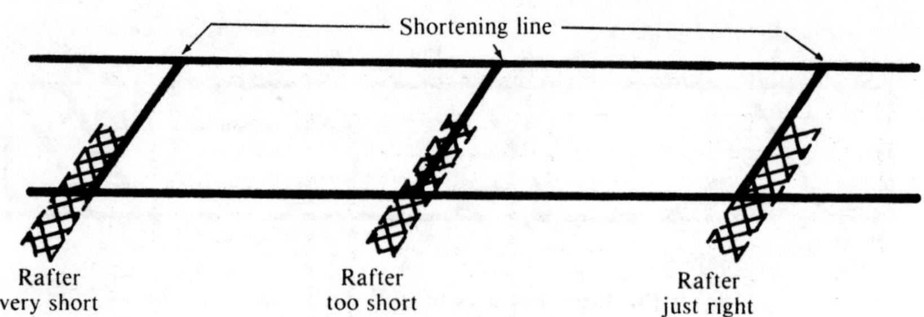

The line and the saw blade
Figure 2-32

perpendicular to the sole plate of the saw (on zero degrees on the angle scale). The saw is set this way for all common rafter cuts.

The ridge line isn't cut. That would be wasted effort. Instead, cut on the outside of the shortening line. See Figure 2-32.

Next, make the fascia cut and the plancher cut. Cut the birdsmouth carefully so you don't go into the material deeper than the lines drawn. Finish the birdsmouth with a hand saw just for pride of workmanship! That finishes the first rafter.

You'll have to cut seven more common rafters for the model. Practice laying out each one carefully to the nearest sixteenth of an inch. Then cut out these rafters.

Place the rafters side by side and see if all lengths and cuts are equal. If not, cut more rafters until you have an equal set.

For the model, you'll probably want to pre-drill holes where nails will be needed. We'll use 8d and 16d green vinyl sinkers. (Look back to Figure 2-30.)

The Gable Ridge

The gable ridge is a straight piece of lumber that extends beyond the building line on each end by the amount of the overhang. See Figure 2-33.

Since the roof overhang is 6'', we'll need a ridge board that's 12'' longer than the garage. Then the 6'' overhang will be uniform along each gable end. Cut a 6 foot gable ridge out of 2 x 6 material.

Notice how the length of the rafter ridge cut is longer than 3½''. If you had used a 2 x 4 for the ridge, the lower end of each rafter would hang below the ridge. The rafters need all the bearing surface available at the ridge. That's why ridge material is at least 2'' deeper than the rafter material on nearly every roof we cut. Be sure to crown the ridge board just the way you crown the rafters.

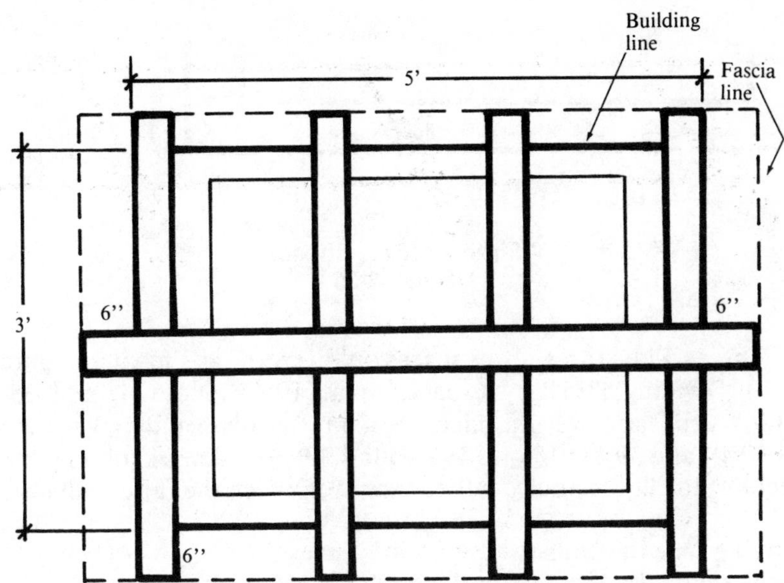

Top view of the 3' x 5' model gable roof
Figure 2-33

Dividing Space
Next, we have to mark where each rafter will be nailed. The rafters are 1½'' wide. So the center line of each rafter is 3/4'' in from each end. The garage model is 5 feet, or 60 inches long. The last rafter on each end will be flush with the exterior of each end wall. See Figure 2-34. The space between the center lines of the two end rafters will be 60 inches less 3/4'' twice. Sixty inches less 1½'' equals 58½''.

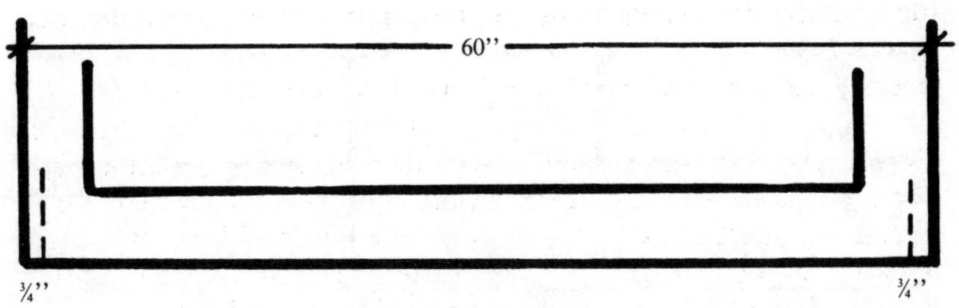

Evenly dividing a space
Figure 2-34

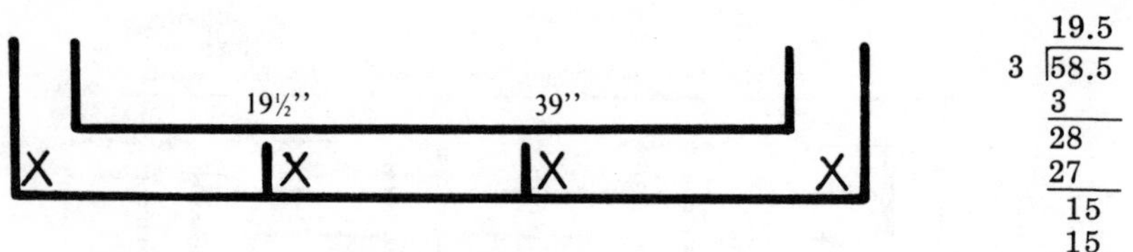

Marking off the divisions
Figure 2-35

There will be three rafter spaces on this roof, as shown in Figure 2-33. Dividing 58½'' by 3 spaces equals 19½''. Note Figure 2-35. Hook your tape over the edge of each rafter plate and make a line at 19½'' and 39'' (19½ plus 19½ equals 39). We'll mark an X to the right of the line to indicate the space over which the rafter will rest.

Laying Out the Ridge

Now mark the top of the ridge board. Hook your tape over the edge of the ridge and mark off 6'' for the overhang. The rafters will be marked at 19½'' and 39'' from the overhang mark. Add 6'' to the 19½'' and 39'' measurements to allow for the overhang. That gives us 25½'' and 45''. Add X's in the right places. See Figure 2-36.

On most roof layouts the rafters will be on 2' centers so the edge of each 4' x 8' plywood sheet falls on a rafter without trimming. Remember to figure 2' centers from the *outer* edge of the barge board. That will be 1½'' beyond the end of the gable ridge.

Roof sheathing at the overhang should be the best sheathing available for this roof. Sheathing on the overhang is exposed both to weather and observation from the ground, therefore it is usually a better quality material than the rest of the roof sheathing.

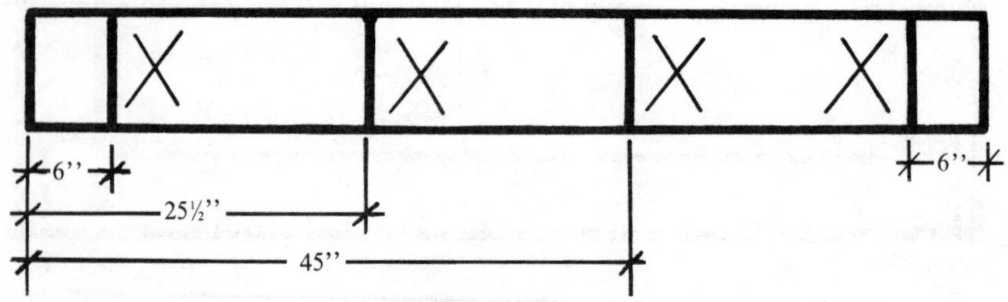

Marking off the ridge
Figure 2-36

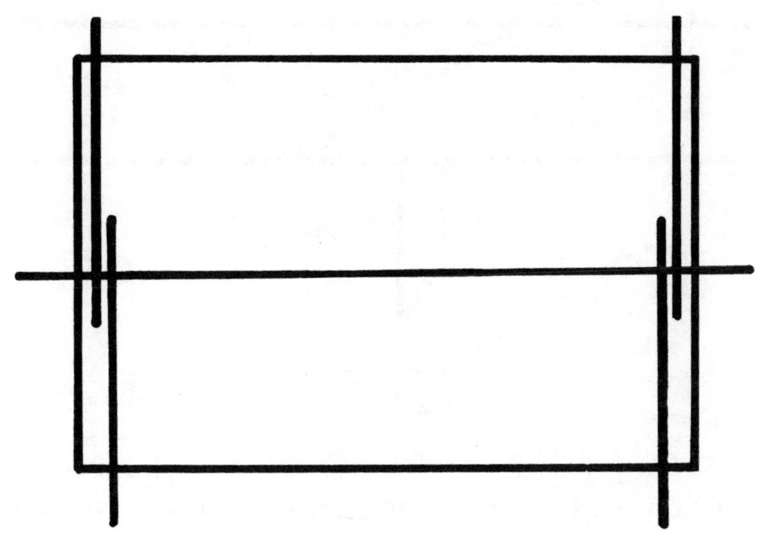

The first step
Figure 2-37

Framing

Framing the roof comes next. We'll put the rafters and ridge in position and nail them in place. It's a good idea to be sure the walls are plumb and true before framing begins. There isn't much you can do to correct walls that are out of square or not vertical. But you should point out the problem to the framers. And you *will* have to make allowances so the roof fits the building as it *is*, not as it should be.

There are many way to erect the rafters. Unless you have a better way, follow my system:

Lay the ridge across the end plates (and ceiling joists). Then lay the common rafters at each end across the ridge and toenail these rafters at the seat cut. See Figure 2-37. Then raise the ridge. That also raises both pairs of common rafters. When the ridge is in the right position, each pair of common rafters then acts like a vice to pin the ridge in place. But don't stand under the ridge until it's nailed off securely!

On a long building, the ridge board is made from two lengths of lumber spliced together. Cut each piece to a convenient angle at a place where a pair of common rafters will fall. See Figure 2-38. You'll need to add bracing directly under the splice so that the ridge remains level and true.

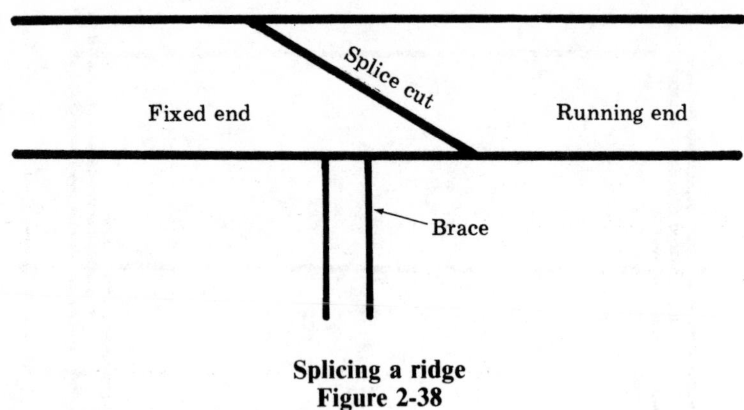

Splicing a ridge
Figure 2-38

After framing the common rafters, nail sway braces to the ridge about every 6'. The braces should run down at a 45-degree angle or more from the ridge to the top of a partition wall or a 2 x 6 nailed across the top of the joists.

Now begin framing the roof on your model. Use 16d and 8d green vinyl sinkers. Be sure to pre-drill a pilot hole so the wood doesn't split. You can refer back to Figure 2-30 for placement of nails.

Here's a tip to prevent splitting. A dull nail won't split wood as easily as one with a sharp point. So flatten off the point by holding the nail, point up, on a hard surface like a concrete slab. Then tap the point with a hammer. The nail then acts like a punch, pushing its way through the wood instead of splitting it.

Congratulations, your model roof is now finished! Things will become a little more complex from this point on, but they won't get any more difficult. If you've come this far, you'll have no trouble becoming a skilled roof cutter.

I'll continue to work step-by-step. If you've done reasonably well so far, you'll have equally good success throughout the rest of the chapters in this book.

Methods of Computation

Calculating the mathematical length of rafters on our model garage was easy. It was one and one-half times the unit rafter length of 14.42''. On larger and more complex jobs you'll have to do more figuring than that. The rest of this chapter will suggest better ways to figure common rafter lengths, including using a powerful carpentry tool: the modern hand-held calculator.

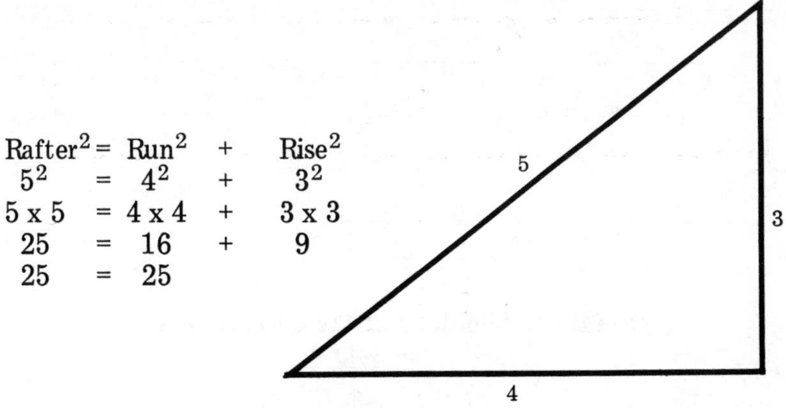

Pythagorean theorem
Figure 2-39

Many roof cutters use rafter length books to find the mathematical length of rafters for each job they have. A book, *Rafter Length Manual,* lists rafter lengths for nearly every span and rise on a regular roof, and can be ordered from Craftsman Book Company, P. O. Box 6500, Carlsbad, CA 92008. But you'll need to figure your own rafter lengths on irregular roofs. We'll cover irregular roofs in the second half of this manual.

Even if you had a book that covered every possible roofing situation, you should know how to figure rafter lengths on your jobs. The book could be wrong or you could lose your book!

Here are two ways to calculate rafter lengths. They both work quite well on any regular or irregular roof.

First Ancient and True Method: Pythagorean Theorem

Look back to Figures 2-3 to 2-5. Notice the basic relationships in a right triangle. Over 2000 years ago the Greeks figured out how to use triangles to make square corners. Today, carpenters use the information to get their buildings square.

A man named Pythagoras first proved that in a right triangle the total run squared (that is: multiplied by itself), plus the total rise squared, would equal the length of the diagonal (rafter), squared.

A carpenter uses this information when he checks the square of a corner: Measure from the corner along one wall 4', from the same corner along the other wall 3', then measure between the 4' and 3' points. See Figure 2-39. If the distance is exactly 5', the two walls are at 90 degrees, or perpendicular to each other.

	3	**4**	**5**
2 x	6	8	10
3 x	9	12	15
4 x	12	16	20
6 x	18	24	30

Useful combinations of the basic triangle
Table 2-40

In a mathematical equation, such as 5 squared equals 4 squared plus 3 squared, you can multiply all numbers on both sides of the equals sign by the same amount without changing the truth of the equation. If we multiplied everything by 2, we would have 10, 8, and 6. Table 2-40 shows the basic 3-4-5 triangle multiplied up to 6 times.

A Practical Use

Notice that on the framing square, on the back side at the heel, there are three numbers written like this: 18/24 30. This is each number of the 3-4-5 triangle multiplied 6 times.

Here's why we use a 3-4-5 triangle multiplied 2 or more times. Let's say that at the back of a house we want to set form boards for a 16' square cement patio. Using a 3', 4' and 5' triangle might not be accurate enough. If any measurement is off by as little as 1/4", the corner would be noticeably out of square. To make it more accurate, multiply all dimensions by 4: use 12', 16' and 20'.

First, stake down the forms near the house, as indicated at the left side of Figure 2-41. See that you have 16' clearance on the inside of the forms. Also, the two side boards must be cut at exactly 16'. Measure along the inside edge of the front corner 16', then measure 12' on one side. Measure between these points and move the front form board from side to side until the tape reads 20', as in the right illustration in Figure 2-41. Now the patio forms are square.

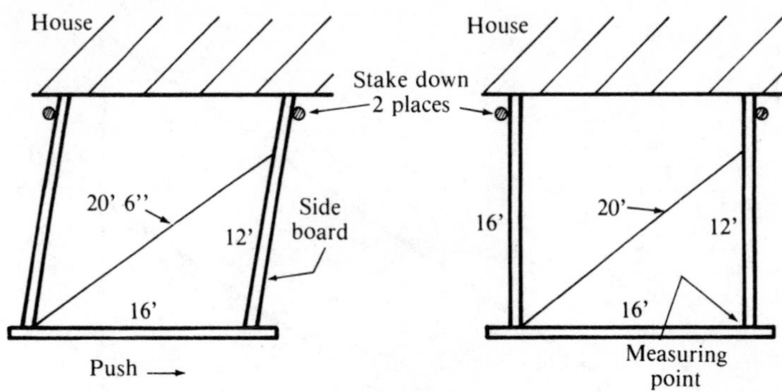

Squaring forms
Figure 2-41

Let's check our triangle with sides of 12', 16' and 20' by the Pythagorean Theorem to see if it's a right triangle:

$$(20)^2 = (12)^2 + (16)^2$$
$$400 = 144 + 256$$
$$400 = 400$$
$$\sqrt{400} = 20 \quad \text{It works!}$$

The science of triangles hasn't changed much in the last 20 centuries and won't change in our lifetime. You can assume that the Pythagorean Theorem can be used to calculate the diagonal (rafter length) of a right triangle.

All rafter tables, either in books or on the framing square, are based on the Pythagorean Theorem. But no book of rafter tables can cover more than the most common rafter lengths and roof spans. Even the best rafter table is like a hammer that will drive only 4d box nails. Sometimes you need more than that.

Let's meet a powerful 20th-century tool that makes rafter calculations really simple.

The Hand-held Calculator

According to the Pythagorean Theorem, the rafter length multiplied by itself is equal to the total run multiplied by itself plus the total rise multiplied by itself. In the case of our model, Figure 2-42, the total run is 1½ feet, or 18''. (These calculations work only if all the dimensions are expressed either in inches or in feet.) Since the rafter is small, let's put everything in inches.

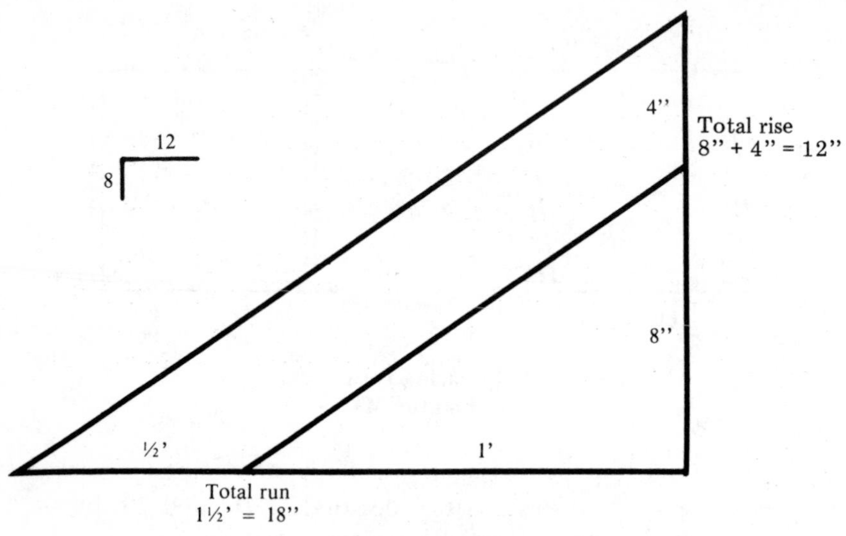

The model rafter
Figure 2-42

Figure 2-42 shows an 8 in 12 roof: it rises 8 inches for every foot of total run. What's the rafter length?

We know the total run is 18". If we can find the total rise, we could apply the Pythagorean Theorem to find the rafter length.

In this example, the total run is one and one-half times the unit run. So the total rise must be one and one-half times the unit rise. 1½ times 8" equals 12". So now we have:

$$\text{Rafter}^2 = 18^2 + 12^2$$

This could be difficult to solve with pencil and paper. But any inexpensive hand calculator with a square root key will make it quite simple. I use a Texas Instruments TI-35 calculator and will use examples based on that machine. Your calculator will give the same answers. Only the sequence of keystrokes may be slightly different.

Turn your calculator on and punch the following buttons: [1] [8]. Now 18 appears in the display window. Punch [x^2]. Now 324 appears. Punch [+] [1] [2]. Now 12 appears. Punch [x^2]. Now 144 appears. Punch [=]. Now 468 appears. Punch [$\sqrt{}$]. Now 21.6333 appears. (The TI-35 calculator will give you answers carried to six places after the decimal point. For simplicity, we'll round the answers to four digits in our calculations.)

This means that the rafter is 21.63'' long. Let's convert this to inches and fractions of an inch to make layout easier.

Subtract 21 inches and we have 0.63'', or 63 hundredths of an inch. Multiplying this by 16 will give the number of sixteenths in 0.63''. (Refer back to the section under "Unit Length on the Framing Square" in this chapter if you don't understand this.)

Here's how to do it on the calculator. With 21.6333 showing, [−] [2] [1] [=] and 0.6333 appears. Punch [×] [1] [6] [=] and 10.1329 appears, meaning that 0.63 inches is 10.1 sixteenths of an inch. Combining 21'' and 10/16'', we find the length we're looking for is 21⅝''.

Look back at Figure 2-17. That's exactly the dimension we marked for the building line on our model. It looks like the calculator gave us the same answer as the framing square. And the advantage is that the calculator can provide far more correct answers than could possibly be printed on a square or in a book of rafter tables, to say nothing of the irregular calculations in the second half of this book.

The run of the overhang (the projection of the roof) is half a foot. Therefore the rise is 4''. See Figure 2-43.

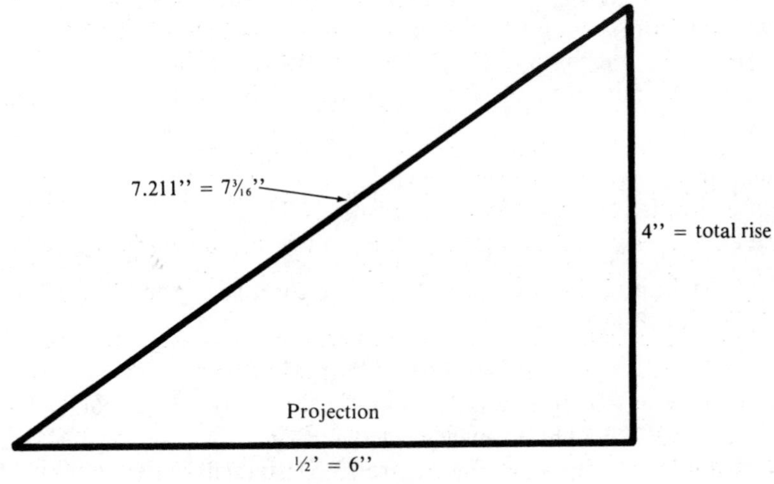

Finding the overhang
Figure 2-43

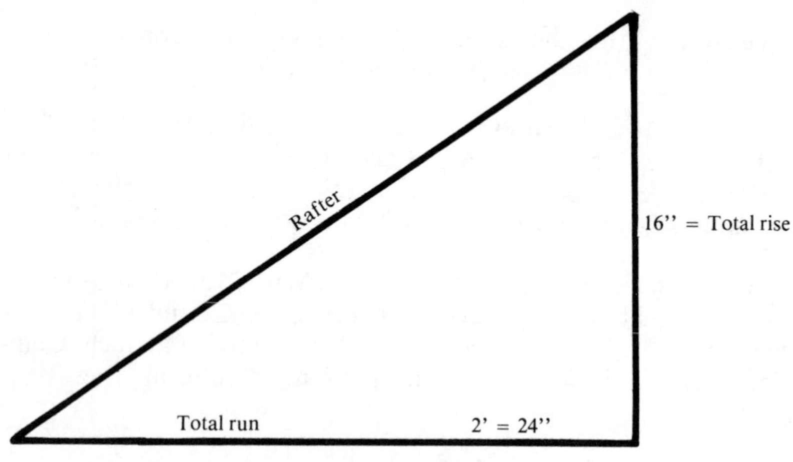

The full rafter
Figure 2-44

The length of the overhang is then:

$$\text{Overhang}^2 = \text{Projection}^2 + \text{Total Rise}^2$$

$$\text{Overhang}^2 = 6^2 + 4^2 \text{ (all in inches)}$$

Here's how to get the answer on the TI-35: Punch [6] [x²] [+] [4] [x²] [=] and 52 appears. Then punch the square root button and 7.2111 appears. The length of the overhang is 7.21 inches. Convert that to inches and a fraction by subtracting the 7 whole inches and multiplying the 0.21 inches by 16. Your answer is 3.37, or 3/16''. The overhang is 7 3/16'' long, from the building line to the fascia cut, measured along the top edge of the rafter.

Adding:

$$7\tfrac{3}{16}'' + 21\tfrac{5}{8}'' = 28\tfrac{13}{16}''$$

That's very close to the number we used for the fascia cut line back in Figure 2-17.

You can also figure out the fascia cut line by taking the total run (including the projection) and the total rise. See Figure 2-44.

The formula would be:

$$\text{Rafter}^2 = 24^2 + 16^2$$

To find the rafter length, punch [2] [4] [x²] [+] [1] [6] [x²] [=] [√] and 28.8444 appears. Punch [−] [2] [8] [=] [×] [1] [6] [=] and 13.51 appears. There are 13.51 sixteenths (let's call it 14 sixteenths, or 7/8 inch) and 28 whole inches. This agrees with the measurement used in Figure 2-17.

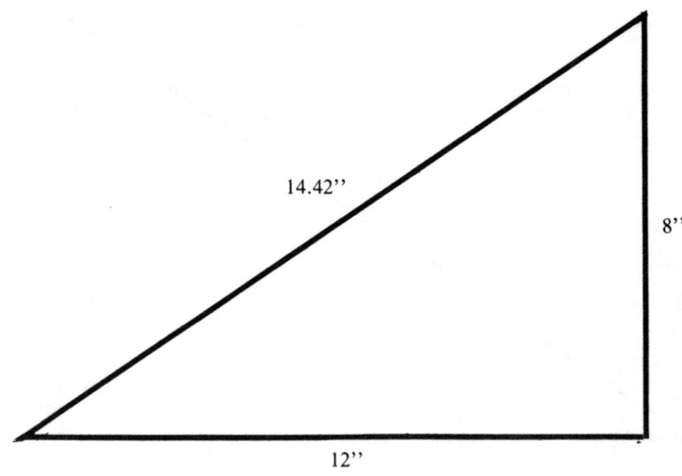

A common rafter unit triangle
Figure 2-45

Framing Square Numbers

In Figure 2-4, we laid out a unit triangle for an 8 in 12 roof and discovered that the unit length of this triangle was 14$\frac{7}{16}$". In Figure 2-7, the number 14.42 is shown printed on the framing square under 8" and along the line of "common rafters per foot run" (which means 12"). See Figure 2-45.

The formula should read:

$$\text{Unit length}^2 = \text{unit run}^2 + \text{unit rise}^2$$

$$\text{Unit length}^2 = 12^2 + 8^2$$

Notice this is all in inches. Punch [1] [2] [x^2] [+] [8] [x^2] [=] [√] and 14.4222 appears. Since the model rafter has a total run of 1½ feet, multiply the 14.4222 by 1.5 and you get 21.6333 inches. Punch [−] [2] [1] [=] and 0.6333 appears. Punch [×] [1] [6] [=] and you get 10.1329. That's 10/16ths or 5/8ths. So again we have arrived at 21⅝ inches for the model roof building line measurement, this time using the unit length.

You should be comfortable with the Pythagorean Theorem now and have some confidence that you can find the rafter run if you know the span and the rise. We'll get a little more practice with it and then go on to another ancient method of finding the third side (rafter length) of a triangle.

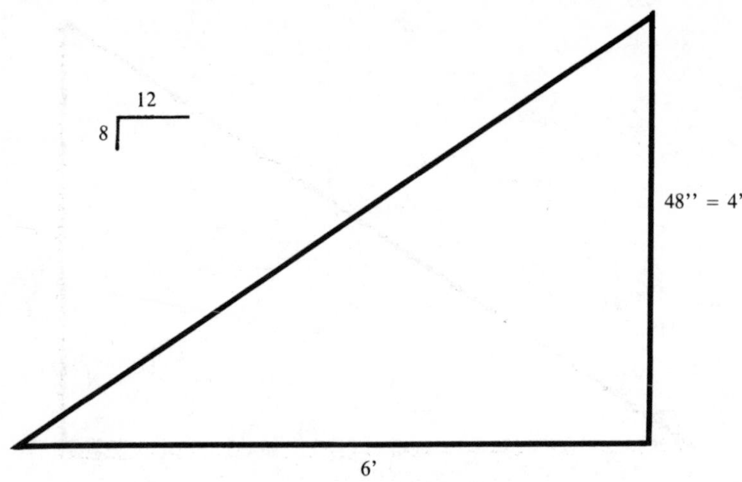

Finding the building line
Figure 2-46

Full Length Calculations
Let's say we're cutting the rafters on our garage job — not the model this time — the actual garage rafters.

The total run is 6'. Therefore the total rise is 6 x 8''. That's 48'', or 4', as in Figure 2-46.

The formula would be:

$$\text{Rafter}^2 = 6^2 + 4^2$$

To solve this problem, punch [6] [x²] [+] [4] [x²] [=] [√]. The window display will show 7.2111.

Because the rafters are considerably longer this time, we changed everything into feet. That makes the answer, including the decimal part, in feet. When we had a decimal part of an inch, we multiplied by 16 to discover how many 16ths there were. Now we must multiply by 12 to discover how many decimal inches 0.2111 is.

Punch [−] [7] [=]. That strips off the whole feet, leaving us with a decimal part of a foot. Now convert that to inches: Punch [×] [1] [2] [=] and 2.5332 appears. That's 2.5332 inches.

We have 7' plus 2.5332 inches. Now strip off the whole inches by punching [−] [2] [=] [×] [1] [6] and 8.5316 appears.

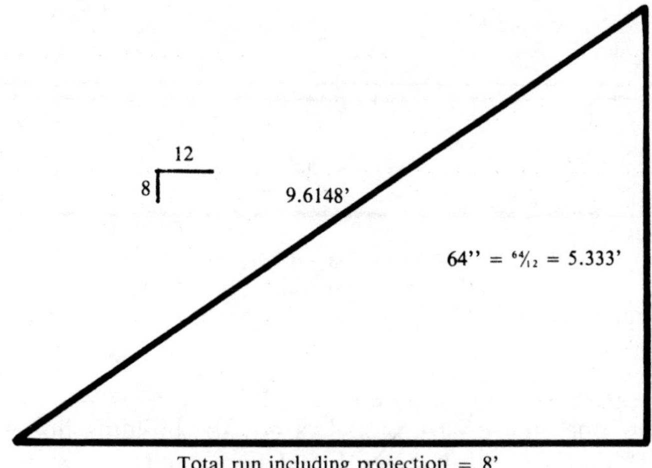

Finding the fascia line length
Figure 2-47

We multiplied 0.5332 by 16, so this is the answer in 16ths. The whole answer in feet, inches and 16ths is 7' 2" and 8.53/16ths. Let's call the rafter length 7'2$\frac{9}{16}$" from the ridge cut to the building line. Here's a summary of how we got that answer:

Rafter length	7.2111	= 7' +
Decimal feet x 12 gives inches	0.2111' x 12	= 7' + 2.5332"
Decimal inches x 16 gives 16ths	0.5332" x 16	= 7' + 2$\frac{9}{16}$"

This would be easier if we were using meters because the metric system is a decimal system. There's no need to convert to inches and 16ths.

Now let's find the fascia cut line: The formula is:

$$\text{Rafter}^2 = 8^2 + 5.333^2$$

Punch [6] [4] [÷] [1] [2] [=] [x^2] [+] [8] [x^2] [=] [$\sqrt{\ }$]. Your answer is 9.6148'. See Figure 2-47.

Convert this to feet, inches and 16ths. Strip off the 9 and find the inches: punch [−] [9] [=] [×] [1] [2] [=] and you have 7.3776". This means that the rafter length is 9' and 7.3776" long.

Strip off the 7 so we can multiply the 0.3776 by 16 to discover how many 16ths there are in 0.3776". Punch [−] [7] [=] [×] [1] [6] [=] and 6.0422 appears. This means 6/16 or 3/8". Therefore, the answer is 9'7$\frac{3}{8}$" from the plumb line to the fascia cut line.

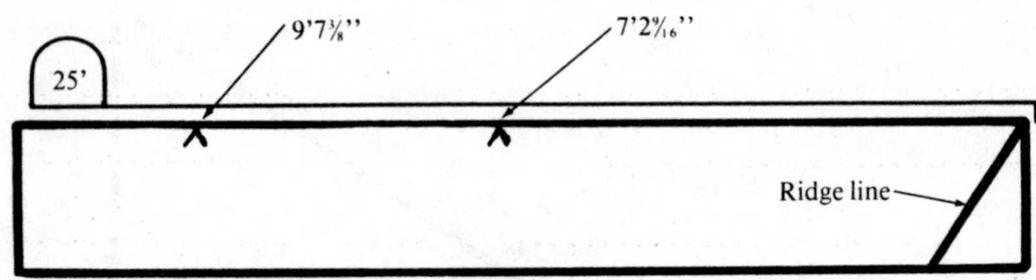

The full-sized rafter
Figure 2-48

Your rafter would look like Figure 2-48.

On our one-quarter-size model rafter, the building line was at 21⅝". If that's multiplied by 4 and converted to feet you get 7'2½". That's very close to the building line measurement of the actual rafter. The 1/16" difference is due to rounding. And so we find this whole thing really works!

Second Ancient and True Method

In the U.S. we usually use roof slope designations from 1 in 12 to 24 in 12. That includes all the half-step increments such as 7½ in 12, 3½ in 12, etc. In Europe there is no such thing as 4 in 12. There, the slope of the roof would be given as 18 degrees 26'. (Read that as eighteen degrees, twenty-six minutes). See Figure 2-49.

This means that the roof makes an 18 degree 26' angle with the horizon. This, then, would be the value of angle A.

When America converts fully to metrics, we'll replace our tape measures and framing squares, of course. But the words 4 in 12 will lose all meaning because 4 in 12 is in inches with 12 inches or 1

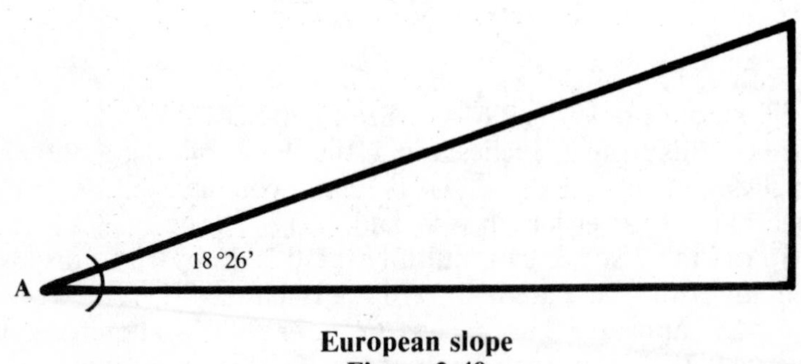

European slope
Figure 2-49

Two different systems
Figure 2-50

foot as the base. In meters, the base will be in tenths and not twelfths. Under the metric system we might refer to a roof as being 3 in 10. That will be the closest to our current 4 in 12 roof. See Figure 2-50.

If we don't change from a 4 in 12 type system to a 3 in 10 system, we could go to a system that refers to slope in degrees of incline. For example, we might have a 16 degree 42 minute roof. This is meaningful, simple, and direct. Slope has been expressed like this in degrees and minutes for over 3,000 years.

The Tangent of a Right Triangle

In a right triangle there are two important relationships for the roof cutter. One is the rise divided by the run (Figure 2-51). On our model roof the rise divided by the run is 8/12.

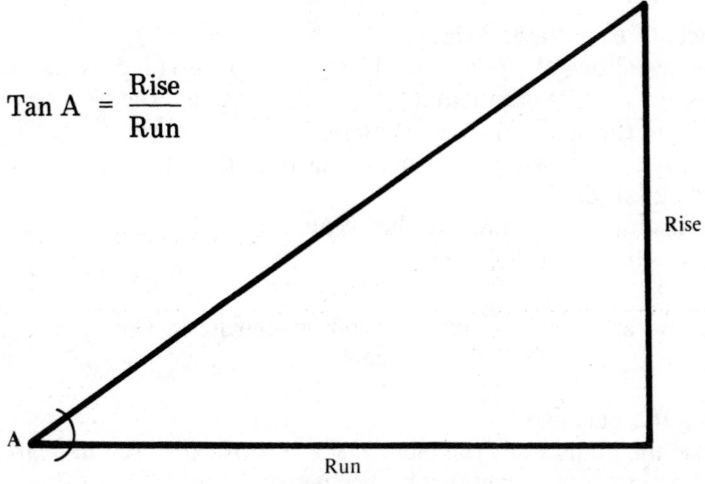

The tangent of angle A
Figure 2-51

If we divide 12 into 8 we get a decimal number. Punch [8] [÷] [1] [2] [=] and 0.6667 appears. This number expresses a relationship between the length of what a mathematician would call the opposite and adjacent (rise and run) sides. It's called the *tangent* and always indicates that the run has been divided into the rise.

If the tangent is known, your calculator will find angle A, the slope of the roof.

To find this angle on your calculator, put 0.6667 in the display. Notice the key [tan]. This is an abbreviation for the word tangent. Before punching [tan], hit the button marked [INV]. Then punch [tan]. The number 33.6900 appears in the display. This means that the angle of the roof is 33.69 degrees.

To go from 33.69 degrees to 0.6667 simply punch [tan]. To go from the tangent to the angle, push [INV] [tan].

The number 33.69 degrees is the decimal form of the angle. Each degree is broken into 60 minutes. To find the minutes, strip off the 33 whole degrees and multiply the 0.6900 by 60. The answer is 41.40 minutes. Therefore the angle is 33 degrees 41'. (Read this as thirty-three degrees forty-one minutes.)

This is indeed the incline of an 8 in 12 roof. The term 8 in 12 is quite meaningless outside the U.S., but 33 degrees 41' is a mathematical reality as solid as gravity.

Notice that the total rise divided by the total run will have the same tangent and ratio and produce the same incline in degrees as the unit rise divided by the unit run.

The Secant of a Right Triangle

The *secant* shows the relationship of the rafter divided by the run. See Figure 2-52. A mathematician would call this the hypotenuse divided by the side adjacent of angle A.

Since the rafter will always be longer than the run, this ratio will always be larger than 1.

Notice that if we multiply the secant times the run, we get the length of the rafter.

$$\text{Secant} = \frac{\text{Rafter}}{\text{Run}} \quad \text{or} \quad \text{Rafter} = \text{Secant} \times \text{Run}$$

Finding the Secant

Suppose the plans show the incline of the roof expressed in degrees at each angle A, and we also know the total run. Can we find the rafter length? We don't know the rise, so the Pythagorean Theorem won't help.

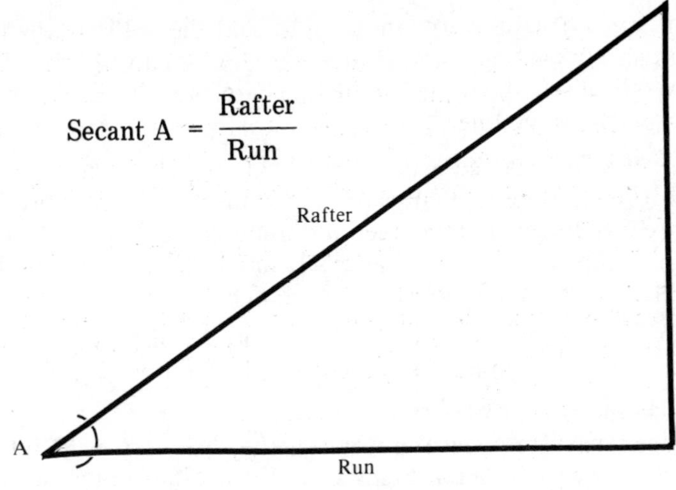

The secant of angle A
Figure 2-52

It's a snap with our calculator. Using the angle at A we can find the secant. See Figure 2-52.

Here's how to do it. Punch [3] [3] [·] [6] [9]. You don't need a degree button.

There's no button that says secant. But you'll notice a [cos] button. The "COS" stands for cosine. A secant is the same as 1 divided by the cosine. There's a [1/x] that divides 1 by any number in the display. We'll use it to change a cosine into a secant.

With 33.69 in the display, punch [cos] and wait a second or two. Then punch [1/x]. Now 1.2018 appears. This is the secant. If we multiply it by the total run, we'll have our rafter length.

Punch [STO]. That stores 1.2018, the secant, in the calculator's memory so we can use it later.

Next, punch [×] [1] [8] which stands for the 18 inches of total run to the building line in our model. Punch [=] and 21.6332 appears. This is the rafter length in inches from the ridge line to the building line.

Punch [−] [2] [1] [=] to strip off the 21 inches so just the 0.6332 is left.

Punch [×] [1] [6] [=] and the display shows 10.1326. This is 10/16ths or 5/8ths of an inch, making our rafter 21⅝".

To find the rafter length from the ridge line to the fascia line, go back to the secant. Punch [RCL] (for recall) and the secant reappears. Punch [×] [2] [4] [=] and 28.8443 appears.

Again, strip off the whole inches and find the 16ths: Punch [−] [2] [8] [=] and 0.8443 appears. Punch [×] [1] [6] [=] and 13.5101 appears. We'll call it 14/16, or 7/8 inch. Therefore, the fascia line is 28⅞'' from the ridge line.

Notice that we started with 33.69 degrees. That was in pure decimal form so we could enter it directly into the calculator. But angles are usually given in degrees and minutes, such as 33 degrees 41'. You would first have to change 41' into a decimal by dividing by 60 since there are 60 minutes in each degree.

Punch [4] [1] [÷] [6] [0] [=] [+] [3] [3] [=] and the display will show 33.6833. Then you would continue as explained above, punching [cos] [1/x] to get the secant.

Suppose you were given the slope of the roof as 8 in 12 rather than as 33.69 degrees. What then? Could you still find the secant?

Of course you could! First, use the unit rise and unit run to find the tangent. Then find the degrees of angle A. From there it's easy to get the secant.

For our 8 in 12 roof, [8] [÷] [1] [2] [=] [INV] [tan] [cos] [1/x] and the secant appears again, 1.2018.

Punch [STO] [×] [1] [8] [=] and 21.6333 appears.

Punch [−] [2] [1] [=] [×] [1] [6] [=] and 10.1329 appears. Here the building line is 21⅝''. Now punch [RCL] [×] [2] [4] [=] and 28.8444 is on the display.

Punch [−] [2] [8] [=] [×] [1] [6] [=] and 13.5105 appears. Therefore, the fascia line is 28⅞''.

Roof Framing Power

We've come through some pretty heavy math in the last few pages. It may be time for you to take a breather before we go on to the next chapter. All these theorems, tangents and secants are a little confusing at times, but they're very powerful tools in the hands of a roof framer.

If you understand tangents and secants, you have an unlimited system that will figure rafter lengths on any regular or *irregular* roof. You're free from the rafter tables that limit the work that most roof framers can do. You're prepared to become a master roof cutter capable of tackling any roof on any job.

But before putting this book down, check your understanding of the points covered in this chapter by doing the problems that follow.

Problems

Solve the following problems: (All answers are in the back of the book.)

1) *How many 16ths are there in 0.383''?*

2) *How many degrees in a 10 in 12 roof?*

3) *Convert 0.3796' into inches and sixteenths.*

4) *What's the secant of an 8 in 12 common rafter?*

Do ***problems 5), 6)*** and ***7)*** using the square root method:

5) *A 7 in 12 roof has a span of 18'. The overhang is 2'. Calculate the common rafter mathematical length of:*

a) *the building line length*
b) *the fascia line length*

6) *A 4 in 12 roof has a span of 15'. The overhang is 18''. Calculate the common rafter mathematical length of* ***a)*** *and* ***b)*** *above.*

7) *A 16' by 23' building has a gable roof with the mathematical height of the ridge at 72''. The overhang is 2'. Calculate the common rafter mathematical length of* **a)** *and* **b)** *above.*

8) *Do* ***problem 5)*** *using the secant method.*

9) *Do* ***problem 6)*** *using the secant method.*

10) *Do* ***problem 7)*** *using the secant method.*

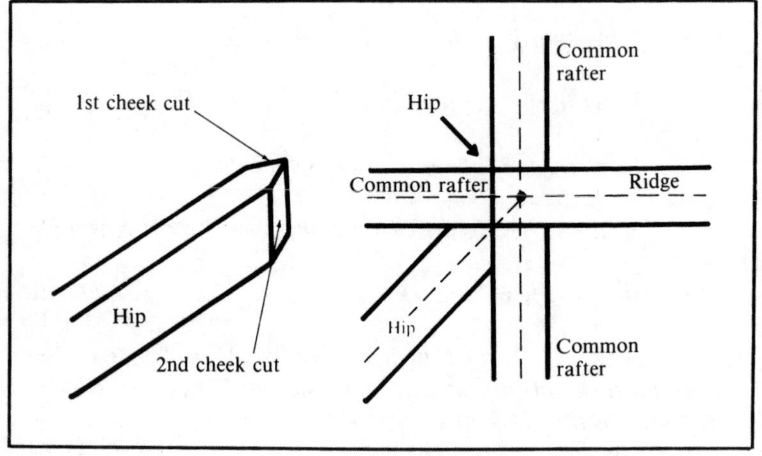

3

Getting the Ridge Length Right

Do you remember from the last chapter how the length of the ridge board is calculated on a gable roof? You'll recall that on a gable roof the ridge is longer than the building length by the amount of the overhang on both ends. Finding the ridge length is a little more complex on a hip roof where we have four sloping sides rather than just two.

Look at Figure 3-1. When a square is cut from corner to corner, the cut is at a 45-degree angle to the sides. Side A and side B are exactly the same length. Keep that in mind as you look at Figure 3-2, the top view of a regular hip roof. You can see that the ridge length has to be calculated very carefully.

A regular hip roof has at least tour sloping sides. The hip rafters go from the corners of the building up to the ridge at a 45-degree angle. This type of hip construction is most common. But there's another type of hip roof with hip rafters at some angle other than 45 degrees. That's called an irregular hip roof and we'll explain it in the second half of this book.

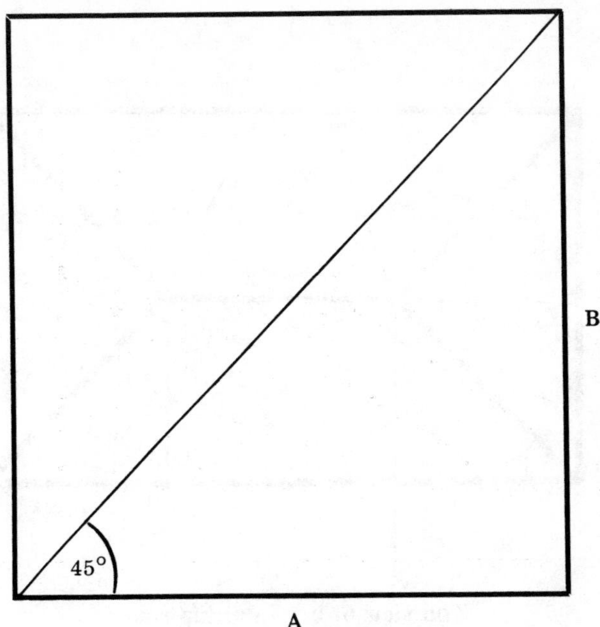

A square cut by a 45° line
Figure 3-1

Figure 3-2 shows that the length of the ridge on a regular hip roof is shorter than the length of the building by the amount, on each side, of half the span. For a hip roof, then:

Mathematical length of ridge = Building length - Building width.

Sometimes you'll hear the mathematical or calculated length referred to as the "apparent length." It just means the length before the width of the lumber is considered.

Test yourself on all this by working the following examples:

a) A 30'-long building is to have a regular hip roof. The span of the building is 10'. What's the *mathematical* length of the ridge?

Apparent ridge = Length - Width
= 30' - 10'
= 20'

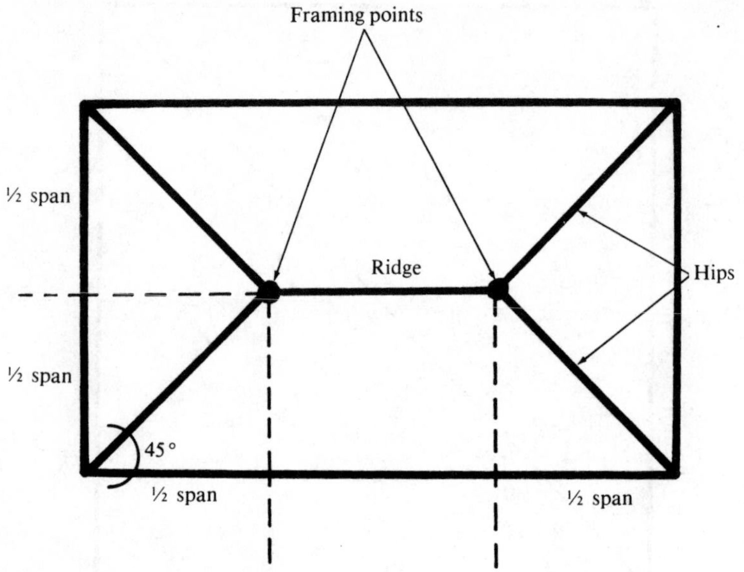

Top view of a regular hip roof
Figure 3-2

b) A 25'-long building has a total run of 8'6". The plans call for a regular hip roof. What's the *apparent* length of the ridge?

Apparent ridge = 25' - 17'
= 8'

The Ridge in a Parallelogram

At times, a ridge will form one side of a parallelogram, as in Figure 3-3. In this case, the mathematical length of the ridge will be the length of the side it is opposite. This is true since any opposite pair of sides in a parallelogram are equal. In Figure 3-3 the ridge length is the same as the side labeled "side of building."

Framing Points Defined

Figure 3-2 indicates two framing points. A framing point is the calculated point where framing members intersect. Notice that the framing point falls at the end of the mathematical length. In Figure 3-2, the distance between the two framing points is the mathematical length of the ridge. But because of the thickness of the framing members, the actual ridge length has to be longer.

In the last chapter we found that the mathematical length of the common rafters shorten by half the thickness of the ridge. Now

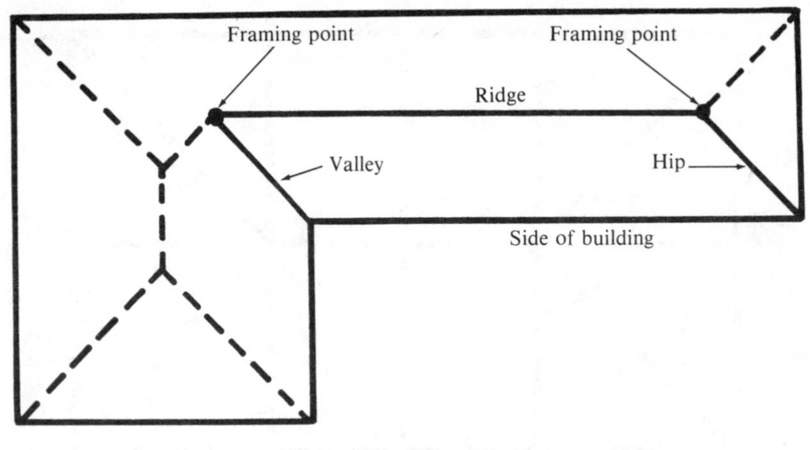

Equal sides
Figure 3-3

we'll see that the mathematical length of ridges at a regular hip end grow according to the type of hip configuration.

There are two types of regular hip configurations: the double cheek cut hip and the single cheek cut hip. Each causes the ridge to grow differently. Let's cover these cheek cuts separately.

Double Cheek Cut Hips

Look at Figure 3-4. This is a top view of a regular hip roof in a double cheek cut configuration. Figure 3-5 is an enlarged view of the portion circled in Figure 3-4. Figure 3-6 shows why the name "double cheek cut hip" applies. The hip rafter joins this intersection with two beveled sides.

A double cheek cut roof is designed so that the common rafters at (a) in Figure 3-4 are centered with the framing points. Figure 3-5 shows the common rafters centered on the dotted line drawn through the framing point.

In a double cheek cut hip, the common rafter at (b) in Figure 3-4 is centered on the ridge and on the framing point.

Notice in Figure 3-5 that the common rafters are shortened (cut short of the framing point) by half the thickness of the ridge. Notice also that the ridge extends beyond the framing point by a distance equal to half the thickness of the common rafter. Since this roof has the same hip configuration on each end, the ridge has to grow (be lengthened beyond the mathematical length) by the same amount on each end.

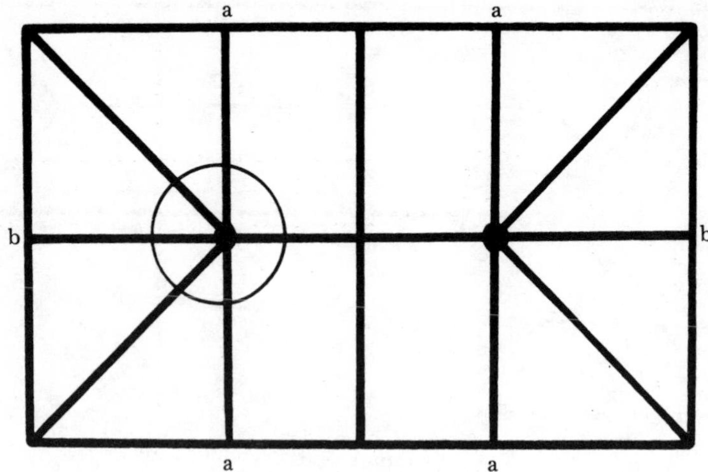

Top view
Figure 3-4

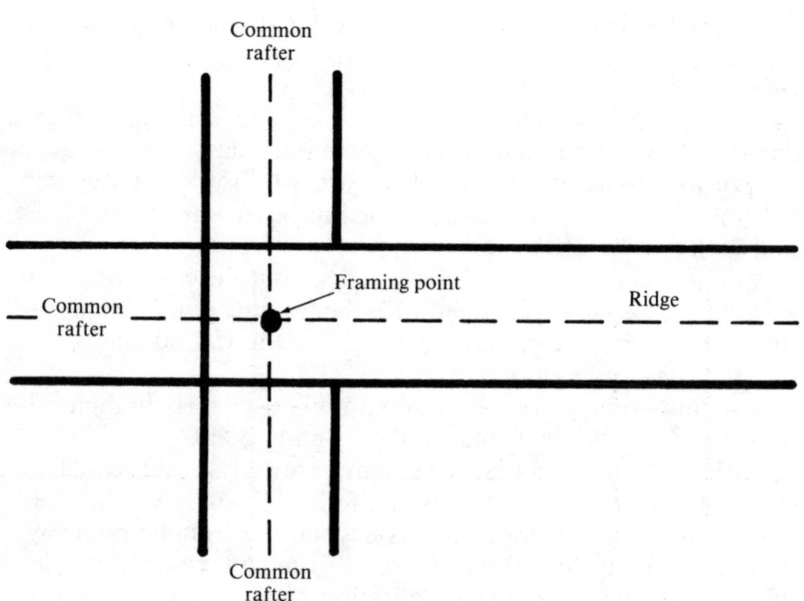

Ridge growth
Figure 3-5

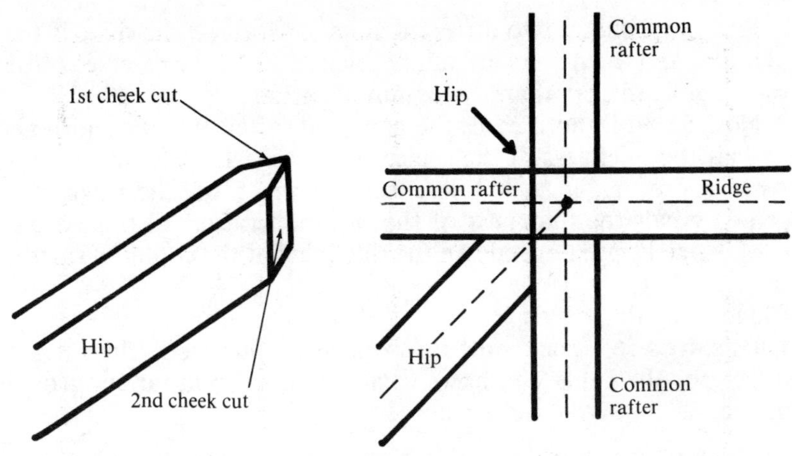

The double cheek cut
Figure 3-6

Here's what you have to remember: For a double cheek cut hip, add to the mathematical length of the ridge one-half the thickness of the common rafter *at each end*. That makes the ridge the right length.

Figure 3-6 shows the double cheek cut on the hip rafter and how that rafter fits into the junction of rafters and ridge.

The Proof

Figure 3-7 should help you see that a double cheek cut hip causes the ridge to grow by half the thickness of the common rafter.

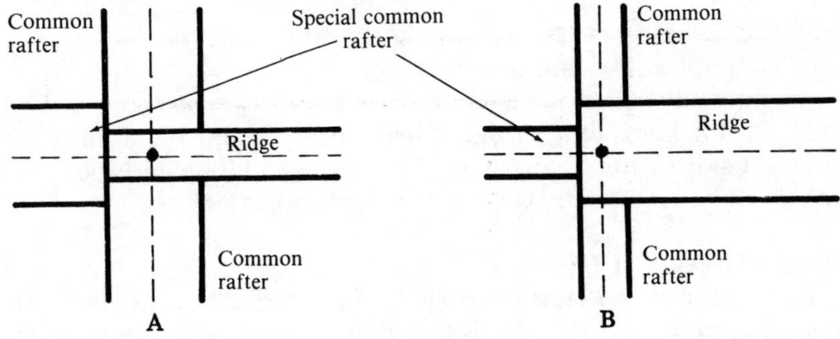

Ridge growth
Figure 3-7

Figure 3-7 shows two different situations. In A, the ridge is thin material and the rafters are much thicker. In B, the reverse is true; the ridge is thicker than the common rafters.

Notice that in both cases the common rafter must be shortened by half the thickness of the ridge, no matter what the thickness is. Also notice that in a double cheek cut hip roof, the ridge must grow by half the thickness of the common rafter. In this type of roof, ridge length depends on the thickness of the common rafters.

Ridge Growth

Let's start a reference table that will help you master the rules of ridge growth. Table 3-8 shows what we know so far about growth of the ridge board.

Type of Hip	Ridge Growth
Double cheek cut	½ thickness of common rafter

Ridge growth
Table 3-8

The Special Common Rafter

When the thickness of common rafters and the ridge are the same, shortening of the end common rafters (marked "special common rafter" in Figure 3-7) and other common rafters will be the same. But when the thickness of the ridge and commons is different, then the end common rafter must be shortened by half the thickness of the common rafter and not the ridge.

Suppose all common rafters in Figure 3-7 had been shortened by half the thickness of the ridge. Then the "special" or end rafter at the left side of the figure would have been too long and the same rafter at the right side would have been too short!

Single Cheek Cut Hips

Figure 3-9 is a top view of a regular hip roof with a single cheek cut. The obvious difference is that only two hip rafters are attached at the end of the ridge board. No common rafters reach these framing points.

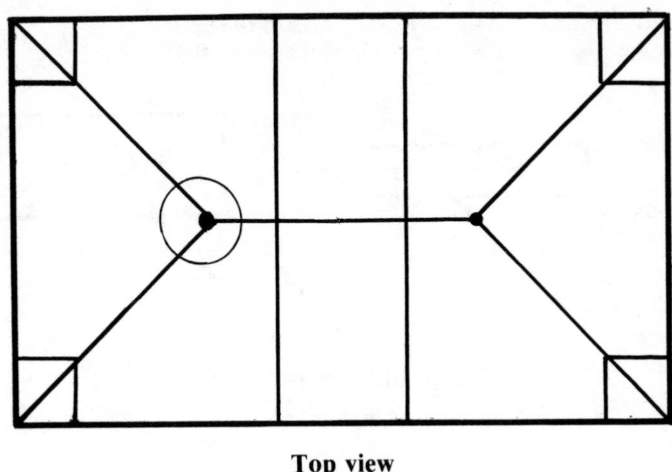

Top view
Figure 3-9

This figure shows the ridge between the framing points, four single side cut hip rafters, four common rafters, and eight hip jack rafters (commonly called jack rafters).

Figure 3-10 shows how a single cheek cut hip rafter joins the ridge. Notice that the ridge extends beyond the framing point. We have to know this distance to calculate the ridge length.

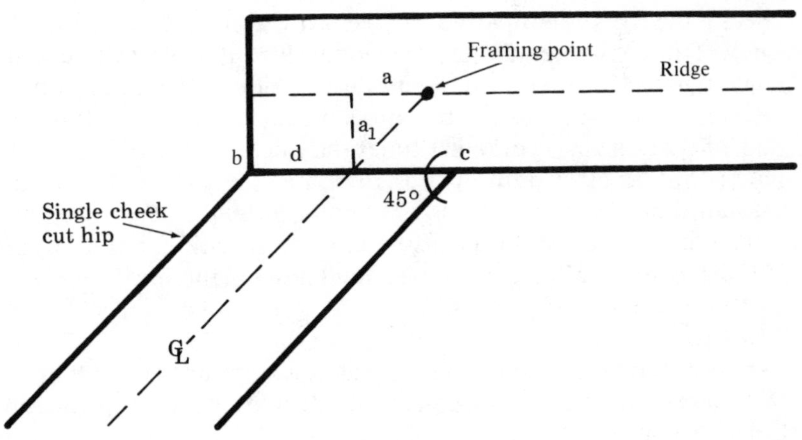

Ridge growth
Figure 3-10

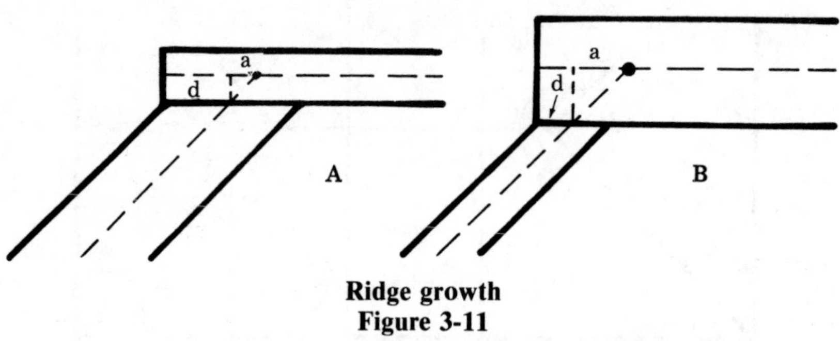

Ridge growth
Figure 3-11

You can see that the distance beyond the framing point is equal to (a) plus (d). All we have to do is find length (a) and then add it to length (d) in Figure 3-10. Here's how to do it.

The center line of the hip is at a 45-degree angle through the framing point. It forms a right triangle with the extended ridge center line (a) and the line (a1) which is half the ridge thickness.

In a 45-degree right triangle, the two sides are equal. Thus (a) must equal (a1). The first part, then, of the ridge growth is equal to half the thickness of the ridge.

Now, if we can find (d), all we have to do is add (a) and (d) together to find ridge growth in a single cheek cut hip roof.

Here's how to find (d). The length (b-c) is the top edge of the cheek cut. This line is at 45 degrees to the sides of the hip. We know from Chapter 2 that the long side of a 45-degree right triangle is 1.414 times one of the shorter sides. So line (b-c) must be equal to 1.414 times the thickness of the hip. Length (d) is one-half of that distance, or 0.707 times the thickness of the hip. If the hip is 1½" thick, length (d) is 0.707 times 1½, or 1.060". That's very close to 1¹⁄₁₆".

Therefore, the second part of the ridge growth is based on the hip width. It's half the 45-degree thickness of the hip.

The Proof

Let's see if any single cheek cut hip causes ridge growth of half the thickness of the ridge plus half the 45-degree thickness of the hip. Look at Figure 3-11.

Figure 3-11 shows two different situations. In A, the ridge is thin and the rafter is thicker. In B, the reverse is true, the ridge is thicker than the common rafter. But in both cases, the ridge for a single cheek cut hip roof grows by half the thickness of the ridge plus half the 45-degree thickness of the hip.

Let's continue building our ridge growth table, Table 3-12.

Type of Hip	Ridge Growth
Double cheek cut	½ thickness of common rafter
Single cheek cut	½ thickness of ridge plus ½ 45° thickness of hip

Ridge growth
Table 3-12

Practice in Framing Points

Now that you know how to figure ridge lengths for single and double cheek cut hips, we can practice laying out more complex roof jobs. And some hip roofs can get pretty complex, as you'll see.

This section isn't necessary for building your model. So you can skip the rest of this chapter if you're not interested in planning a complex roof. But a master roof cutter should be ready to lay out and build just about any roof job that comes along. To do that, you'll need to know what I'm going to explain in the rest of this chapter.

Many times you'll have to design the roof yourself. If the building architect or designer doesn't know how to plan a complex equal pitch roof, it becomes your job.

Planning begins with paper and pencil.

Your plan should show the hips, valleys, ridges and broken hips. (A broken hip is the short connection needed to join a lower ridge and a higher ridge when the span of two roof sections are unequal.) When this is completed, the framing points appear. Finally, add common rafters and various jack rafters to complete the design. When your plan shows all this detail, begin calculating mathematical lengths and then subtract shortenings and setbacks to establish the finished product.

Let's follow this order and start by working up the hips, valleys, ridges and broken hips.

At this point you may want to scan Chapter 4 on hips and Chapter 5 on valleys if you need some background information on these parts of the roof.

Here's how to plan a roof: Let's work in 1/4'' scale. Each quarter of an inch on your paper is equal to one foot. Buy some 8½'' x 11'' graph paper that has four squares to an inch. Lay out the building outlines carefully on your graph paper, using a sharp pencil. See Figure 3-13.

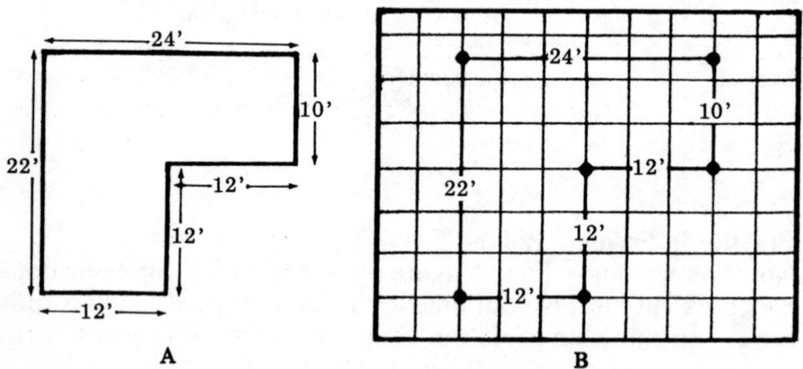

Transferring to graph paper
Figure 3-13

Begin by placing dots on the paper where building corners fall. Write the dimensions as in Figure 3-13. Connect the dots along the lines of the graph paper.

Drawing Equipment
Lines on the graph paper make it easy to draw straight and 45-degree lines. But if you're going to make a living as a roof cutter, invest a few dollars in some inexpensive drafting tools:

- A small drawing board about 11½'' x 13''
- A 6'', 45-degree drafting triangle
- A plastic 12'' T-square

Place the graph paper toward the top of the drafting board and tape the lower two corners only.

General Layout Rules

Here are some general rules that apply whenever you're planning a roof.

1) Begin drawing hips from the span or narrow side. When there are several spans, generally start with the smaller span, then move to the larger spans.

2) All valleys run full length from an inside corner of the rafter plate, and must end at one end of a ridge. There are two exceptions to this rule: *1)* The shortened valley rafter; *2)* A supporting valley rafter that hits a ridge at mid-point. These are explained in Chapter 5.

3) Ridges are generally connected by broken hips that go between two ridges. Occasionally two ridges connect at 90 degrees under Construction Rule 4 below.

Construction Rules Listed

These are the rules that we'll follow when making roof layouts. They apply on all equal pitch regular hip roofs. I'll state the rules and then explain each in detail on the following pages. Don't worry if you don't understand the rules on first reading. But remember that you can't break any rule while applying another.

Rule 1, Centered Ridge Rule— Ridges are always centered between two walls or wall sections.

Rule 2, Double Hip Rule— a) The near valley terminates the ridge; b) The far valley intersects the far side of a second ridge if both ridges are parallel.

Rule 3, Triple Hip Rule— a) The near valley intersects the broken hip. (A new ridge follows from that broken hip, but not as a Y.); b) The far valley intersects the far side of a second ridge if both ridges are parallel.

Rule 4, Ridge-to-ridge Rule— Two ridges will intersect at 90 degrees if: a) the addition span is equal to the main building span and the rise is equal; b) an opposite hip (or broken hip) and valley pass through the intersection.

Rule 5, Broken Hip Rule— When a valley terminates a ridge, this generates a broken hip to the valley side.

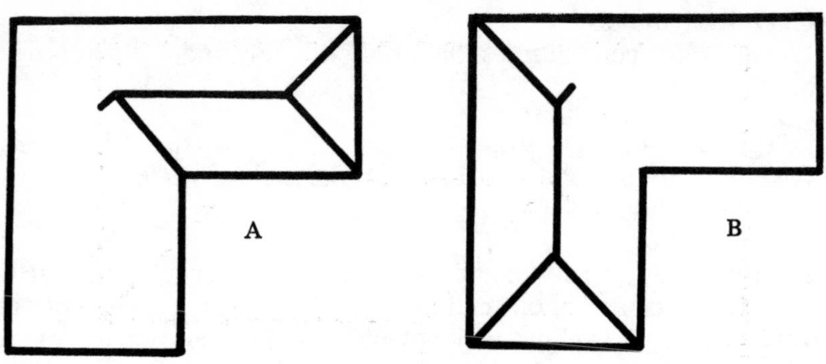

Two and three hip ridges
Figure 3-14

Here's how to figure the ridge length under Construction Rules 2 through 4. Again, don't be concerned if you don't understand these right off the bat. I'll give plenty of examples.

Under Rule 2a, ridge length is equal to the length of the near offset wall.

Under Rule 2b, ridge length is equal to the length of the far valley offset minus the length of the near valley offset.

Under Rule 3a, ridge length is equal to the wall length minus the span.

Under Rule 3b, ridge length is equal to the far offset wall less the near offset wall less the length of the Rule 3a ridge.

Under Rule 4a, and Rule 4b, ridge length is equal to the length of the respective offsets.

Five Construction Rules Illustrated

Different types of ridges are possible on complex regular hip roofs. By far the most common is the ridge shown in Figure 3-14 A. It's called a two hip ridge because it has only two full hips connected to it. Figure 3-14 B shows a three hip ridge, so-called because it has three full hips connected to it.

Rule 2 applies to two hip ridges and Rule 3 applies to three hip ridges.

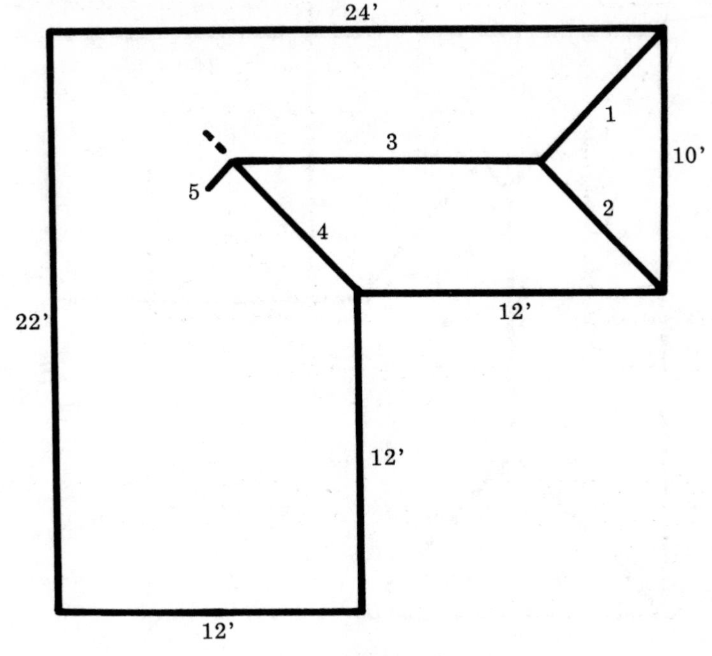

Rule 2a
Figure 3-15

Rule 2a - Double Hip Rule
Look at Figure 3-15. Rule 2a states that if there are only two full hips, line (1) and (2), to a ridge (3), then the nearest valley (4) will terminate that ridge. You have drawn the outline of Figure 3-15 on graph paper. Now draw the four roof members.

Rule 5 - Broken Hip Rule
When a valley terminates a ridge, a broken hip will be generated on the valley side of the ridge. The broken hip (line 5) in Figure 3-15 is correct. The dotted line shows the wrong position. A valley can terminate a ridge without generating a broken hip, but only in the case of Rule 4b.

Rule 3a - Triple Hip Rule
Rule 3a states that for a three hip ridge (lines 6, 7, 8 and 9 in Figure 3-16), the near valley (4) intersects the broken hip (5). If we had started planning our roof from this side of the building, we would next apply Rule 5 in reverse and pencil in the dotted line ridge.

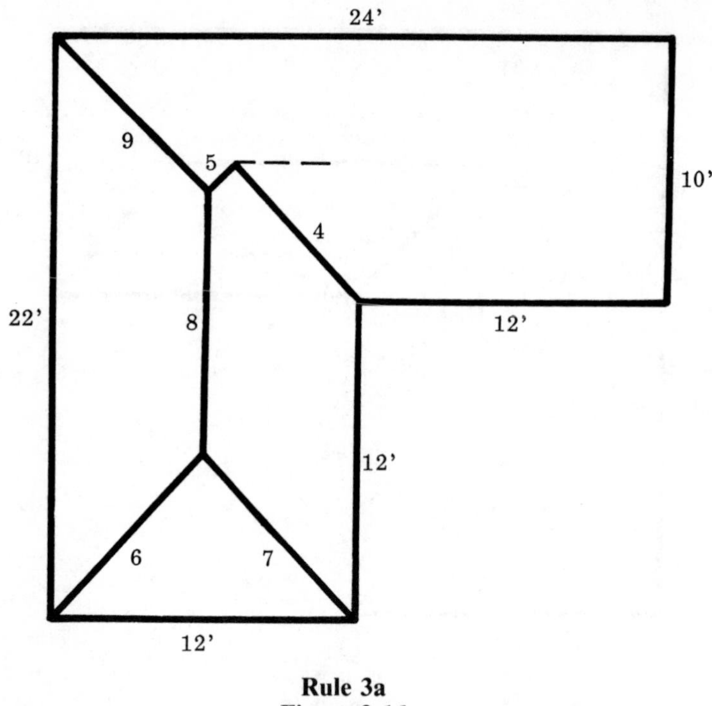

Rule 3a
Figure 3-16

Figure 3-17 shows the rest of the framing points for the roof in Figure 3-16. Next, you would fill in the two sets of common rafters and all the jack rafters. Then you can begin calculating the length of each type rafter.

Rule 4b - Ridge-to-ridge Rule

Now let's change the building outline to see how framing points are affected by building span. Suppose we had two 10' spans instead of a 10' and 12' span.

Draw the outline of Figure 3-18 on graph paper. Begin with Rule 3 by drawing lines (1) through (4). Under Rule 3, a broken hip would be next. But the valley (5) is in a direct line with the hip (4). This signals that we can use Rule 4b. Ridge (6) must be drawn at 90 degrees to the ridge (3) intersect.

Notice that ridge (3) is the same length as the side marked 10' (a) and ridge (6) is the same length as the 12' side. You remember that under Rule 4b, the ridge length is equal to the length of the parallel offset.

Finally, draw the two remaining hips of Figure 3-18.

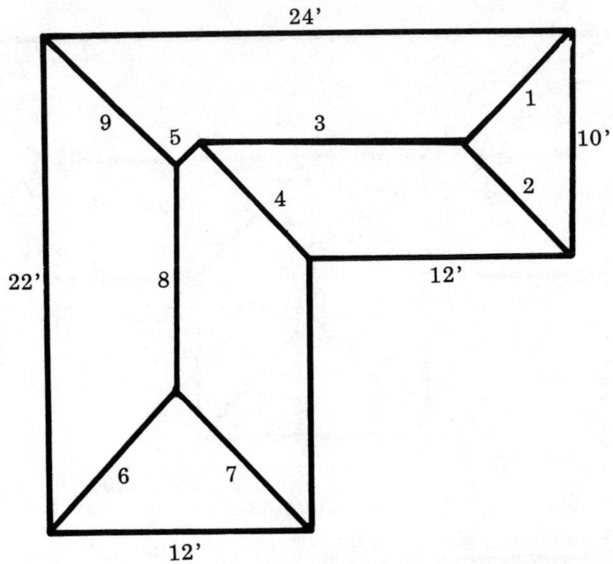

The complete regular hip roof
Figure 3-17

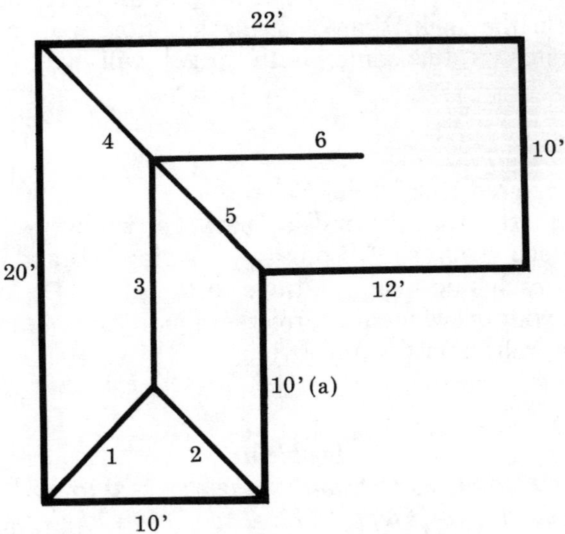

Rule 4b
Figure 3-18

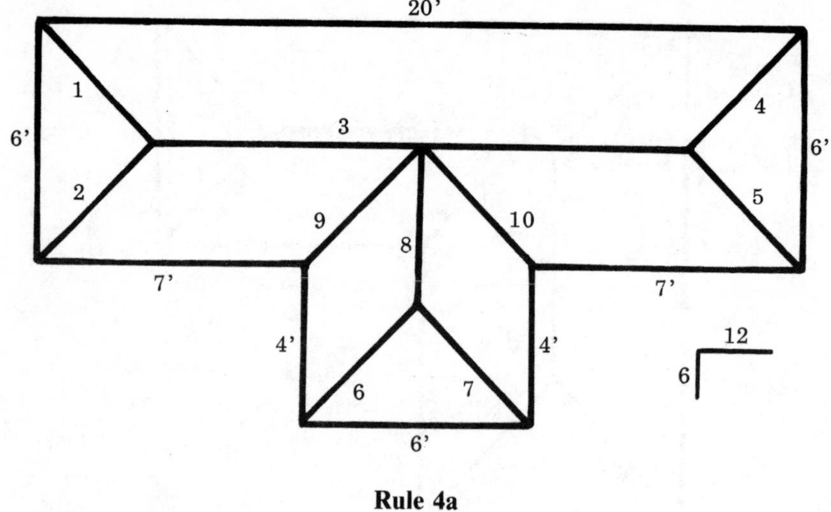

Rule 4a
Figure 3-19

Rule 4a - Ridge-to-ridge Rule
Draw the outline of Figure 3-19 on graph paper. Under Rule 2, draw lines (1) through (5). Also under Rule 2, draw (6), (7) and (8). The valleys (9) and (10) meet ridge (8) and all three touch ridge (3). Because the main span and the addition span are equal, and the pitch is the same, both ridges will have the same height.

Rule 1 - Centered Ridge Rule
The ridge must always be midway between two opposite outside walls in a regular hip roof. Look at your drawings and count the squares on each side from the ridge to the plate. The ridges are centered if your drawings are correct. On more complicated plans, the outside walls won't be so obvious.

Problems

1) Transfer Plans 1, 2, 3, 4, and 5 (Figures 3-20 to 3-24) to graph paper as you did for Figure 3-13. Then, using the Construction Rules, plan the roof for each. If you have trouble, refer to the solutions at the back of the book.

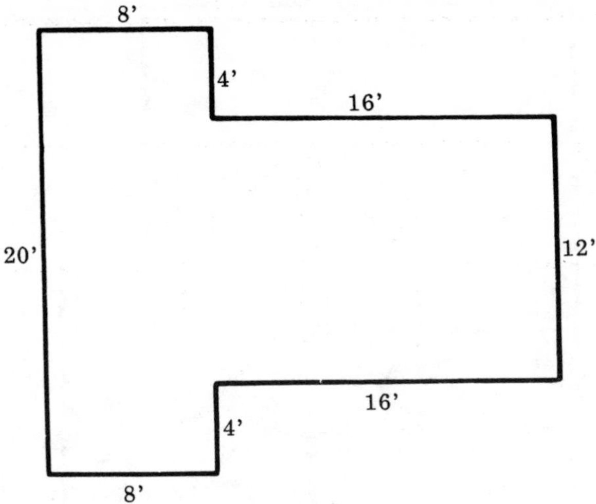

Plan 1 presented
Figure 3-20

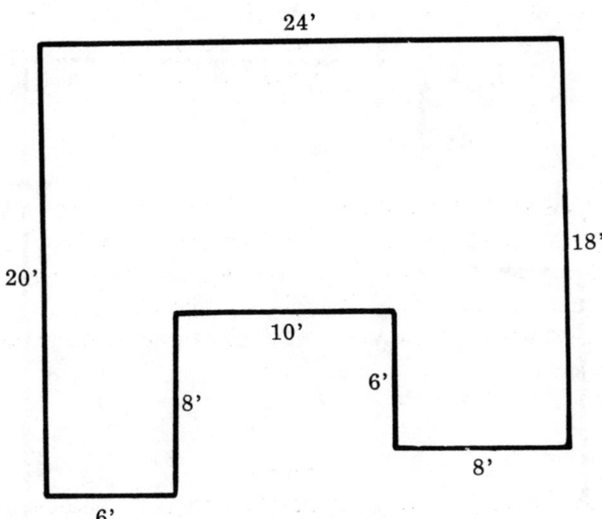

Plan 2 presented
Figure 3-21

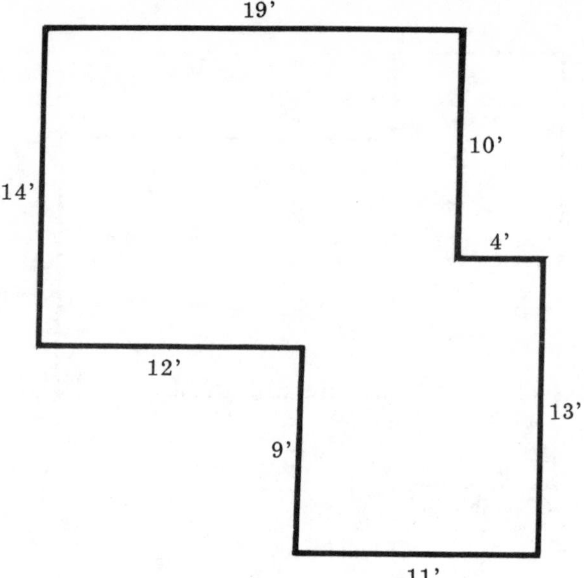

Plan 3 presented
Figure 3-22

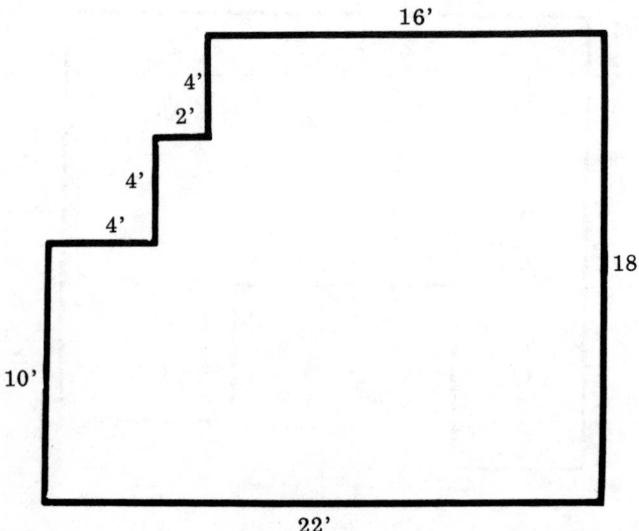

Plan 4 presented
Figure 3-23

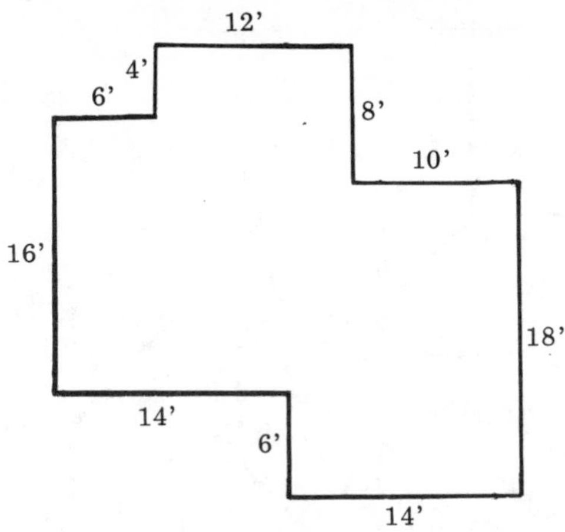

Plan 5 presented
Figure 3-24

Rule 2b - Far Valleys
Transfer the outline of Figure 3-25 to graph paper. This is a complex roof worthy of a master roof cutter. It's a good test of roof planning and cutting skill. We'll use this drawing to apply Rules 2b and 3b. In Figure 3-26 we've started to plan this roof job.

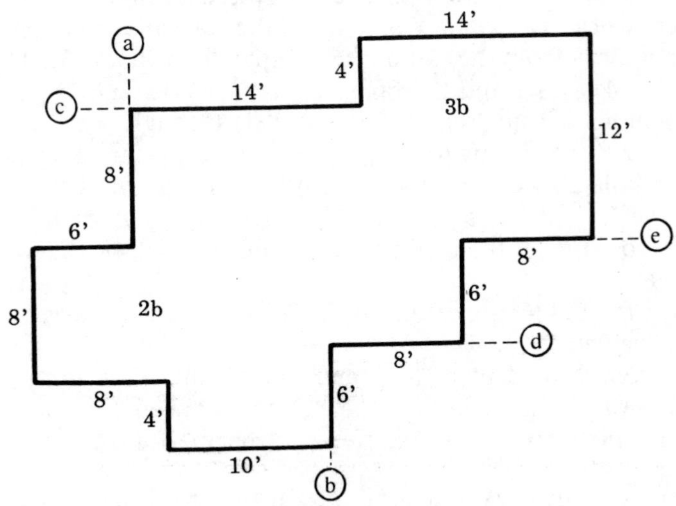

Rule 2b and 3b
Figure 3-25

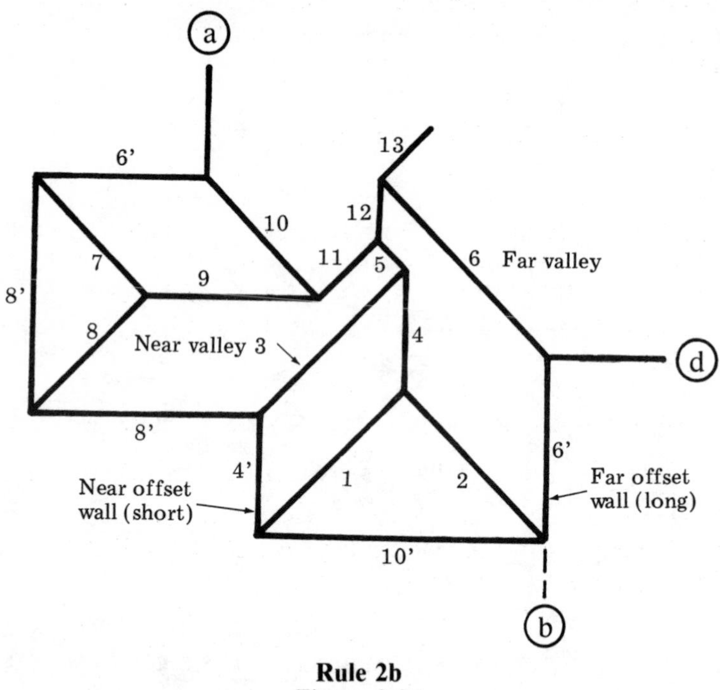

Rule 2b
Figure 3-26

Valley (3) in Figure 3-26 is called the near valley because it's closest to the 10' span. Valley (6) is the far valley because it's 2' further away from the end of the 10' span than valley (3). The near valley is 4' away, and the far valley is 6' away.

Begin with Rule 2a and draw lines (1) through (4). Broken hip (5) is drawn according to Rule 5. Since there is a far valley present, under Rule 2b we'll draw valley (6) in anticipation of its intersecting the far side of a second ridge. The near side of the second ridge would meet the broken hip (5) and the two ridges would be parallel.

Now (5) and (6) are extended, but we don't know the location of the ridge yet.

On complicated plans like this, work from different outer sections toward the center of the building. When (5) and (6) are drawn, move to another section and begin working toward the center again.

Following Rule 2a, draw (7), (8), (9) and (10). Line (11) is drawn by Rule 5. It locates the framing point for ridge (12). Extend valley (6) until it meets ridge (12). Rule 2b is complete if ridge (12) is centered. Ridge (12) is centered between wall sections (a) and (b). Add line (13) under Rule 5 and move to another section of the building.

Rule 2 - The Two Mathematical Ridge Lengths

You'll remember that under Rule 2a, ridge length is equal to the length of the near valley offset wall. In Figure 3-26, the length of ridge (4) is 4', the same as the near valley offset wall. Here's why. Sides (1), (3), (4), and the short (near) offset wall form a parallelogram. In a parallelogram, the length of opposite sides has to be equal. That's why ridge (4) is equal to the length of the near offset wall.

In this case, ridge (4) has a mathematical length of 4'. Of course, the final actual length depends on how you decide to frame at the framing points. I'll cover this later in the chapter.

For review: Under Rule 2a, ridge length is equal to the length of the near offset wall.

Now, how long is ridge (12) in Figure 3-26? It's a Rule 2b ridge.

Under Rule 2b, the ridge is terminated by the far valley and has a length equal to the difference of the far offset and near offset. This is true only if the two ridges are parallel. The far offset of 6' minus the near offset of 4' equals 2'. In Figure 3-26 ridge (12) is 2' long.

Rule 3b - Far Valleys

Under Rule 3a, draw lines (14) through (19) as in Figure 3-27. Line (20) is drawn under Rule 5. Rule 3b is very simple. The far valley (21) must intersect the far side of ridge (20) if (20) is parallel with ridge (16). Line (22) is added under Rule 5. Notice that ridge (20) is midway between (c) and (e).

Completing the Figure

Broken hips (22) and (13) are pointing toward the center of the building, but we don't know yet where they end.

Finishing the roof is simple. Figure 3-41, near the end of this chapter, shows the last two hips (23) and (24). Both of these intersect their respective broken hips at mid-span. Ridge (25) is midway between (c) and (b). That does it! *Congratulations.* You've completed a complicated regular hip roof.

From Rule 3 - The Two Mathematical Ridge Lengths

The ridge length for a Rule 3a section is the length of that section minus the width of that section.

The length of the ridge (16) in Figure 3-27 is 14' minus 12', or 2'.

Ridge (20) is a Rule 3b ridge, the far valley ridge. Figuring its length is a little more complex.

Look at Figure 3-27. Subtract the near offset wall (8') from the far offset wall (14') and then subtract the length of the first ridge. This answer is the length of Rule 3b ridge (20) when (16) and (20) are parallel.

The length of a Rule 3b ridge equals the length of the far offset wall minus the length of the near offset wall minus the length of the 3a ridge.

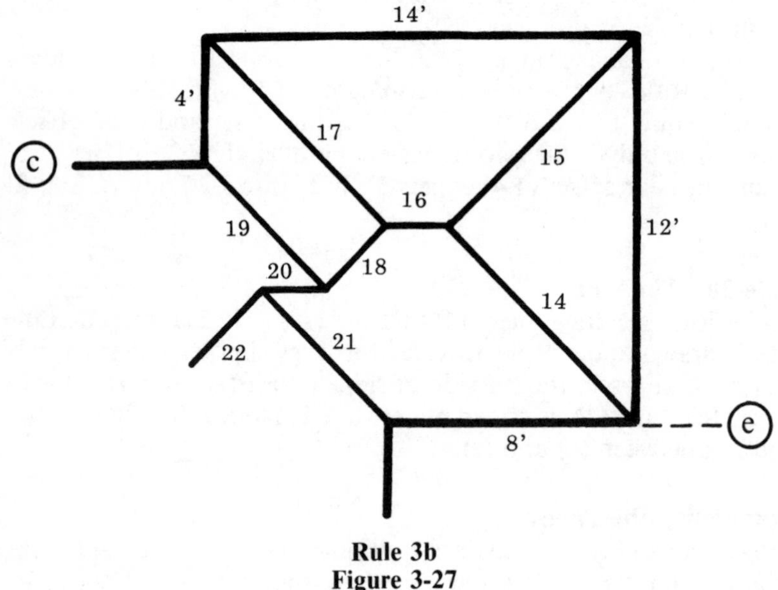

Rule 3b
Figure 3-27

Problems

2) Transfer Figures 3-28 to 3-33 (Plans 6 to 11) onto graph paper as you did for Figure 3-13. Then draw in the roof of each. The solutions are at the back of the book with the other answers.

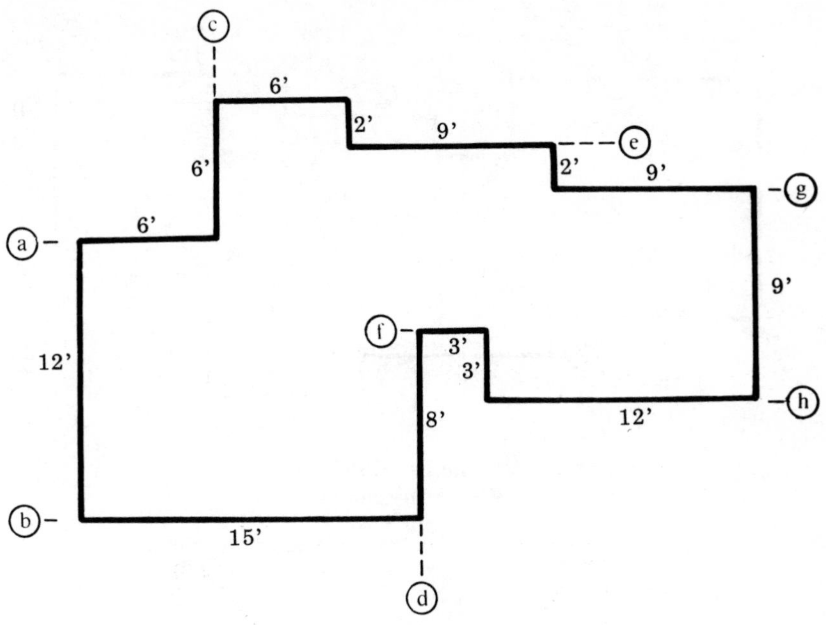

Plan 6 presented
Figure 3-28

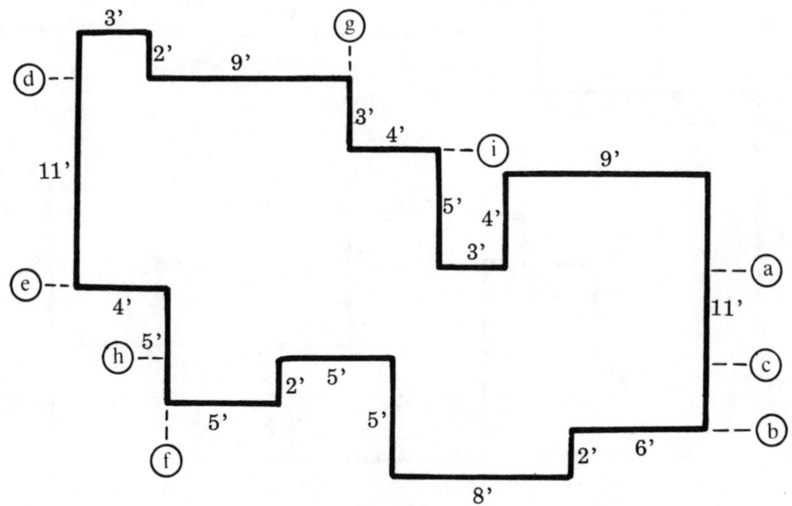

Plan 7 presented
Figure 3-29

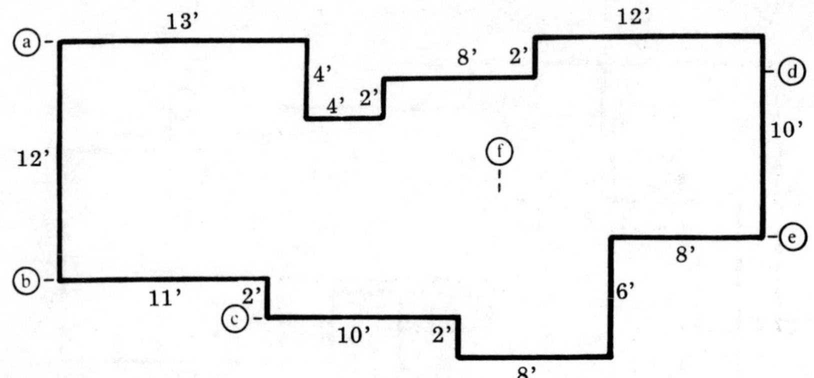

"The Invisible"
Plan 8 presented
Figure 3-30

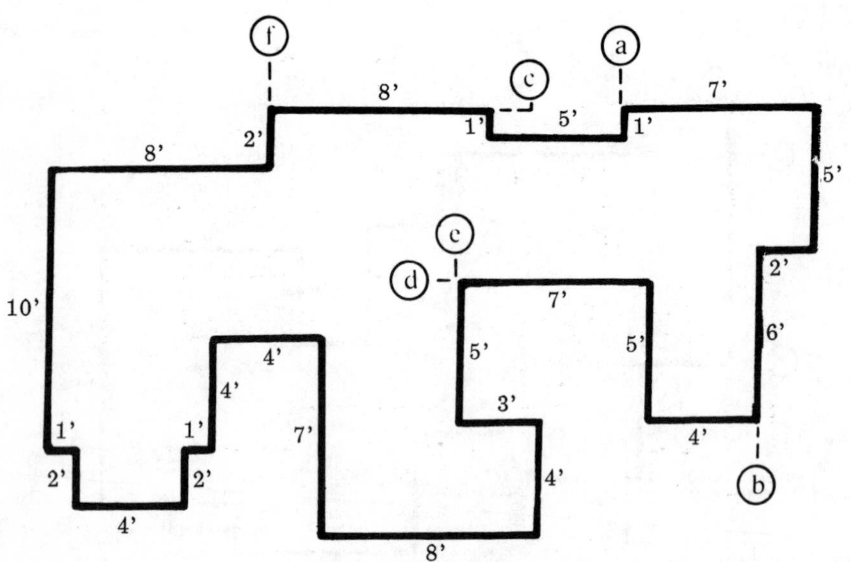

"The Bushwhacker"
Plan 9 presented
Figure 3-31

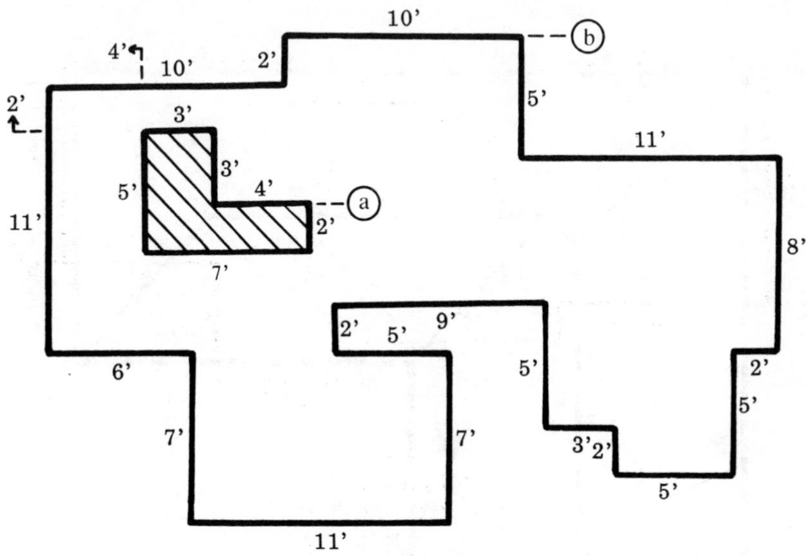

"The Long Nighter"
Plan 10 presented
Figure 3-32

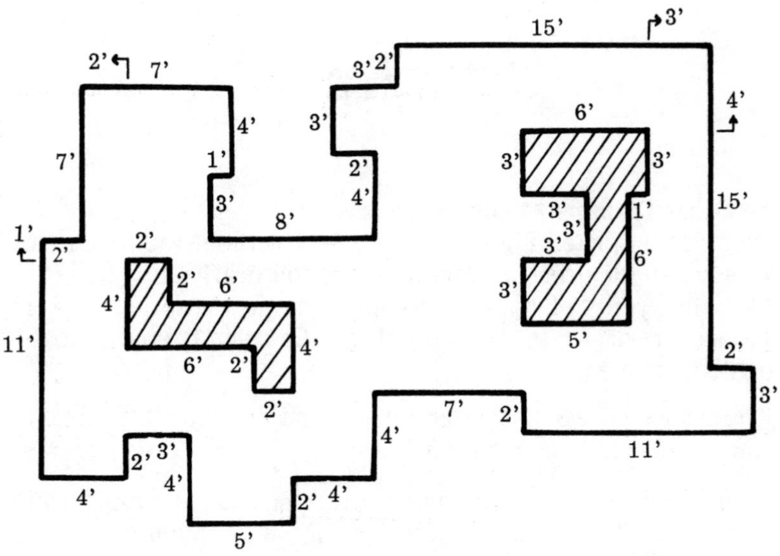

"The Stumbler"
Plan 11 presented
Figure 3-33

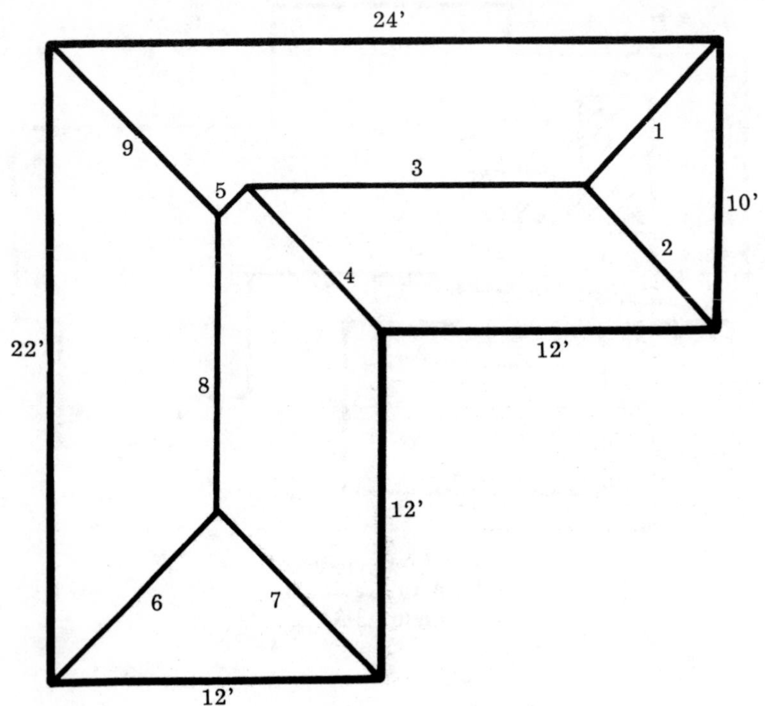

Plan from Figure 3-11
Figure 3-34

The Other Framing Members

If you've completed Figures 3-28 to 3-33 without too much trouble, you can identify the framing points on nearly any regular hip roof. But your plan isn't complete until the other framing members are drawn in. Let's use the outline plan from Figure 3-13 for this purpose.

Transfer Figure 3-13 to graph paper and draw in all the framing point members, as in Figure 3-34.

Start by drawing in the common rafters from the hip framing point of the 12' span. See Figure 3-35. Draw common rafters at 2' intervals. That would be at every other blue line on the graph paper.

Figure 3-36 shows rafters (a), (b), (c) and (d) drawn properly.

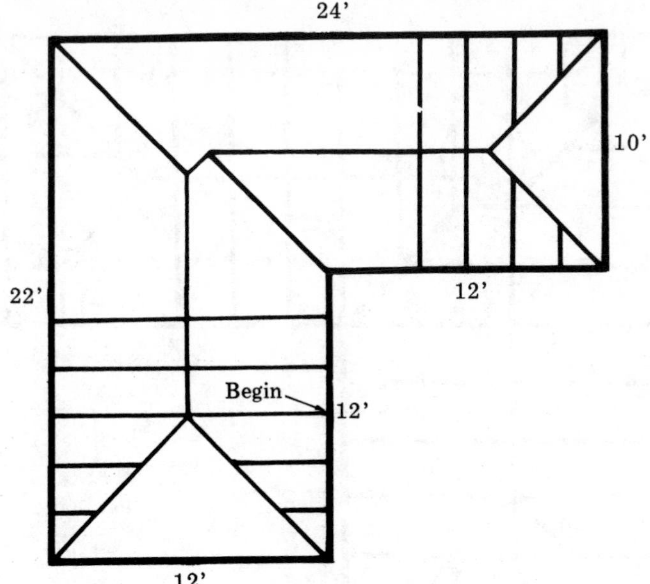

The plan continued
Figure 3-35

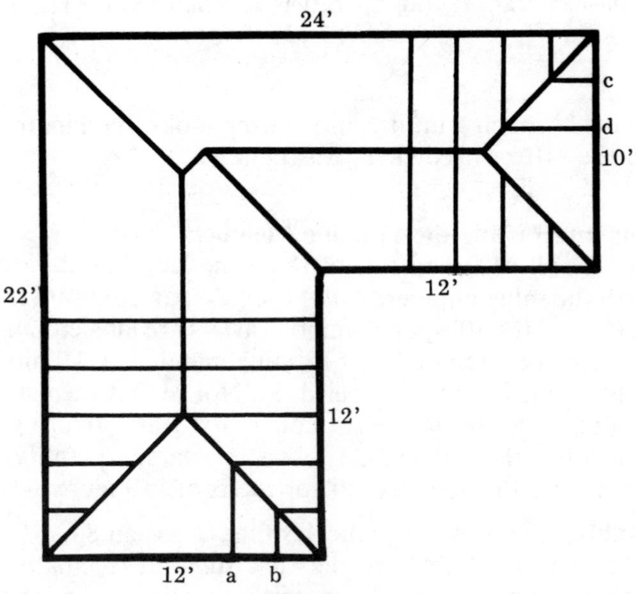

Adding more rafters
Figure 3-36

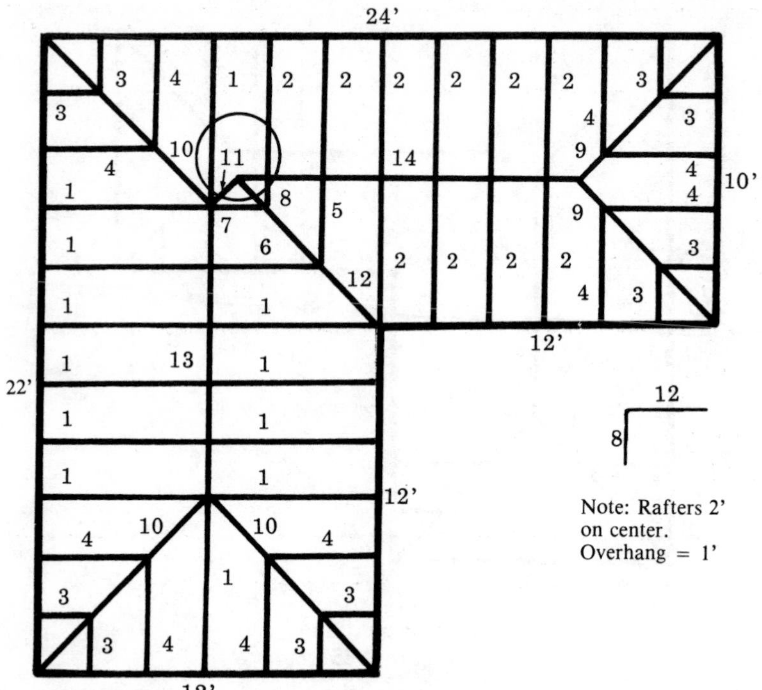

The finished drawing with the rafters re-numbered for listing
Figure 3-37

Continue the 2' spacing until your drawing looks like Figure 3-37. That's all the rafters needed for this plan.

Numbering and Listing the Framing Members

In Figure 3-37 all of the rafters of the same length and kind are labeled with the same number. All of the 12' span common rafters are called (1). All the 10' span common rafters are labeled (2). (See Figure 1-16 for the names of roof framing members.) All hip jack rafters of the same length are labeled (3). Notice that there are ten rafters called (3). Now we're going to start figuring the mathematical lengths. Table 3-38 shows the name, quantity, and apparent run of all rafters needed for the roof in Figure 3-37.

The symbol in Figure 3-37 indicates that this is an 8 in 12 roof. The overhang is 1'. These are the first steps in calculating the mathematical rafter lengths. The second is to keep in mind that the runs in Table 3-38 are *apparent* runs.

No.	Kind	Quantity	Apparent run	Math length @ building	Math length @ overhang
1	Common	12	6'-0"		
2	Common	10	5'-0"		
3	Hip jack	10	2'-0"		
4	Hip jack	10	4'-0"		
5	Valley jack	1	3'-0"		
6	Valley jack	1	4'-0"		
7	Valley jack	1	2'-0"		
8	Valley jack	1	1'-0"		
9	Hip	2	5'-0"		
10	Hip	3	6'-0"		
11	Broken hip	1	1'-0"		
12	Valley	1	5'-0"		
13	Ridge	1	10'-0"		
14	Ridge	1	12'-0"		

Listing the framing members
Table 3-38

With the framing square method, use the apparent run to calculate all the members. If you use the secant method, the apparent run of the 45-degree members must be changed to *actual* run.

For the secant method, the apparent run of (9), (10), (11) and (12) in Figure 3-37 must be multiplied by the square root of 2, which is 1.414. The result is the *actual* run, which is discussed in Chapter 4. Or you can use the secant times $\sqrt{2}$ column from Figure 4-65. See also Figure 4-29. For the other framing members, the apparent run is equal to the actual run. The calculations are given in the back of the book. The run for jack rafters is in Chapter 6.

Choosing a Method of Framing

The roof in Figure 3-37 has two ridges. That gives you four ridge ends to frame. Ridge (13) will be framed on both ends with double cheek cut hips. The ridge (14) hip end will be framed with single side cut hips. On the last ridge end, made up of (11), (12) and (14), you must decide on the framing method.

Figure 3-39 is an enlarged view of the circled portion of Figure 3-37. It shows two ways to connect the framing members. Each drawing has its own shortening and setback measurements. By the end of Chapter 4 you'll know how to make these calculations.

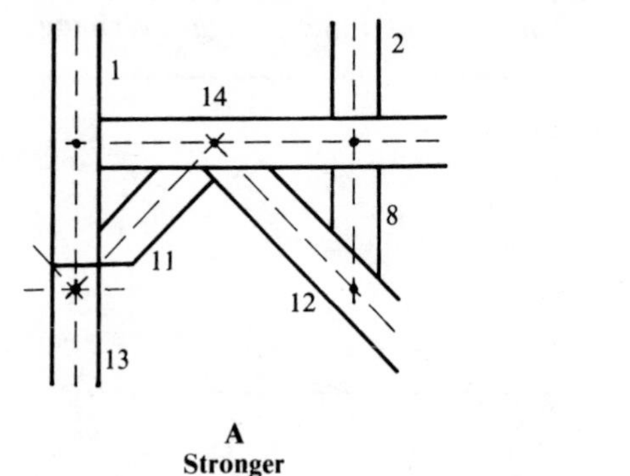

A
Stronger

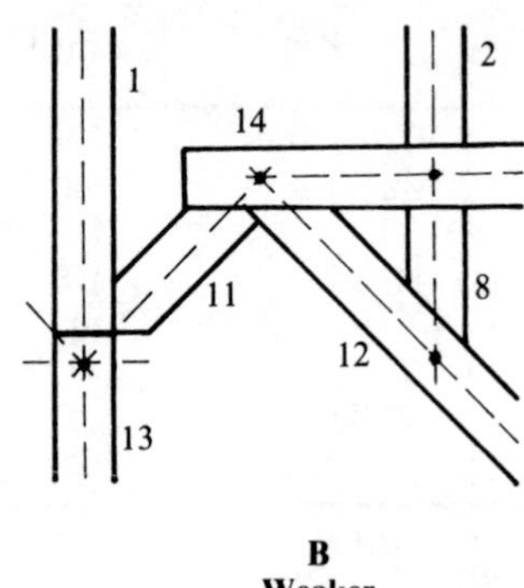

B
Weaker

Choices in framing
Figure 3-39

Drawing Elevations

Up to this point all of our planning has been done with plan views — looking down on the roof from above. Sometimes it will be useful to have an elevation view of the roof. It's fairly easy to convert a plan view to an elevation.

Once again, transfer Figure 3-13 onto graph paper. But this time draw it only half the former size. Let each 1/4'' square be equal to 2'. Make the height of the rafter plate 8'.

Look at Figure 3-40. Project dotted lines outward at each framing point and wall edge for each side of the building. Be sure these lines are straight. To find the correct roof heights, multiply the unit rise by the total run.

When complete, you have four elevation views of the house and roof that's to be framed.

Second Elevation

Lay out Figure 3-25 on graph paper, making one 1/4'' square equal 2'. Center the drawing on your graph paper. Draw in the roof members. Then extend dotted lines outward from all the framing points and walls until your drawing looks like Figure 3-41.

Notice that new numbers have been given to the framing members. These new numbers indicate the sequence for constructing these elevation views.

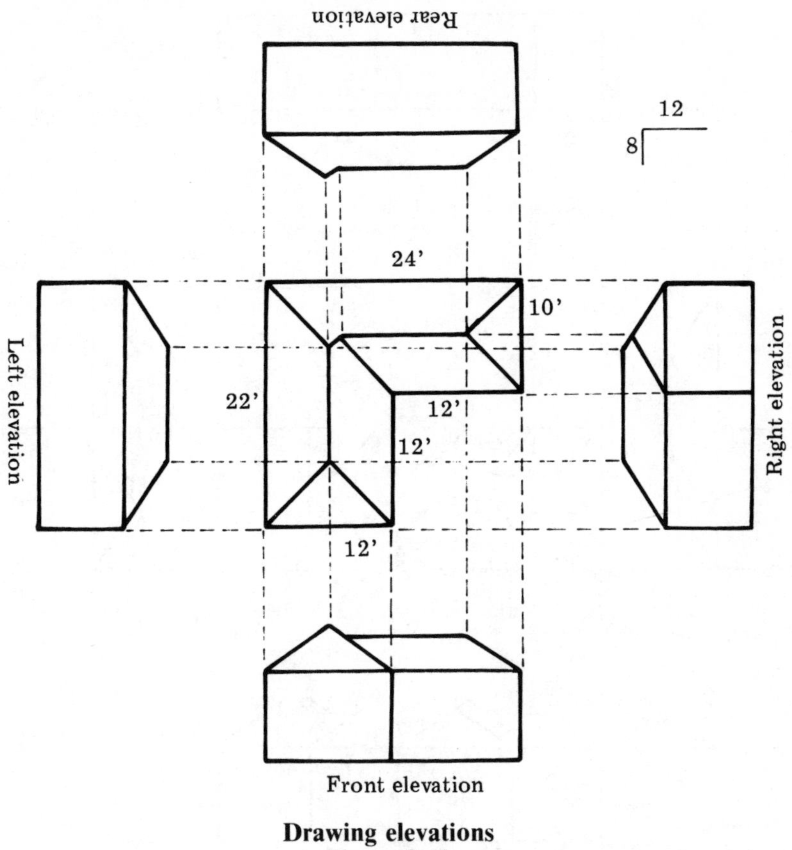

Drawing elevations
Figure 3-40

Follow the dotted lines (c), (g), and (k) and draw in the building edges (1), (2) and (3). Locate the 32'' ridge height on the drawing between (b) and (d). The ridge must be 32'' high because this is an 8 in 12 roof with an 8' span and a 4' total run. Four times eight inches equals 32''. Draw ridge (4) between (b) and (d).

Line (5) connects the framing points between (a) and (b). Locate the 40'' height on line (f). Connect lines (6) and (7). The extension of (7) above the ridge represents the broken hip that goes between the 40''-high ridge and the 48''-high ridge.

Line (8a) is the broken hip that goes between the 32''-high ridge and the 48''-high ridge. It's drawn between (d) and (e). Line (8b) is the cripple hip that goes between the 48''-high ridge and the 56''-high ridge. It's drawn between (e) and (f).

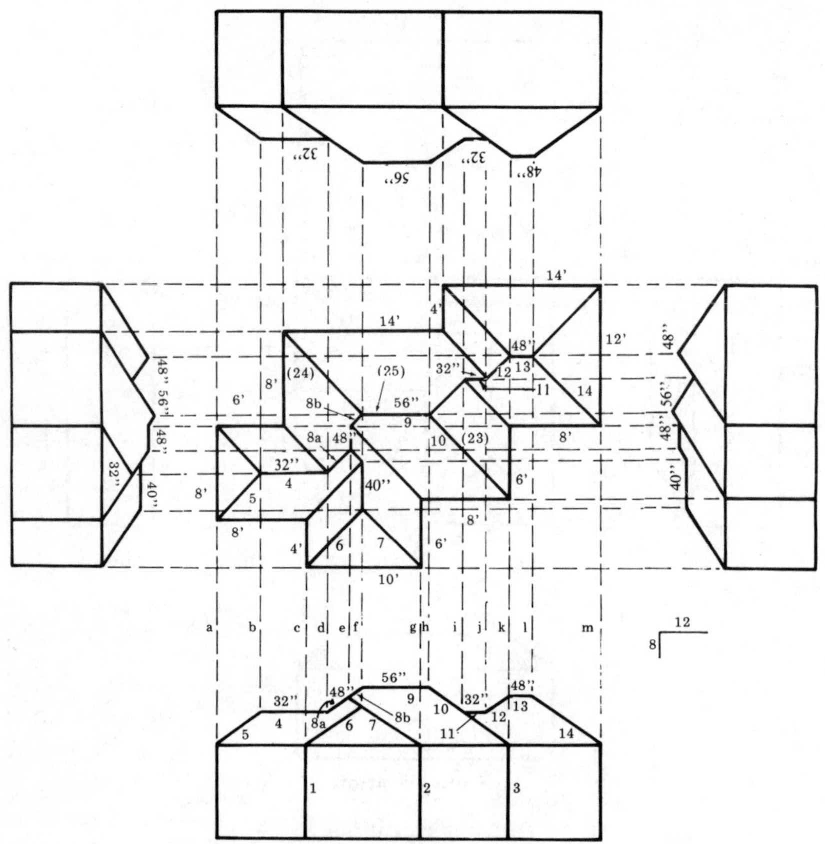

The building from Figure 3-25
Figure 3-41

The height of ridge (9) is 56''. Draw it in between (f) and (h). Draw line (10) between (h) and (k). Find the height of ridge (11) and draw it between its framing points (i) and (j). Find the height of ridge (13) and draw it between (k) and (l). Connect line (12) between these two framing points.

Complete the front elevation by putting in line (14). Continue this process for the other three views. You may want to get bigger drafting paper and lay this out in a larger scale with an architect's ruler.

Actual Ridge Height

In two cases you'll want to know the actual ridge height. The first is when you are checking the ridge height for correct placement.

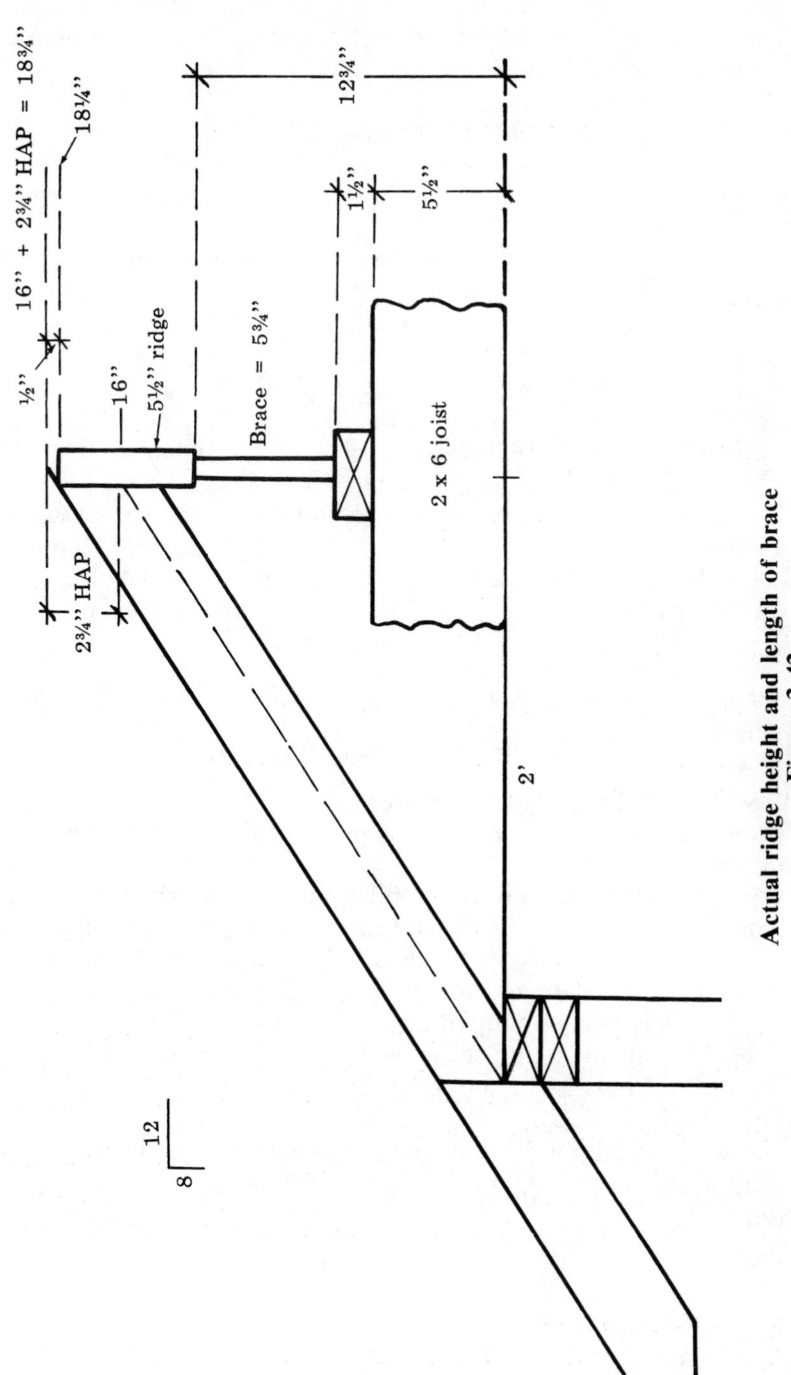

Actual ridge height and length of brace
Figure 3-42

The second is when you're cutting a brace during the framing of the ridge. Here's how to find the actual ridge height.

Case 1 - The Actual Height of the Ridge

Let's find the actual ridge height on an 8 in 12 model roof with a 2' total run. The mathematical height of a ridge is the total run times the unit rise.

Look at Figure 3-42. The mathematical height of the triangle in Figure 3-42 will be 16''. On the actual model, ridge height depends on two factors:

1) HAP (Height Above Plate): Since the HAP is a measurement along the vertical building line, the HAP must be added to ridge height. This model HAP is 2¾''. 16'' plus 2¾'' equals 18¾''.

2) Loss At Top: Because the triangle does not continue on top of the ridge, there is a slight loss in the measured height at this point. For an 8 in 12 roof with a 1½''-thick ridge, the loss will be 1/2''. (If you're curious about this calculation, look ahead to Figure 7-23 in Chapter 7.) 18¾'' minus 1/2'' equals 18¼''.

This is the actual height of this ridge from the top of the rafter plate.

Case 2 - The Length of the Brace

There are two things to consider here: The ridge itself and the ceiling joist.

Say the ridge is a 2 x 6. After milling, the measured width of the board is 5½''. This must be subtracted from the actual height of the ridge. 18¼'' minus 5½'' equals 12¾''. This is the distance between the rafter plate and the bottom of the ridge board.

If a 2 x 4 is to lay on top of 2 x 6 ceiling joists, then the width of these two members must be subtracted. This would remove another 5½'' plus 1½'' from the remaining space. 12¾'' minus 5½'' minus 1½'' equals 5¾''.

Cut a brace about 5'' long and cut two wedges so the ridge can be adjusted exactly. Add the brace as close as possible to a partition wall.

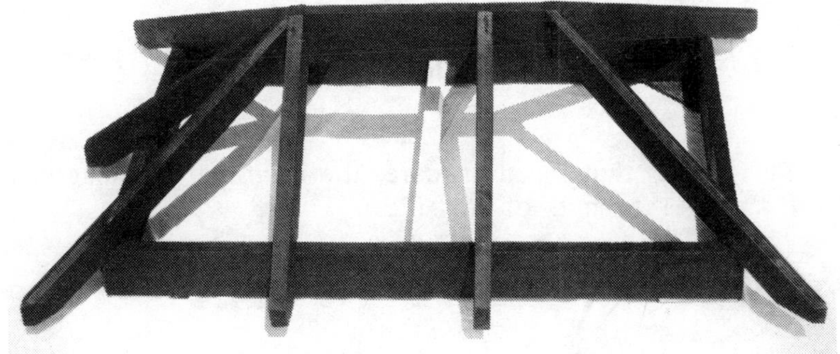

4

Cutting Hip Rafters

Common rafters run parallel to the span. Hip and valley rafters are different. They run at 45 degrees to the span. That means that hip and valley rafters have to be longer than common rafters on the same span.

Unit Run of the Hip

Fortunately, there's a fixed mathematical relationship between the unit run of common rafters and the unit run of hip and valley rafters. For common rafters the unit run has been established as 12''. For hip and valley rafters the unit run is 16.97''.

Remember near the end of Chapter 2 when we discussed the Pythagorean Theorem? Figure 4-1 shows a special case of the theorem when both the run and the rise are equal and therefore the angle is 45 degrees. If the number 1 was used for both the run and rise, the theorem would look like this:

$$c^2 = a^2 + b^2 \quad \text{(the theorem)}$$
$$c^2 = 1^2 + 1^2 \quad \text{(filling in for a and b)}$$
$$c = \sqrt{1^2 + 1^2} \quad \text{(taking the square root of each side)}$$
$$c = \sqrt{1 + 1} \quad \text{(squaring the two numbers)}$$
$$c = \sqrt{2} \quad \text{(doing the addition)}$$
$$c = 1.414 \quad \text{(taking the square root)}$$

This means that in any square, the length of the diagonal is 1.414 times the length of any side.

$$\text{side a} \times \sqrt{2} = 1.414$$
$$\text{(or)} \quad 1 \times 1.414 = 1.414$$

Look at the right half of Figure 4-1. In this country the unit run is always 12''. Since each common rafter forms one side of the square, the diagonal (or hypotenuse) is equal to 12'' times the square root of 2. Twelve inches times 1.414 equals 16.97''.

Because we use 12'' for the common rafter unit run, the hip rafter unit run (the diagonal) is fixed at 16.97''.

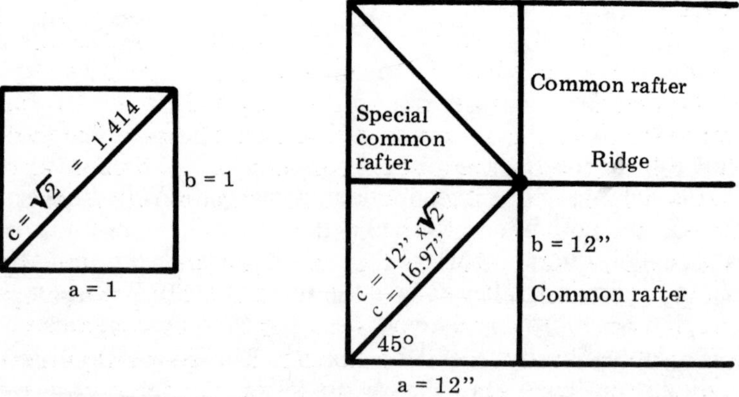

Diagonal of a square
Figure 4-1

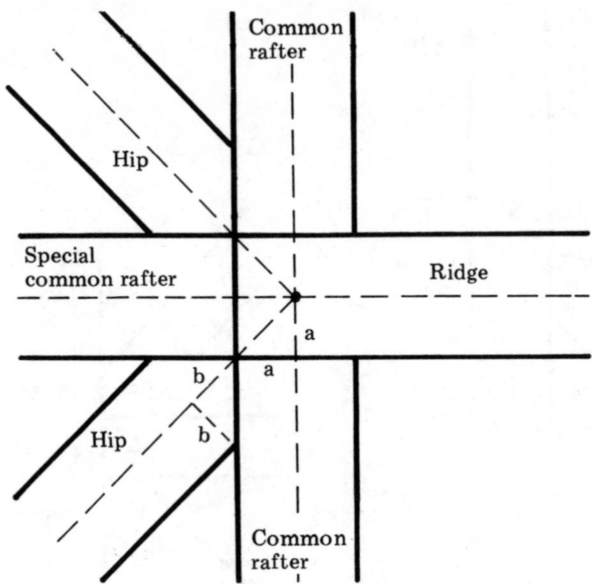

Hip shortening and setback
Figure 4-2

Hip Shortening and Setback

In Chapter 3 we were concerned only with the growth of the ridge at the framing point where the hip met the ridge. Now let's look at those drawings again, but this time we'll see how the hip is affected.

Double Cheek Cut Hips

Figure 4-2 is the same drawing as Figure 3-5. But now the hips are drawn in. Since (a) is half the thickness of the common rafter, it's obvious that the hip shortens along the diagonal from the framing point to the tip of the hip rafter. This distance is spoken of as one-half the 45-degree thickness of the common rafter. It could also be stated as one-half the thickness of the common rafter times the square root of 2.

The term *setback* refers to the proper distance to move back from the shortening line to begin the cheek cut. Since (b) in Figure 4-2 is equal to half the thickness of the hip, and the diagonal is the cheek cut, you must move back from the shortening line the distance (b). That's equal to one-half the thickness of the hip rafter. All regular hip roof cheek cuts have a setback equal to one-half the thickness of the hip material itself.

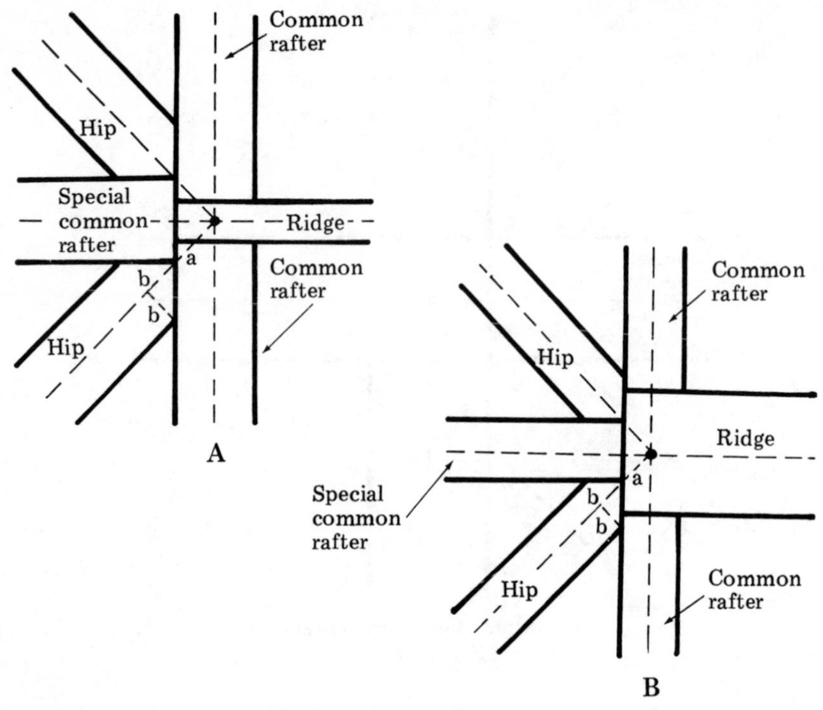

Relationships
Figure 4-3

Setback is often placed toward the head or top end of the rafter. Because of this, it's sometimes called *set forward.* They are both the same distance, which is half the thickness of the material, but set forward is laid out toward the ridge while setback is laid out toward the fascia.

The Proof

Let's see if a double cheek cut hip always shortens by one-half the 45-degree thickness of the common rafter and really has a setback of one-half the thickness of the hip.

Figure 4-3 shows two different situations. In 4-3 A, the ridge is a thin material and the rafters are much thicker. In 4-3 B the reverse is true. Despite these differences, notice that in both cases the shortening of this hip rafter depends only on the thickness of the common rafter. The setback depends only on the thickness of the hip itself. We can now expand Table 3-12 in Chapter 3 to include this valuable information. See Table 4-4.

Type of Hip	Common Rafter Shortening	Ridge Growth	Hip Shortening	Setbacks (for the Cheek Cut)
Double cheek cut	½ thickness of ridge *	½ thickness of common rafter	½ 45° thickness of common rafter (not hip or ridge)	½ thickness of hip
Single cheek cut	½ thickness of ridge	½ thickness of ridge + ½ 45° thickness of hip		

*See the section "The Special Common" in Chapter 3 (pages 67 and 68)

Effects of the framing members
Table 4-4

Single Cheek Cut Hips

Figure 4-5 is the same drawing as Figure 3-10. But this time we'll focus on what happens to the hip rafter. Since the center line of the hip passes through the framing point at 45 degrees, Line (a) must be one-half the 45-degree thickness of the ridge. This, then, is the shortening of the single cheek cut hip.

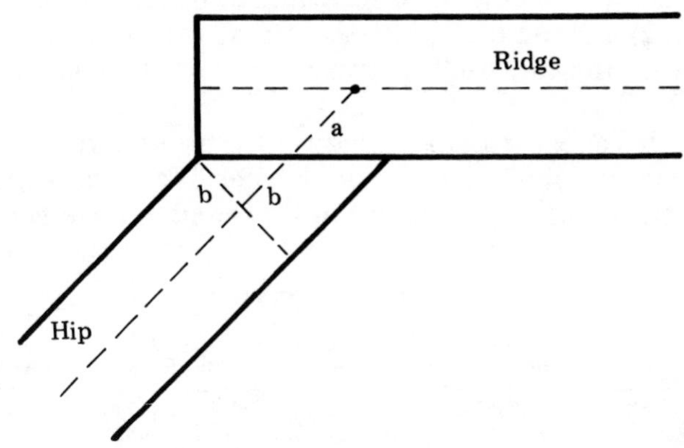

Hip shortening and setback
Figure 4-5

The setback, again, depends on the thickness of the hip material itself. Therefore the setback is equal to one-half the thickness of the hip.

The Proof

Here's a test to see if any single cheek cut hip really shortens by one-half the 45-degree thickness of the ridge and has a setback of one-half the thickness of the hip. Look at Figure 4-6.

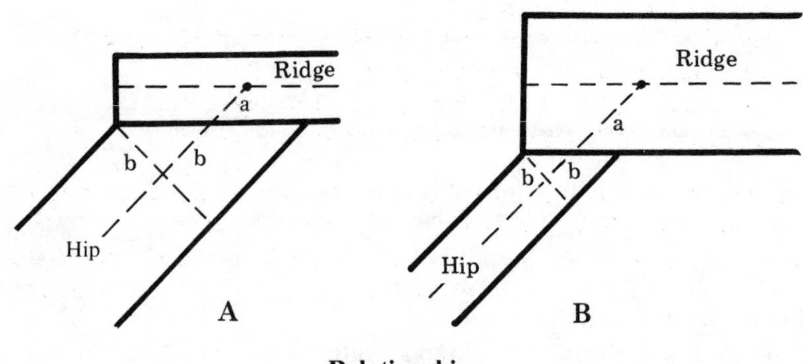

Relationships
Figure 4-6

Figure 4-6 shows two different situations. In 4-6 A, the ridge is thin and the hip is much thicker. In 4-6 B, the reverse is true. You can see that in both cases the shortening of this rafter depends only on the thickness of the ridge; the setback depends only on the thickness of the hip itself. Let's expand Table 4-4 to look like Table 4-7.

Notice in Table 4-7 that common rafter shortening and hip setback remain the same, but the double cheek cut and single cheek cut hips affect ridge growth and hip shortening differently.

Type of Hip	Common Rafter Shortening	Ridge Growth	Hip Shortening	Setbacks (for the Cheek Cut)
Double cheek cut	½ thickness of ridge *	½ thickness of common rafter	½ 45° thickness of common rafter (not hip or ridge)	½ thickness of hip
Single cheek cut hip	½ thickness of ridge	½ thickness of ridge + ½ 45° thickness of hip	½ 45° thickness of ridge	½ thickness of hip

* See the section "The Special Common" in Chapter 3 (pages 67 and 68)

Effects of the framing members
Table 4-7

Problems

Here are some problems to test your understanding of these important points. Answers are at the back of the book.

1) A 7 in 12 hip roof uses double cheek cut hips on each end of a 24' x 36' building. The framing members are nominal 2" wide stock and actually measure 1½".

a) What will the shortening of the common rafters be?
b) What's the actual length of the ridge?
c) How much does the hip rafter shorten?
d) How much should be allowed for the setback?

*2) A 6 in 12 exposed beam hip roof has single cheek cut hips on each end of a 44' x 64' lodge building. The ridge is 6" x 16' rough sawn material a full 6" thick. The hips and common rafters are rough sawn 4 x 10's. Answer **a), b), c),** and **d)** from **Problem 1** above.*

Finding the Length of Hip Rafters

There are four ways to find the length of hip and valley rafters. You can use printed tables or slide rules, the framing square tables, the square root method or the secant method. We'll look at them one at a time.

Printed Tables or Special Slide Rules

The length of selected regular hips can be found on a rafter slide rule or in a book of rafter length tables.

Rafter Length Manual, published by Craftsman Book Company, 6058 Corte del Cedro, Carlsbad, California 92008, shows the total length of the rafter for many spans. You just find the correct total span under the appropriate unit rise. All the calculations are done for you.

For each pitch and total run in the table, there will be two rafter lengths listed. In Figure 4-8A, the table is for common rafters and will be based on a 12" run. Figure 4-8B is for the hip or valley rafter and will be based on a run of 12" times the square root of 2, or 16.97".

Framing Square Tables

Figure 2-7 in Chapter 2 shows common rafter lengths on the top line of the rafter scale. The second line begins with ditto marks under the word length, then the words "hip or valley," and then three more sets of ditto marks. This second line should be read as "length hip or valley per foot run."

9 Foot 0 Inch Run — Common Rafter Lengths

Run -	9' 0"			9' 0 1/4"			9' 0 1/2"			9' 0 3/4"		
Pitch	Ft	In	16th"	Ft	In	16th"	Ft	In	16th"	Ft	In	16th"
1 IN 12	9'	0"	6	9'	0"	10	9'	0"	14	9'	1"	2
2 IN 12	9'	1"	8	9'	1"	12	9'	2"	0	9'	2"	4
2.5 IN 12	9'	2"	5	9'	2"	9	9'	2"	13	9'	3"	1
3 IN 12	9'	3"	5	9'	3"	9	9'	3"	13	9'	4"	2
3.5 IN 12	9'	4"	8	9'	4"	12	9'	5"	0	9'	5"	5
4 IN 12	9'	5"	13	9'	6"	2	9'	6"	6	9'	6"	10
4.5 IN 12	9'	7"	6	9'	7"	10	9'	7"	14	9'	8"	2
5 IN 12	9'	9"	0	9'	9"	4	9'	9"	9	9'	9"	13
5.5 IN 12	9'	10"	13	9'	11"	1	9'	11"	6	9'	11"	10
6 IN 12	10'	0"	12	10'	1"	0	10'	1"	5	10'	1"	9
6.5 IN 12	10'	2"	13	10'	3"	2	10'	3"	6	10'	3"	11
7 IN 12	10'	5"	1	10'	5"	5	10'	5"	10	10'	5"	14
8 IN 12	10'	9"	13	10'	10"	2	10'	10"	6	10'	10"	11
9 IN 12	11'	3"	0	11'	3"	5	11'	3"	10	11'	3"	15
10 IN 12	11'	8"	9	11'	8"	15	11'	9"	4	11'	9"	9
11 IN 12	12'	2"	8	12'	2"	14	12'	3"	3	12'	3"	8
12 IN 12	12'	8"	12	12'	9"	1	12'	9"	7	12'	9"	13
13 IN 12	13'	3"	4	13'	3"	10	13'	3"	15	13'	4"	5
14 IN 12	13'	9"	15	13'	10"	5	13'	10"	12	13'	11"	2
15 IN 12	14'	4"	14	14'	5"	5	14'	5"	11	14'	6"	1
16 IN 12	15'	0"	0	15'	0"	7	15'	0"	13	15'	1"	4
17 IN 12	15'	7"	4	15'	7"	11	15'	8"	2	15'	8"	9
18 IN 12	16'	2"	11	16'	3"	2	16'	3"	10	16'	4"	1
19 IN 12	16'	10"	4	16'	10"	11	16'	11"	3	16'	11"	10
20 IN 12	17'	5"	15	17'	6"	6	17'	6"	14	17'	7"	6
21 IN 12	18'	1"	11	18'	2"	3	18'	2"	11	18'	3"	3
22 IN 12	18'	9"	9	18'	10"	1	18'	10"	9	18'	11"	2
23 IN 12	19'	5"	8	19'	6"	0	19'	6"	9	19'	7"	2
24 IN 12	20'	1"	8	20'	2"	1	20'	2"	10	20'	3"	3
25 IN 12	20'	9"	9	20'	10"	2	20'	10"	12	20'	11"	5

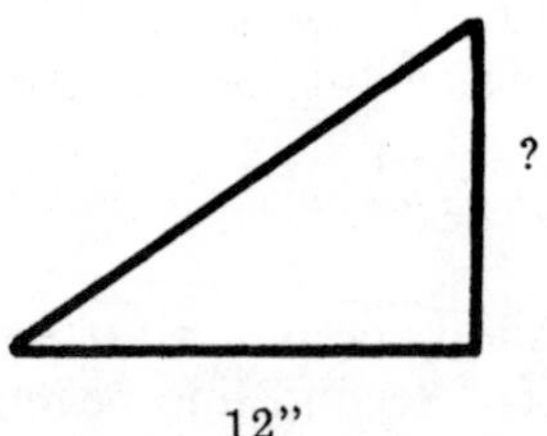

Rafter length manual
Figure 4-8A

9 Foot 0 Inch Run — Hip Or Valley Rafter Lengths

Run -	9' 0"			9' 0 1/4"			9' 0 1/2"			9' 0 3/4"		
Pitch	Ft	In	16th"	Ft	In	16th"	Ft	In	16th"	Ft	In	16th"
1 IN 12	12'	9"	0	12'	9"	6	12'	9"	11	12'	10"	1
2 IN 12	12'	9"	13	12'	10"	2	12'	10"	8	12'	10"	14
2.5 IN 12	12'	10"	6	12'	10"	12	12'	11"	2	12'	11"	7
3 IN 12	12'	11"	2	12'	11"	7	12'	11"	13	13'	0"	3
3.5 IN 12	12'	11"	15	13'	0"	5	13'	0"	11	13'	1"	1
4 IN 12	13'	0"	15	13'	1"	5	13'	1"	10	13'	2"	0
4.5 IN 12	13'	2"	0	13'	2"	6	13'	2"	12	13'	3"	2
5 IN 12	13'	3"	4	13'	3"	10	13'	3"	15	13'	4"	5
5.5 IN 12	13'	4"	9	13'	4"	15	13'	5"	5	13'	5"	11
6 IN 12	13'	6"	0	13'	6"	6	13'	6"	12	13'	7"	2
6.5 IN 12	13'	7"	9	13'	7"	15	13'	8"	5	13'	8"	11
7 IN 12	13'	9"	3	13'	9"	10	13'	10"	0	13'	10"	6
8 IN 12	14'	0"	14	14'	1"	4	14'	1"	10	14'	2"	0
9 IN 12	14'	4"	14	14'	5"	5	14'	5"	11	14'	6"	1
10 IN 12	14'	9"	4	14'	9"	11	14'	10"	2	14'	10"	8
11 IN 12	15'	2"	0	15'	2"	7	15'	2"	14	15'	3"	4
12 IN 12	15'	7"	1	15'	7"	8	15'	7"	15	15'	8"	6
13 IN 12	16'	0"	6	16'	0"	13	16'	1"	5	16'	1"	12
14 IN 12	16'	6"	0	16'	6"	7	16'	6"	15	16'	7"	6
15 IN 12	16'	11"	14	17'	0"	5	17'	0"	13	17'	1"	4
16 IN 12	17'	5"	15	17'	6"	6	17'	6"	14	17'	7"	6
17 IN 12	18'	0"	3	18'	0"	11	18'	1"	3	18'	1"	11
18 IN 12	18'	6"	10	18'	7"	3	18'	7"	11	18'	8"	3
19 IN 12	19'	1"	4	19'	1"	13	19'	2"	5	19'	2"	14
20 IN 12	19'	8"	1	19'	8"	10	19'	9"	3	19'	9"	11
21 IN 12	20'	3"	0	20'	3"	9	20'	4"	2	20'	4"	11
22 IN 12	20'	10"	1	20'	10"	10	20'	11"	4	20'	11"	13
23 IN 12	21'	5"	4	21'	5"	14	21'	6"	7	21'	7"	1
24 IN 12	22'	0"	9	22'	1"	3	22'	1"	12	22'	2"	6
25 IN 12	22'	7"	15	22'	8"	9	22'	9"	3	22'	9"	13

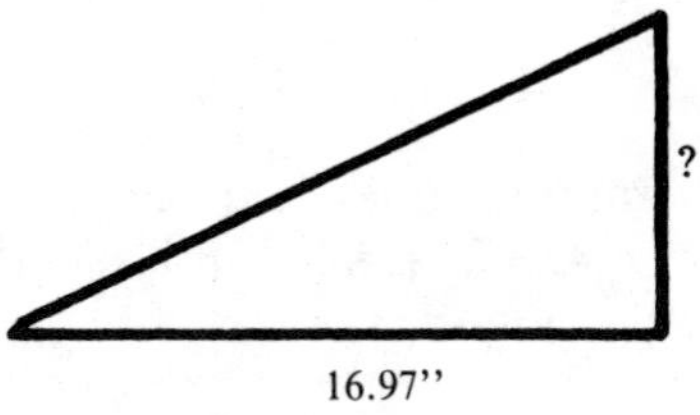

Rafter length manual
Figure 4-8B

All numbers on this second line are unit lengths of both hip and valley rafters per foot of run. They're based on 16.97'' and not on 12'' as is the line above, which is for common rafters. These first two lines are the same as the two sections of the *Rafter Length Manual* or the two sections of the rafter slide rule.

There's a difference, of course, between framing square tables and rafter length books. The framing square covers only unit lengths for each unit run and unit rise. You have to multiply the factor shown by the total run and convert the answer to feet, inches and fractions of an inch. The rafter length book, on the other hand, does this for you.

Let's try an example. Look at Figure 4-9. On the 3' x 5' rafter plate used in Chapter 2, let's build a regular hip roof with double cheek cut hips on one end and single cheek cut hips on the other end. The roof will again have a pitch of 8 in 12.

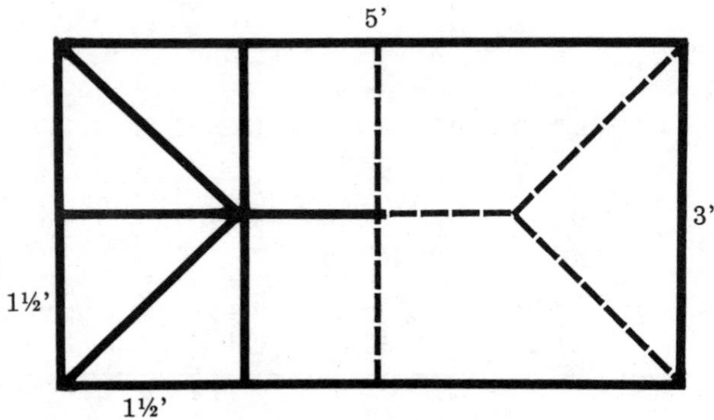

Run of the model hip
Figure 4-9

Under 8'' on the framing square on the first line of the rafter tables is the number 14.42. That's the unit length for each foot of common rafter run. See Figure 4-10.

Since our model common rafter run was 1½', we multiply the 14.42'' by 1½ to get the mathematical length of the common rafter from the ridge to the building line. We calculated this number at 21⅝''.

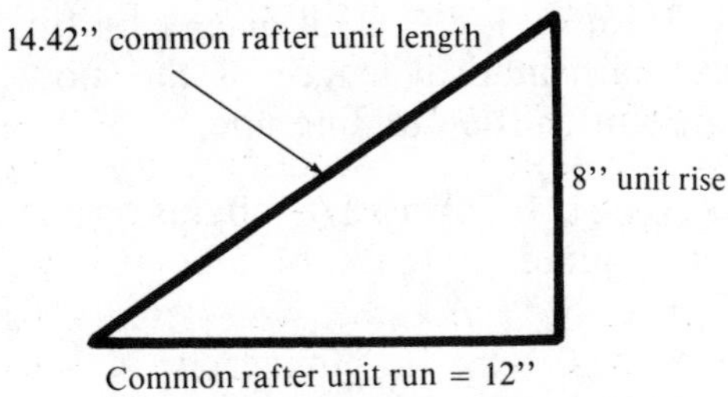

The common rafter unit length
Figure 4-10

The second line on the framing square also shows unit lengths. But now the base of the triangle is 16.97'' for each unit of rise. See Figure 4-11. Notice that the numbers are all slightly larger than on the line above. That's because of the extra distance that the run of the hip or valley rafter needs to go diagonally, on the plan view, from the corner of the building to the center of the span.

Using the Framing Square Tables for the Model Hip

Here's how to find the mathematical length for the model hip rafter. Look on the framing square rafter table under 8''. That's the unit rise for this roof. Look down to the second line, which is the unit length of the hip rafter. The number found is 18.76''. Multiply 18.76'' by half the building span (or the run of the common rafter, which is also called the apparent run of the hip.)

Since the model has a span of 3', half the span is 1½'. Multiplying 18.76 by 1½ gives 28.14''. Change 0.14'' into sixteenths, so it can be read on your tape measure. Multiply 0.14 by 16. Round off

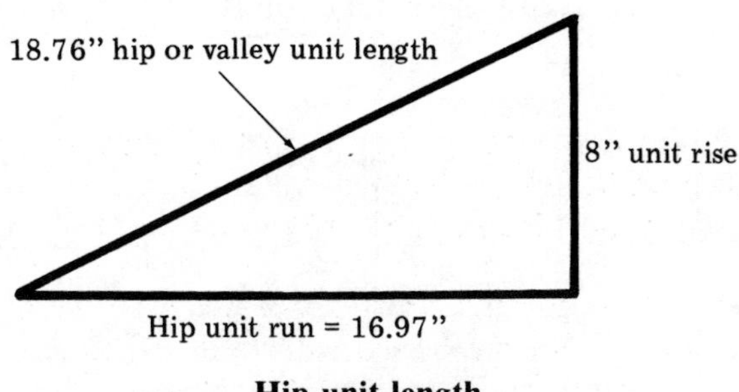

Hip unit length
Figure 4-11

2.24 sixteenths to 2/16, or 1/8''. That gives us a final answer of 28⅛'', the correct mathematical length of the model hip rafter from the framing point to the building line.

You could also convert 0.14'' to 1/8'' by using the hundredths scale on the framing square. Note the 14 indicated by the arrow in Figure 4-12 B.

Notice that we worked from the run of the common rafter, not the run of the hip. We could do this because the unit length 18.76'' is already a function of the hip run, since the hip unit triangle uses 16.97'' rather than 12'' as one of the sides.

The hip rafter lays at a much lower angle than the common rafter because its base is stretched out further, being 16.97'' instead of 12''. This causes the angle of the hip to be 25.24 degrees, as opposed to 33.69 degrees for a common rafter on an 8 in 12 roof.

Of course, this hip rafter still fits into the 8 in 12 common rafter plane of the roof. So it must be considered to begin and end its run in the space between points (a) and (b) in Figure 4-13. That distance is the same as the run of the common rafter. The actual run of the hip is the length of the diagonal, or "a-b" times the square root of 2. See Figures 4-1, 4-13, 4-29, and 7-7 in Chapter 7.

A

.14	decimal part of an inch
16	(16 ths.)
84	
14	
2.24	the number of 16ths.

B

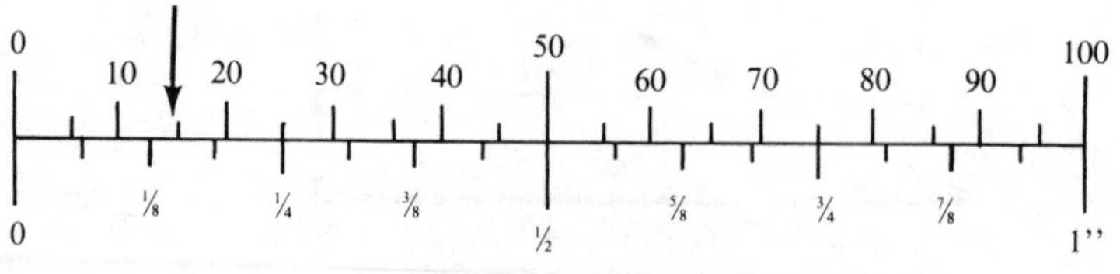

0 10 20 30 40 50 60 70 80 90 100

0 ⅛ ¼ ⅜ ½ ⅝ ¾ ⅞ 1''

Finding 16ths
Figure 4-12

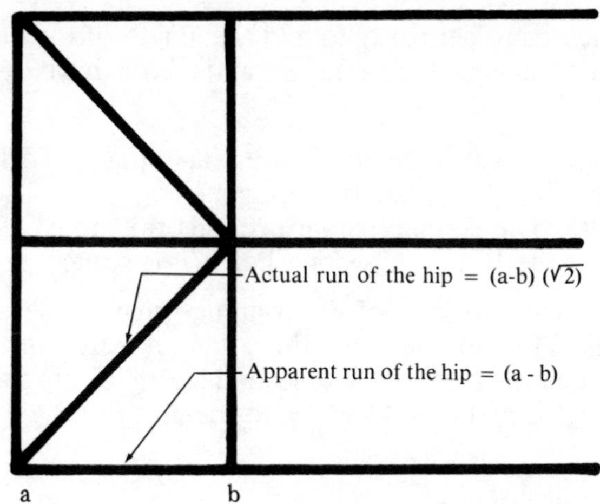

Apparent and actual run of hip
Figure 4-13

On the framing square, all of the numbers on the second line are unit length numbers. Of course, you can find these numbers on your own by solving for a right triangle using 16.97 as the base and any inch number as the pitch.

For example, for a 10 in 12 pitch, the number printed on the second line under 10 is the same as the diagonal in Figure 4-14.

Here's how to check this on your calculator. Punch in [1] [6] [.] [9] and [7] to represent the unit run on a hip or valley rafter. Then square that number by hitting [x^2]. The answer in the display will be 287.9809.

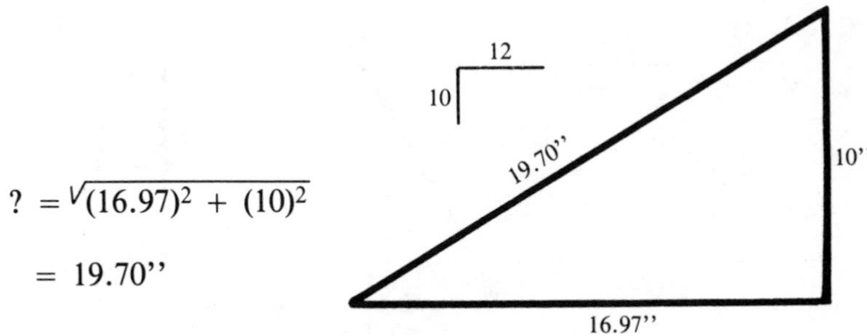

The unit triangle for regular hips and valleys
Figure 4-14

Then punch ⊞ to get ready to add the length of the other side, the unit rise. Punch [1] [0] and [x^2] and you'll have 100 in the display.

Add the two answers together with ⊟. That puts 387.9809 in the display. Then find the square root by punching [√]. The answer will be 19.697. The framing square rounds this up to 19.70 and prints it under the 10 on the second line of the scale.

And so the first two lines of the framing square tables are unit length figures. The top line is for the common rafter unit lengths and uses 12'' and the unit rise. The second line is for hip and valley rafter unit lengths and uses 16.97'' and the unit rise.

The Square Root Method

Say you're working on a 20 in 12 pitch roof. Unfortunately, the framing square numbers only go up to 18. You could simply solve the two right triangles (Figures 4-15 and 4-16) by the square root (Pythagorean) method as we have just done with the other unit triangles and then proceed with the framing square method.

The common rafter unit length triangle for a 20 in 12 pitch roof is:

$$? = \sqrt{(12)^2 + (20)^2}$$

$$= 23.32''$$

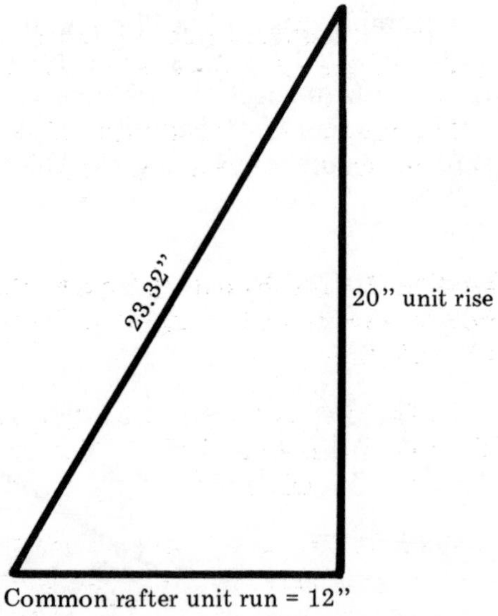

Common rafter unit length
Figure 4-15

Say that the building is to be 18' by 30'. The common rafter run would be half the span, or 9'. To find the mathematical length of a 20 in 12 pitch rafter for this roof, simply multiply the unit length of 23.32'' by 9' to get 209.88'', or 17.49'. That's about the same as 17'5⅞'', which is the mathematical distance for this common rafter from the center line of the ridge to the building line.

Just for the sake of review, let's use our calculator to change 209.88'' into feet, inches and fractions of an inch. Punch [2] [0] [9] [·] [8] and [8]. Then punch [÷] [1] [2] and [=] to convert inches to feet. Write down that number (17.49) and then subtract out the 17: [−] [1] [7] and [=]. You now have 0.49 left in the display.

Punch [×] [1] [2] and [=] and 5.88 appears. Write down the 5, the number of whole inches, and subtract it out: [−] [5] and [=]. Just 0.88 is left in the display.

Convert 0.88'' to sixteenths of an inch: [×] [1] [6] [=] and you have 14.08 sixteenths of an inch. We'll round that to 7/8ths of an inch for 17'5⅞''.

To use the square root method by itself to find the common rafter, use the total run of 9' and the total rise in feet of 15' (20'' divided by 12 times 9').

Nine feet squared plus 15 feet squared equals 306 square feet, whose square root, once again, is 17.49 feet.

The roof cutter usually doesn't use this method. He uses the framing square method, or the secant method which follows.

Finding the Hip Rafter by Square Root

Figure 4-16 shows the hip or valley rafter unit triangle for a 20 in 12 pitch roof.

To find the mathematical length of the hip rafter, multiply the hip unit length 26.23'' by 9'. The answer is 236.06'', or 19'8 1/16'', from the framing point on the ridge to the building line.

Once again we've used the common rafter run for the hip calculation. That's perfectly acceptable because we used 12 times the square root of 2 = 16.97'' in the unit run figure. The square root of 2 factor was built into the unit run. This is an important difference from the secant method which follows.

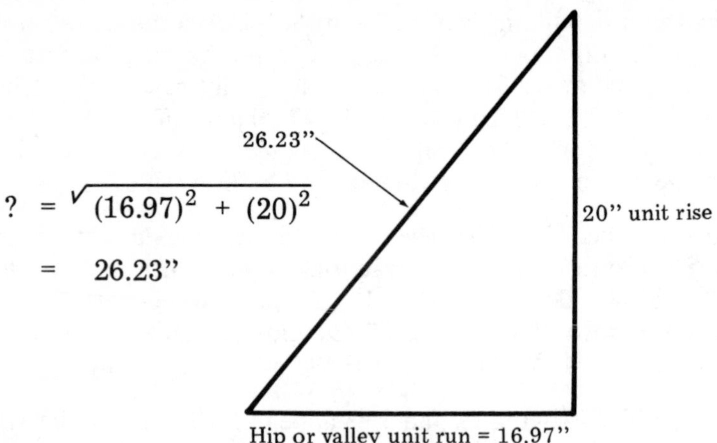

Hip or valley unit length
Figure 4-16

The Secant Method

Toward the end of Chapter 2 we used the secant to find the mathematical length of a common rafter. Remember that the length of the rafter was equal to the secant times the actual run of the rafter.

Let's use the secant method to find the common rafter length for the 20 in 12 pitch roof. Look at Figure 4-17 and follow along as I go through it step by step.

You must first find angle (A). Since the unit run and the unit rise are known, angle (A) can be found by dividing 20" by 12". This particular trigonometric function is called the tangent. On your calculator press [2] [0] [÷] [1] [2] [=] and 1.6667 appears. Now, what angle is this the tangent of? If trigonometry tables were available, you could look through the table under Tangents until you found 1.6667. Reading across you would find 59 degrees and 2'.

But let's use the calculator to find the tangent. With 1.6667 in the display, you press [INV] and then [tan] and 59.0362 appears. Your angle is 59.0362 degrees. Let's convert the 0.0362 to minutes of a degree.

Multiplying 0.0362 by 60 (because each degree is broken into 60 minutes) you have 2.17 minutes, or 59 degrees 2'.

If you have a trigonometry table, look at secant under 59 degrees 2' and you'll find 1.9436. This is the secant of the angle. The secant times the common rafter run will give you the mathematical length of the rafter.

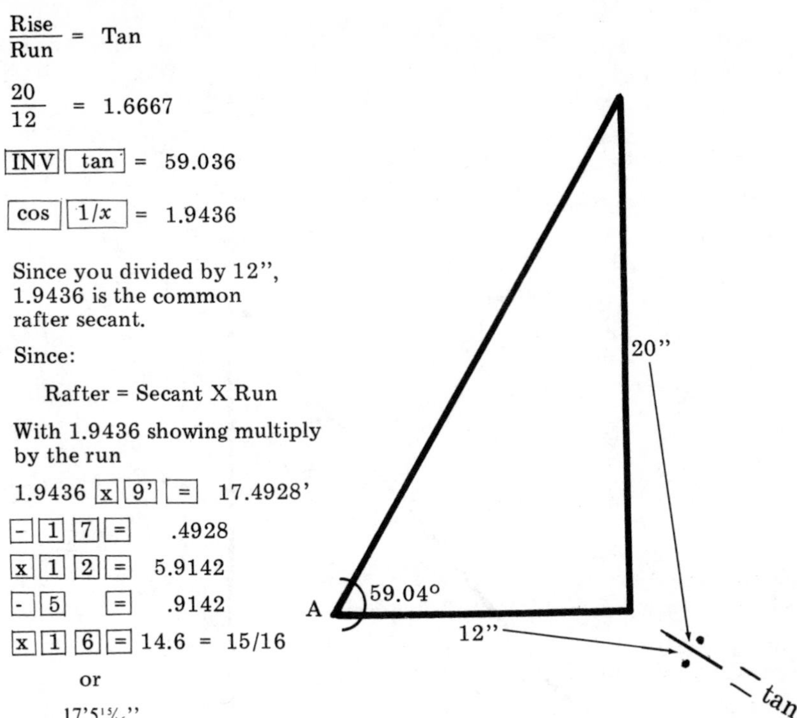

Rise / Run = Tan
20 / 12 = 1.6667
INV tan = 59.036
cos 1/x = 1.9436
Since you divided by 12", 1.9436 is the common rafter secant.
Since:
Rafter = Secant X Run
With 1.9436 showing multiply by the run
1.9436 x 9' = 17.4928'
- 1 7 = .4928
x 1 2 = 5.9142
- 5 = .9142
x 1 6 = 14.6 = 15/16
or
17'5 15/16"
20"
59.04°
A
12"
tan

Tangent of angle A
Figure 4-17

The run was 9'. So 1.9436 times 9' equals 17.4928', or 17'5 15/16". That's the same answer you got using the framing square method.

Here's how to get the same answer on the calculator. Put 59.036 degrees in the display. Punch cos and 0.5144 appears. Press 1/x and 1.9436 appears. Since there is no secant button on this calculator, we use the two buttons cos followed by 1/x to get the secant. Multiplying the secant by 9' gives us the answer, 17'5 15/16".

Finding the Hip Rafter with Secant

You can find the length of a hip or valley rafter with only a slightly different procedure.

Remember that the angle of the hip is lower than the angle of the common rafter. So the secant of the hip is a lower number.

On the calculator, press [2] [0] [÷] [1] [6] [·] [9] [7] and [=]. The number 1.1785 will appear. This is the tangent of angle (A) in Figure 4-18.

In a trigonometry table, under the tangent 1.178, you would find the angle to be 49 degrees and 41 minutes.

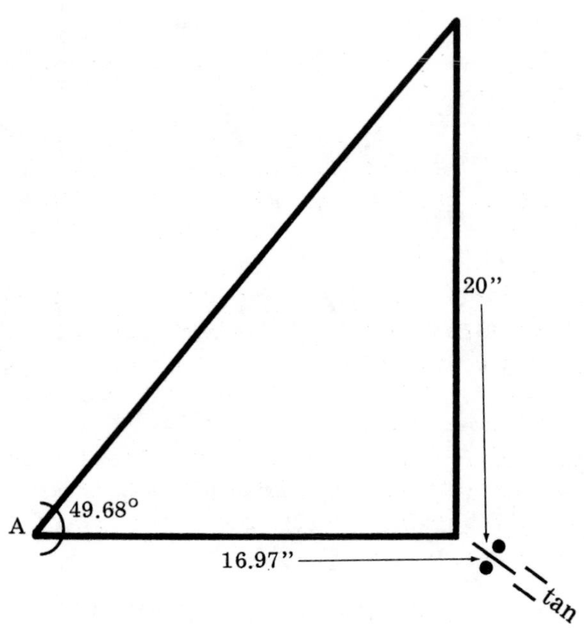

Tangent of angle A for a hip rafter
Figure 4-18

On the calculator, with 1.1785 showing, press [INV] then [tan] and 49.6853 appears. Press [cos] [1/x] and 1.5456 appears. This is shown in Figure 4-65 under secant.

Up to this point, our procedure has been the same as if we were finding a common rafter length. But now it gets just a little different.

Remember in the secant method that the length of a rafter is equal to the secant times the run of that rafter. Also remember that the actual run of the hip or valley rafter is equal to the run of the common rafter (the apparent run) times the square root of 2.

The formula is:

Hip rafter = Secant x actual run of hip

That's the same as:

Hip rafter = Secant x common rafter run x$\sqrt{2}$

Plugging in our numbers:

Hip Rafter equals 1.5456 times 9' times the Square Root of 2 equals 19.6723', or 19'8$\frac{1}{16}$''.

This is the same answer we got with the framing square method and from the *Rafter Length Manual,* Figure 4-8.

Remember, when using the secant method, that the common rafter run must be multiplied by the square root of 2 to get the actual run of the hip.

Since the roof cutter usually works with the apparent run of hip and valley rafters, it would be more convenient to put the square root of 2 factor into the secant, as in the fourth column of Figure 4-65. Now we can use this new secant with the usual apparent run. From the tables, then, 2.1858 x 9' = 19.6723'.

If a rafter length table is available, I'll agree that it's much easier to look up the length in the table. But for the irregular roofs in the second part of this book there are no published tables. You'll have to know how to do the calculations.

Building a 3' x 5' Hip Roof

Let's cut some hip rafters for our 3' x 5' model roof. We'll build a regular hip roof at 8 in 12 with a single cheek cut hip on one end and a double cheek cut hip on the other end. Figure 4-19 shows what we want to do.

The Model Ridge

The mathematical length of the ridge on this model is 5' minus 3', or 2'. Write this down so we can sum up adjustments to the mathematical length.

Now look back to Table 4-7 to see how much is added to the ridge for each type of hip.

The table shows that the ridge growth for a double cheek cut hip is one-half the thickness of the common rafter. Since the ridge and the hips will be nominal 2'' material that's actually 1½'' thick, our growth is 3/4''. Write this number down as in Figure 4-20.

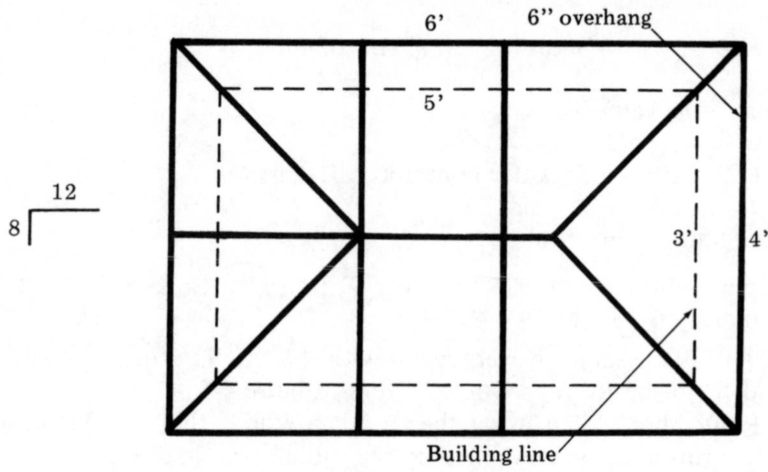

The model 3' x 5' building
Figure 4-19

The ridge growth for a single cheek cut is, first, half the thickness of the ridge. Since the ridge will be 1½" thick, we add 3/4". Then we must add half the 45-degree thickness of the hip.

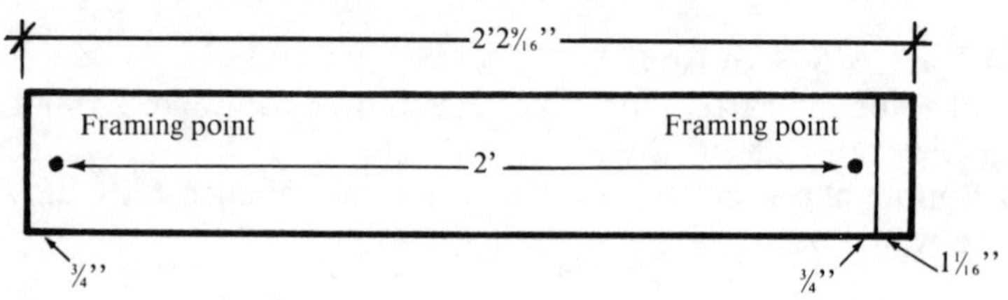

2' + 0"	Mathematical length
+ 3/4"	Double cheek cut end growth
+ 3/4"	Single cheek cut end growth
+ 1-1/16"	
2' 2-9/16"	

Total ridge length
Figure 4-20

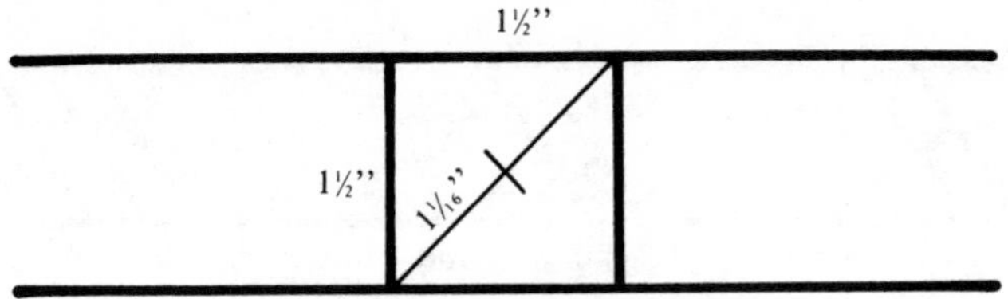

Finding ½ the 45° thickness
Figure 4-21

Figure 4-21 shows the 45-degree distance of 1½'' thick material. To get the diagonal, use the square root method: the thickness times the square root of 2.

45° distance	=	$\sqrt{(1½'')^2 + (1½'')^2}$	1½'' x $\sqrt{2}$
45° distance	=	$\sqrt{2.25 + 2.25}$	1.5 x 1.414
45° distance	=	$\sqrt{4.5}$	=
45° distance	=	2.1213	2.1213

According to the chart, we need only half that distance, 1.06''. Convert that to sixteenths: 16 times 0.06 equals 0.97/16, or 1/16. So for 1½''-thick milled lumber, half the 45-degree thickness is equal to 1$\frac{1}{16}$''. Add this to the ridge growth calculation.

On this model the ridge has grown by a total of 2$\frac{9}{16}$''. Cut a piece of 2 x 6 to a length of 26$\frac{9}{16}$''.

Figure 4-19 shows that you'll need five common rafters. You already have these from the previous model.

The Model Hip Rafters

Figure 4-22 shows the lines for each hip rafter. Each is labeled for easy reference. The lines described below are numbered to correspond with the lines in Figure 4-22.

1) The Ridge Line: Crown a piece of 2 x 4 that's at least 44'' long. Lay the crown away from you. Set your stair gauges and framing square as shown in Figure 4-23. Don't forget to set the stair gauge on the body to 16.97'' rather than 12''.

On the tongue, set 8'', because this is another 8 in 12 roof. Then draw the ridge line as indicated from the corner of the rafter.

2) The Building Line: We have found the model rafter mathematical length by the framing square method to be 28⅛'' from the

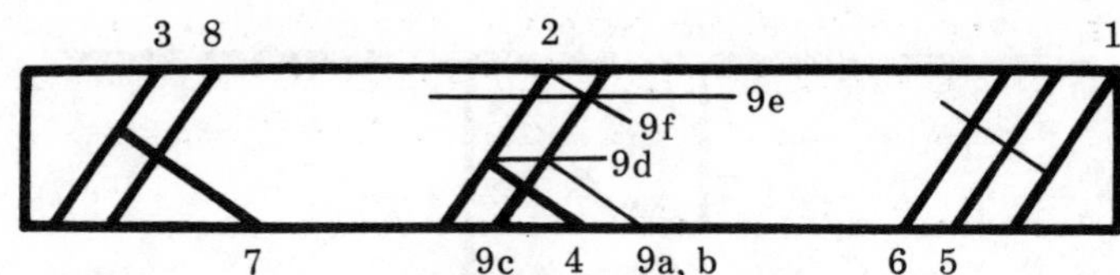

The lines for a hip rafter
Figure 4-22

framing point to the building line. Measure along the top edge of the rafter 28⅛" from the ridge line. See Figure 4-24.

Through this point draw the building line just as you did for the common rafter. See Figure 4-25.

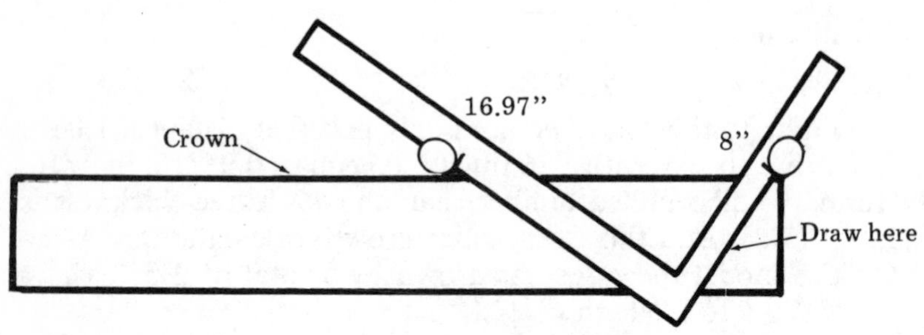

Drawing the ridge line
Figure 4-23

If the rafter stock you are using is too short, rotate the framing square 180 degrees as in Figure 4-26.

Now place the stair gauges against the lower edge of the stock and again draw along the tongue.

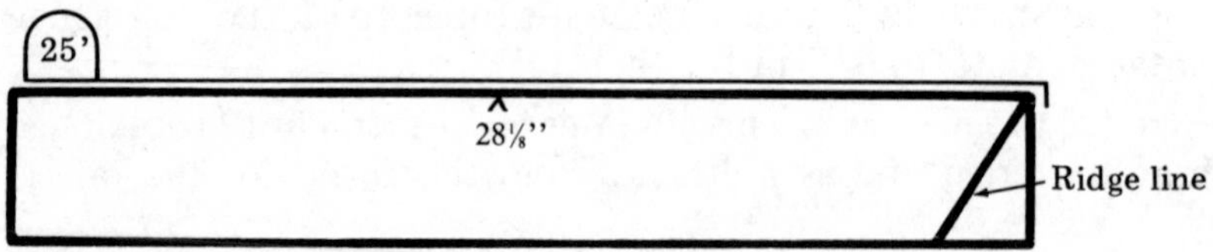

Marking off the mathematical length
Figure 4-24

3) Figuring the Fascia Line: Now we must calculate the overall length of the hip rafter from the framing point to the corner of the overhang. Let's use the framing square method first.

Length of hip = Unit length x Apparent run

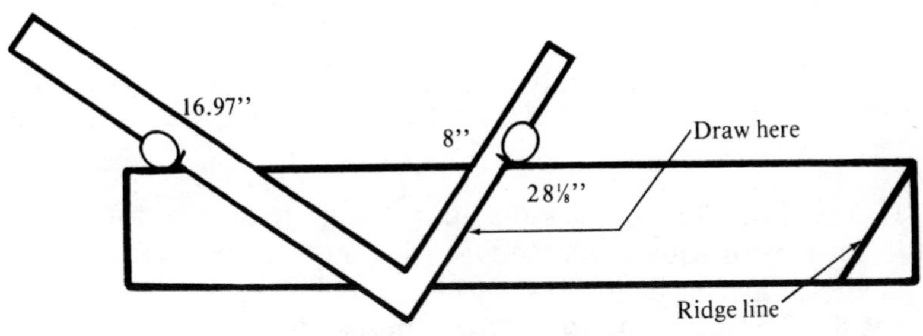

Drawing the building line
Figure 4-25

If the unit length can be found on the framing square, simply write it down and multiply by the *common rafter total run.* Look under 8'' on the rafter table of your framing square. Move down to the second line, which gives the unit lengths for hip or valley rafters. The number there is 18.76. See Figure 4-28.

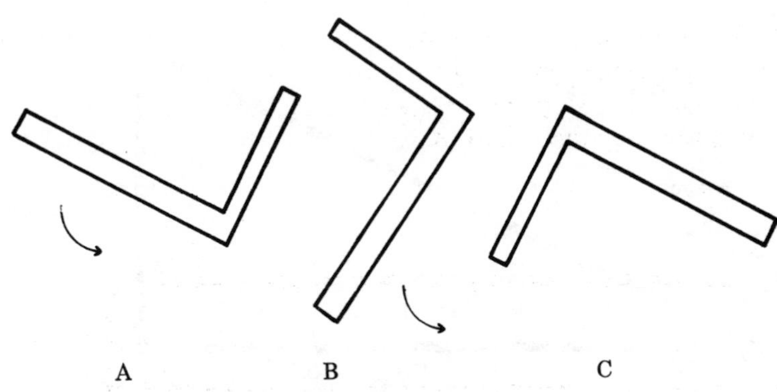

Rotating the framing square
Figure 4-26

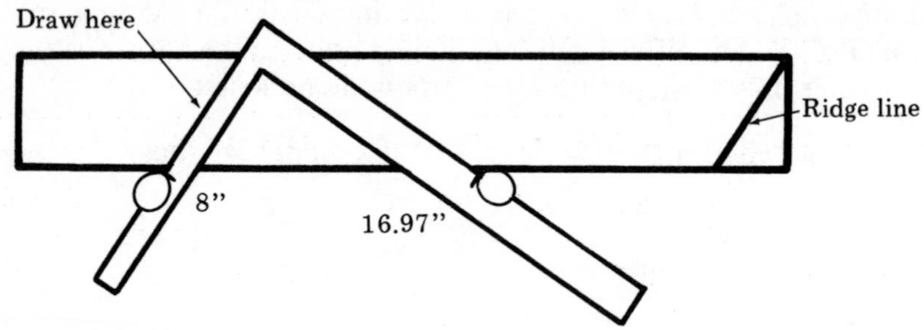

The framing square reversed
Figure 4-27

If you don't have tables on your framing square, but have a calculator with square and square root buttons on it, it's easy to find the unit length. Multiply 16.97 times 16.97 and add it to 8 times 8. Then push the square root button to get 18.76.

Multiply this hip unit length by the common rafter run (which, for the hip rafter, is called the apparent run) to find the mathematical length of the hip. Length of hip equals 18.76" times 2, or 37.52".

Now let's use the two forms of the secant method.

Length of Hip = Secant x Actual Run

The hip secant for the 8" rise roof from Figure 4-65 is 1.1055 times 2' times the square root of 2 equals 3.1269', or 37.52" again.

Length of Hip = (Secant) ($\sqrt{2}$) x Apparent Run

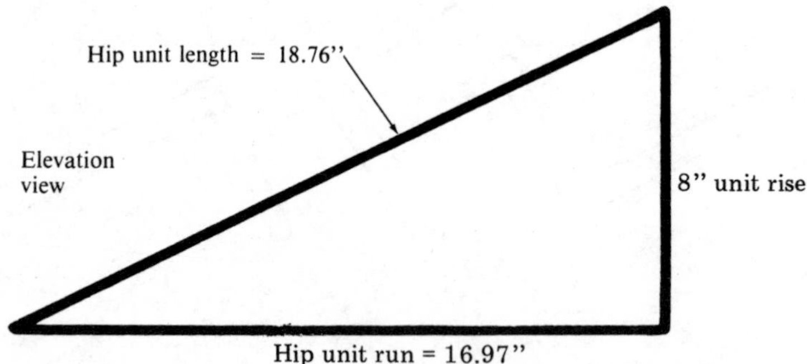

Hip unit length
Figure 4-28

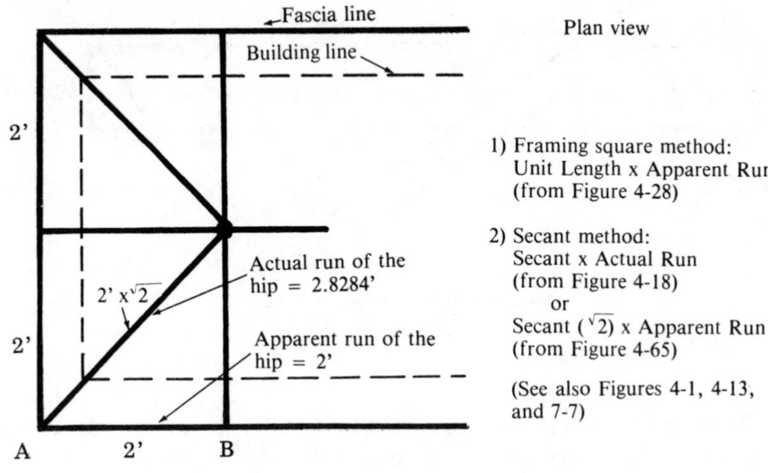

Methods of calculating the hip rafter
Figure 4-29

The secant times the square root of 2 for an 8'' rise roof (from Figure 4-65) is 1.5635 times 2' equals 3.1269' again, or 37½'' from the framing point to the fascia line.

To lay out the fascia line, make a mark 37½'' from the ridge line along the top edge of the rafter. See Figure 4-30.

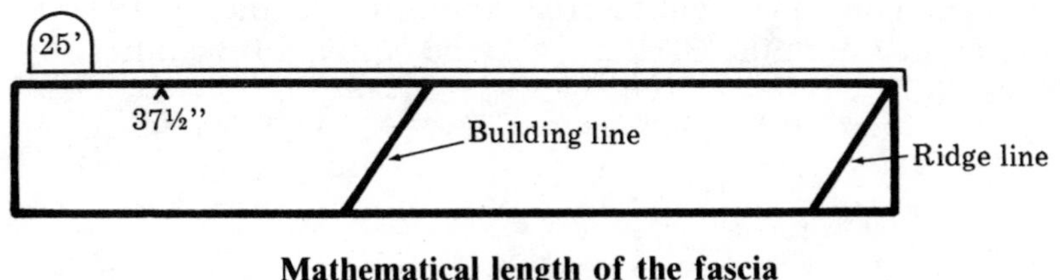

Mathematical length of the fascia
Figure 4-30

Slide the framing square into the position shown and draw the fascia line as in Figure 4-31.

4) The Seat Cut: Remember from Chapter 2 that you must determine the HAP before drawing the seat-cut line, and that there can be only one HAP for any regular roof. The HAP for this model is 2¾''.

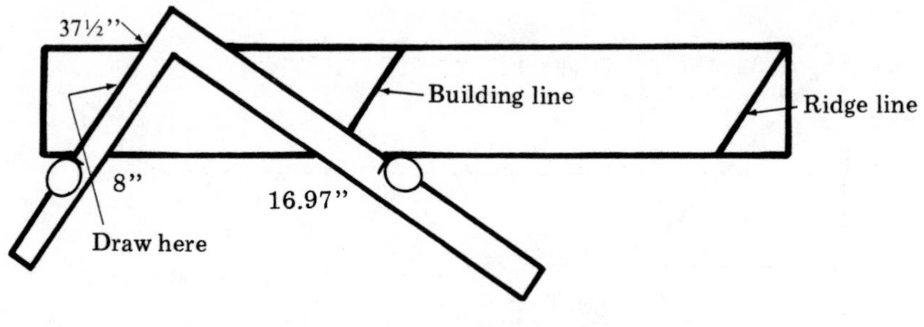

Drawing the fascia line
Figure 4-31

Draw in the HAP mark by measuring 2¾" from the top down and along the building line, as shown in Figure 4-32.

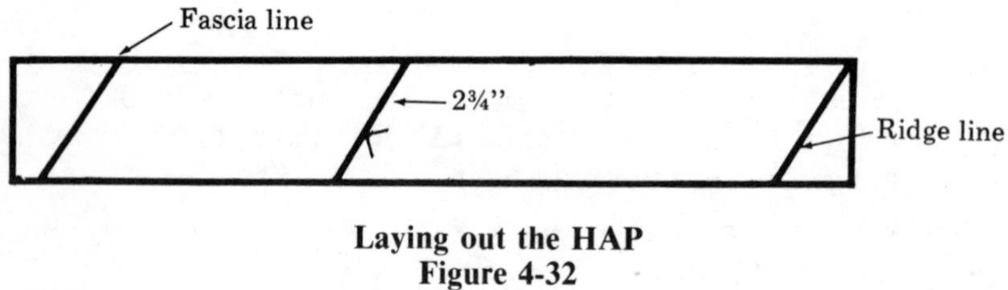

Laying out the HAP
Figure 4-32

Position the square on the HAP mark so that you can draw the seat-cut line from the HAP mark to the bottom of the rafter. See Figure 4-33.

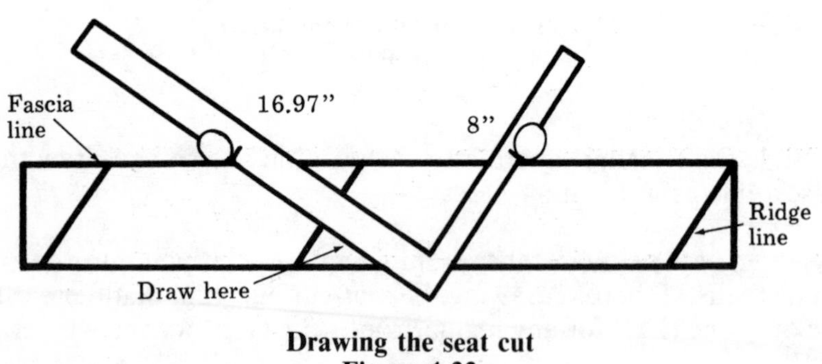

Drawing the seat cut
Figure 4-33

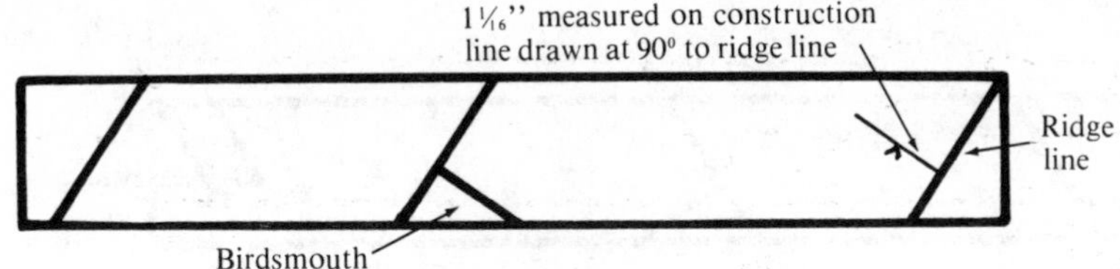

Marking off the shortening
Figure 4-34

5) The Hip Shortening: The mathematical length of the building line and the fascia line are measured along the top edge of the rafter. Their respective lines are drawn through the points you marked. Now, measurements for shortening and cheek cut lines will be made perpendicular to these lines.

Draw a construction line perpendicular to the ridge line. Measure along this line from the ridge line the amount of shortening, as in Figure 4-34.

We'll cut the double cheek cut hip first. Table 4-7 tells us that the shortening for a double cheek cut hip is one-half the 45-degree thickness of the common rafter. Since we are using 2 x 4 material for the common rafter, the shortening will be 1$\frac{1}{16}$".

For the single cheek cut hip, Table 4-7 shows the shortening to be one-half the 45-degree thickness of the ridge. Since we are using 2 x 6 material for the ridge, the shortening will be the same, 1$\frac{1}{16}$". Note that this is true only because the ridge and common rafter are the same thickness, 1½".

Slide the framing square to the position shown in Figure 4-35 and draw the shortening line.

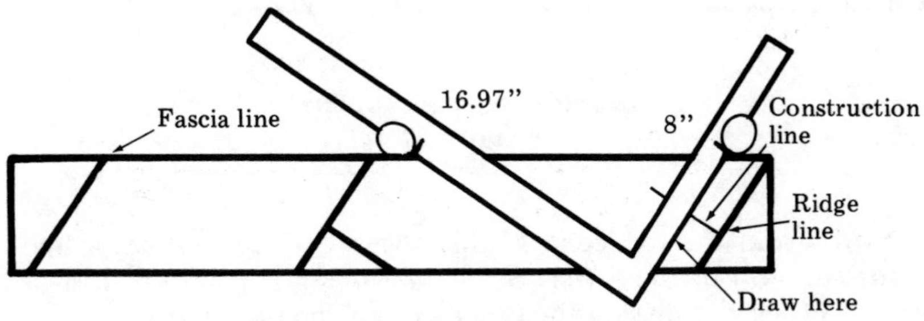

Drawing the shortening line
Figure 4-35

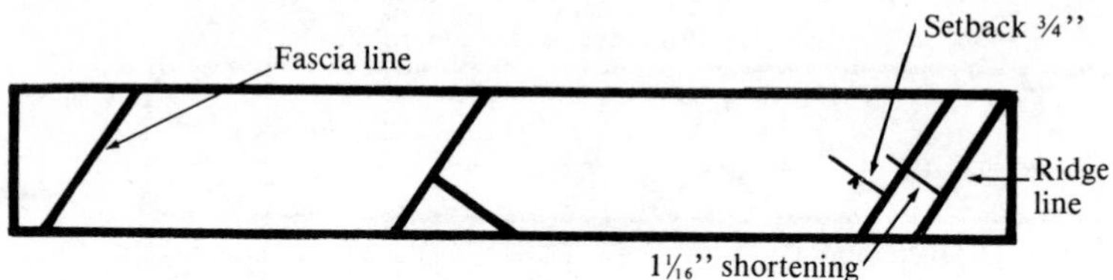

Drawing in the setback
Figure 4-36

6) The Cheek Cut Line: Setback is measured perpendicular from the shortening line for the beginning of the cheek cut. Since the setback distance is half the thickness of the hip material itself, the setback for both the double cheek cut hip and the single cheek cut hip will be the same distance. But there will be a difference in the cutting of each hip.

Extend the old construction line or begin a new one perpendicular to the shortening line, as shown in Figure 4-36. The hip is 2'' material so it's 1½'' thick. The setback is half the thickness of the hip. So you measure 3/4'' along the new construction line from the shortening line.

Draw the cheek cut line as indicated in Figure 4-37.

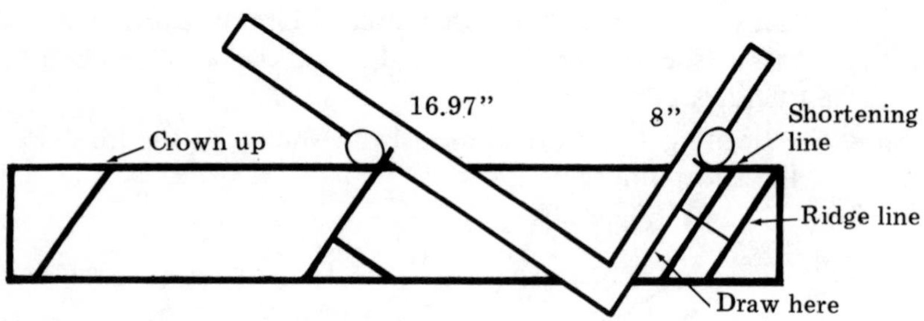

Drawing the cheek cut line
Figure 4-37

Now, from the top edge of this cheek cut line, square a line across the top of the rafter. See Figure 4-38. This lets us draw a similar cheek cut line on the other side of the board.

Move the framing square to the far side of the rafter and draw another cheek cut line as indicated by the dotted line in Figure

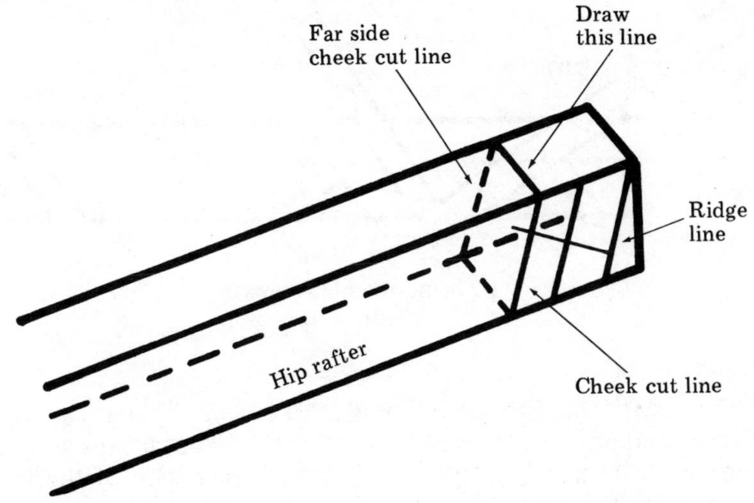

Square the line across the top
Figure 4-38

4-38. To check yourself, square another line across the bottom of the rafter and see if it meets with the first cheek cut line.

7) The Plancher Cut: Just as there's only one HAP measurement for any regular roof, so too there's only one plancher cut measurement. Since we used 2" on the common rafters, we'll use 2" on the hip rafters. Measure down from the top of the rafter 2" along the fascia line and make a mark, as in Figure 4-39.

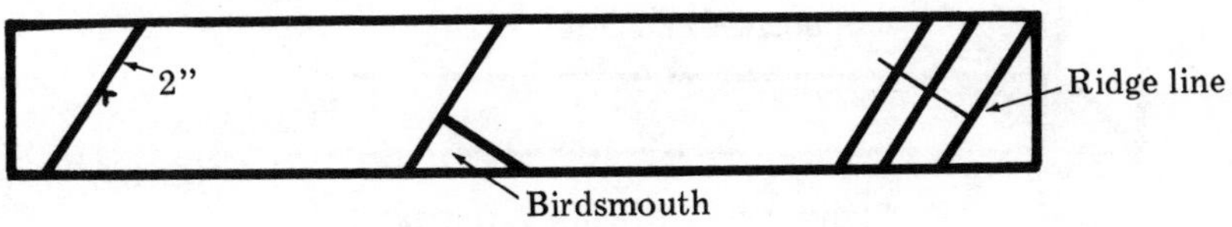

Locating the plancher cut
Figure 4-39

Position the square as shown in Figure 4-40. The plancher cut is a level cut and therefore it must be drawn on the 16.97" side. Refer back to Figure 2-12 in Chapter 2 if you need some review on this point.

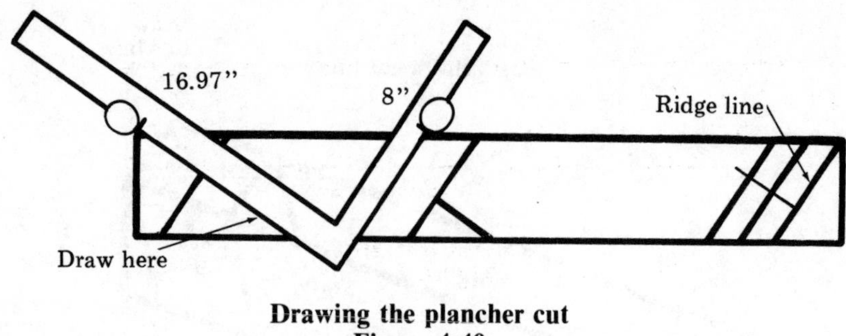

Drawing the plancher cut
Figure 4-40

8) The Tail Cut: The fascia board is to have a mitered corner, so that its end grain isn't visible when finished. That means you need a double cheek cut on the hip to receive each side of the fascia board. See Figure 4-41.

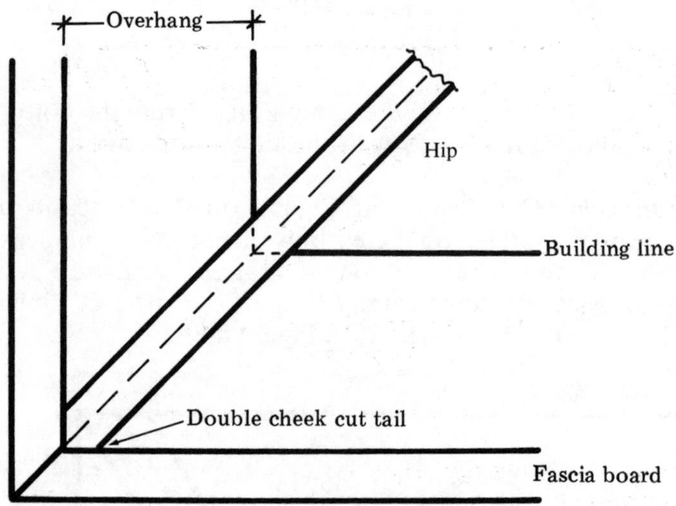

Plan view of the hip rafter tail
Figure 4-41

Since the plancher cut line is perpendicular to the fascia line, we can use this as our construction line. Measure forward half the thickness of the hip, or 3/4" in this case, as shown in Figure 4-42.

Place the framing square as shown in Figure 4-43 and draw the tail cut line. Since this requires a double cheek cut, you'll have to

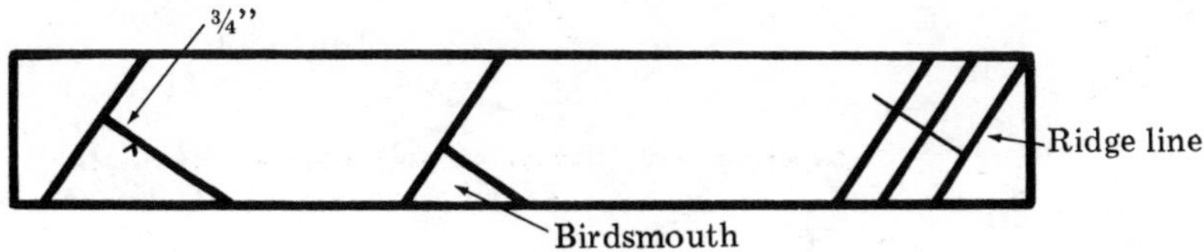

Marking the tail cut
Figure 4-42

square across the top of the rafter and draw a similar tail cut line on the other side. The procedure is the same as in Figure 4-38.

Rather than "setback," this is called "set forward" because you're moving toward the forward part of the rafter from the reference line.

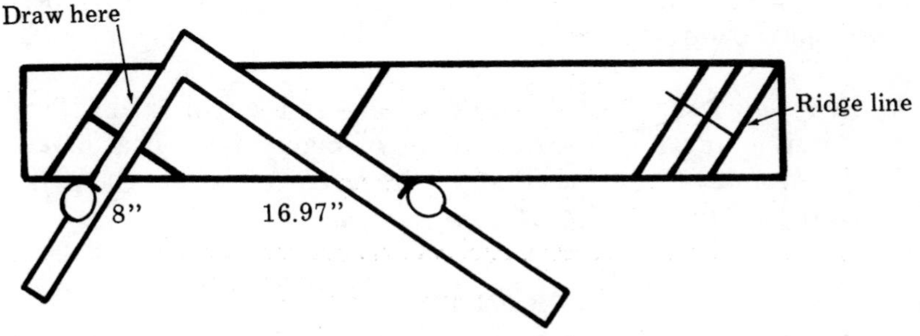

Drawing the tail cheek cut line
Figure 4-43

Backing or Dropping the Hip

If we were to cut the rafter out at this point, there would be one problem. Can you guess what that would be? Look at Figure 4-44.

The line between the framing points of the hip rafter runs along the center of the top edge of the rafter. See Figure 4-44 A. From that center line, the planes of the roof break downward toward each side, as shown in B. Notice that the edges of the hip rafter stick up above the plane of the roof. No master roof cutter wants his roofs to look like that.

Here's one way to remedy this problem: Cut a little bit off both edges of the hip, as in Figure 4-44 C. The cut is called a chamfer

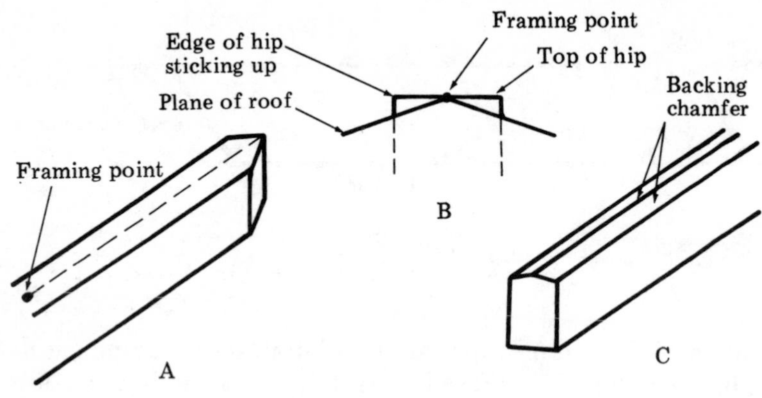

The hip edges
Figure 4-44

cut. This process is called *backing the hip.* But it takes too much time and is seldom done. There's another way (not always the best way) called dropping the hip.

9a) Dropping the Hip: Instead of cutting on the seat cut line, cut the rafter as shown by the dotted line of Figure 4-45. That makes the hip drop lower, so the edges of the rafter fall within the plane of the roof. This is called *dropping the hip.*

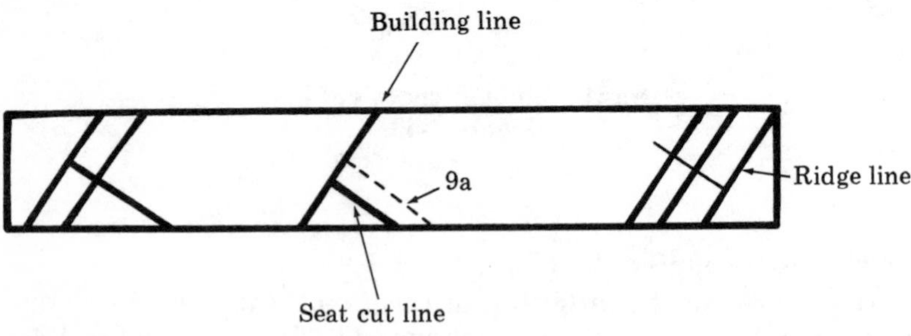

Changing the seat cut
Figure 4-45

The dropping can be figured out mathematically, or laid out on the rafter or on paper. Since the hip must be either backed or dropped, we can proceed to calculate the right amount by any of these methods.

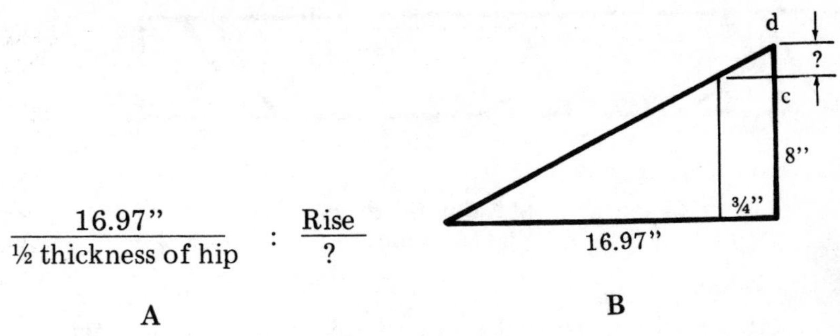

The backing ratio
Figure 4-46

9b) Dropping the Hip Using Math: Here's how to figure it out by the math method. Figure 4-46 A says: As 16.97" is to the rise, so 3/4" is to the backing or dropping. The two dots between the left and right parts of the equation in Figure 4-46 indicate that the problem is expressed as a ratio.

The backing or dropping can be found by multiplying the rise in inches by half the thickness of the hip and then dividing by 16.97. Note that this formula is good only for regular hips.

$$\text{Backing or Dropping} = \frac{(\text{Rise})\ (\tfrac{1}{2}\ \text{thickness of the hip})}{16.97''}$$

Figure 4-46 B shows an 8" by 16.97" triangle. Another line is drawn at 3/4" inside the 8" side because 3/4" is half the thickness of our hip rafter. Where this line hits the hypotenuse, draw a perpendicular projection as shown, and measure the distance from (c) to (d). That's the amount of backing or dropping needed.

Our model has a rise of 8" and a hip that's 1½" thick.

$$\text{Dropping} = \frac{8'' \times .75''}{16.97''} = \frac{6''}{16.97''} = .3535''$$

Multiplying 0.3535" by 16 equals 5.65/16, which rounds to 5½ sixteenths. This is the amount of backing needed.

Using the building line as the construction line, measure from the seat cut line up 5½ sixteenths and make a mark as in Figure 4-47.

Place the framing square in the position shown in Figure 4-48 and make this second seat cut line that passes through the mark. The dropping line is always drawn above the seat cut line.

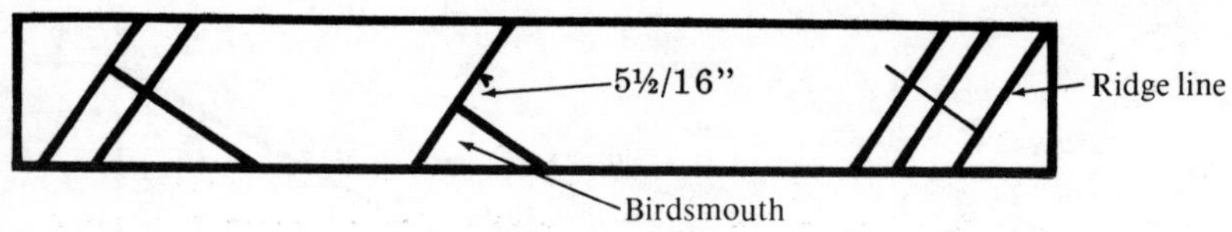

Marking the drop
Figure 4-47

Another formula for calculating backing or dropping will be found at the end of Section 9e, Backing the Hip.

9c-d) Dropping the Hip Using Layout: Using the seat cut line as a construction line, measure forward half the thickness of the hip and make a mark. See Figure 4-49.

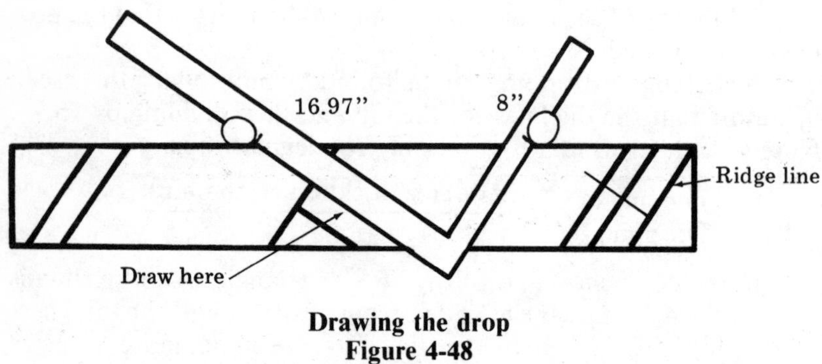

Drawing the drop
Figure 4-48

Position the framing square as in Figure 4-50 and draw a set forward line through this mark.

From the corner of the building line and the seat cut line at point (a) in Figure 4-51, draw line (a-c) parallel to the top edge of the

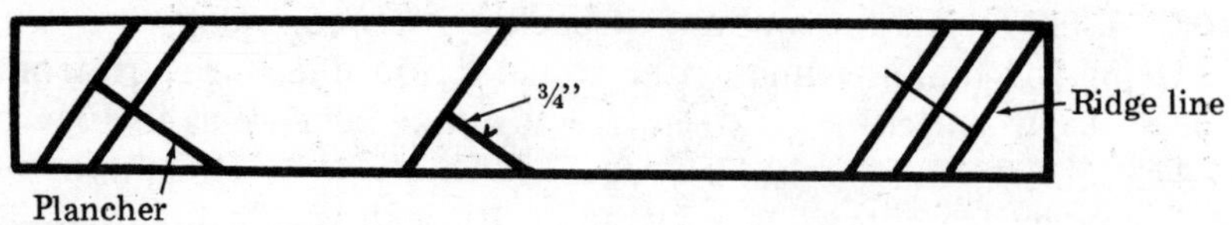

Laying out the dropping
Figure 4-49

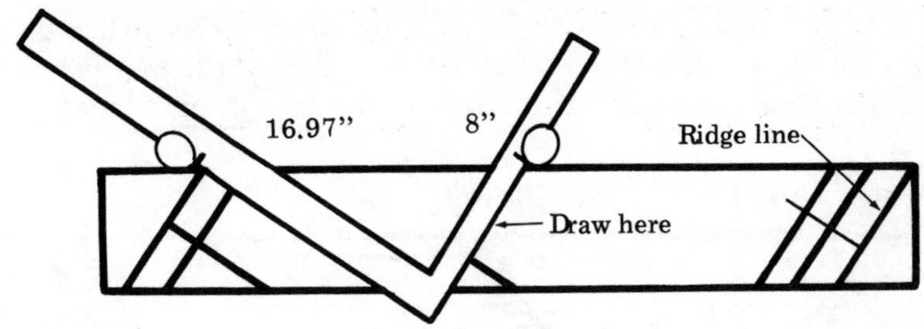

The set forward line
Figure 4-50

board. The line will go from the building line forward, toward the set forward line. Use a combination square to draw this line.

Position the framing square to draw a new seat cut line through point (b). This will be the cut line for the proper dropping.

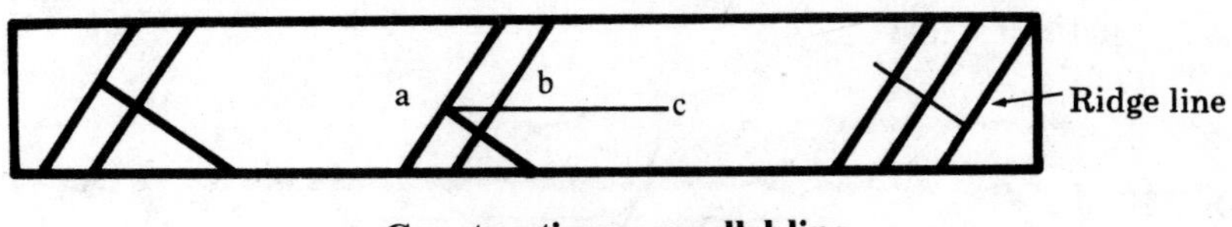

Constructing a parallel line
Figure 4-51

9e) Backing the Hip: If you decide to cut a backing chamfer rather than drop the hip, the amount to chamfer is the same as the formula in 9b.

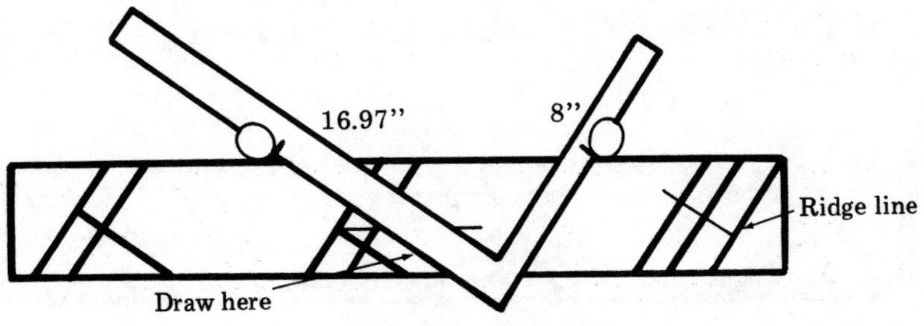

Drawing the new seat cut
Figure 4-52

Measure down perpendicular from the top edge 5½ sixteenths to find point (e) in Figure 4-53. Draw a line through this point parallel to the top edge of the rafter. Do this on both sides.

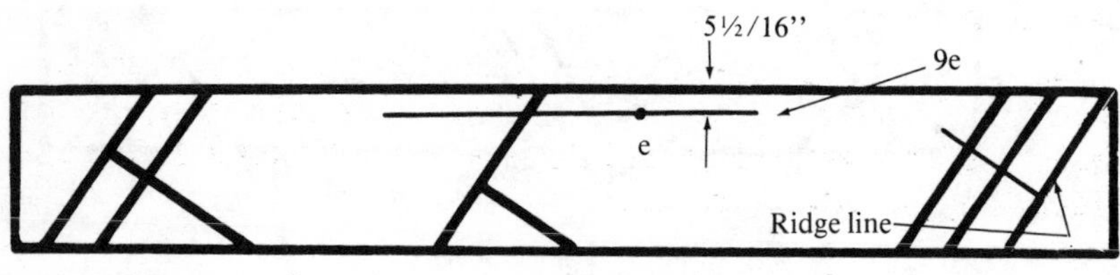

Calculated backing
Figure 4-53

9f) Backing Using the Framing Square: To locate point (e) using the framing square, draw the set forward line as in Figure 4-49 and 4-50. From point (d) in Figure 4-54 draw line (d-f) perpendicular to the building line as shown. Through point (e) draw a line parallel to the top edge of the material. Also draw this line on the other side of the rafter.

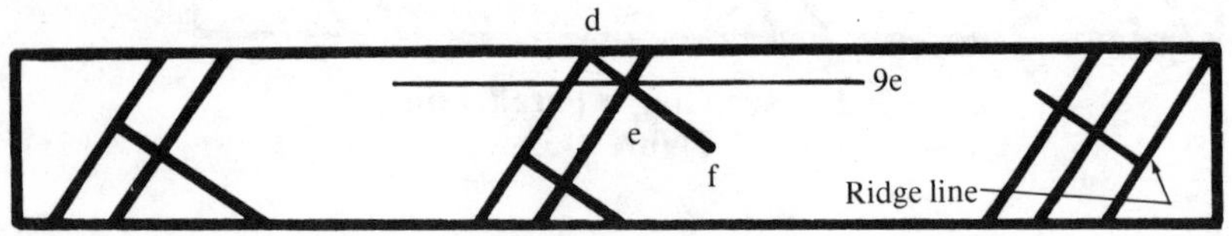

Constructed backing
Figure 4-54

Before cutting out the rafter, label all the lines as in Figure 4-55. This makes a mistake less likely and makes checking your work much easier.

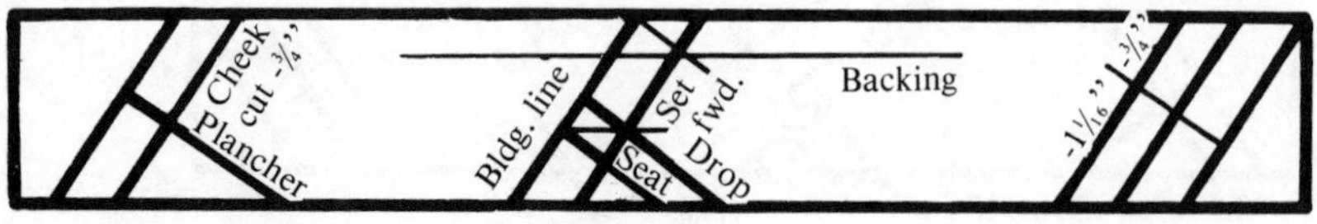

Labeling the rafter
Figure 4-55

Cutting Out the Rafter

When seen from above, hips are at 45 degrees to the span. You'll have to set your saw at 45 degrees for the cheek cuts and tail cuts. Check your saw against some standard to see if the blade is truly at 45 degrees.

For regular hips, use 45 degrees for these cuts whether the roof is an 8 in 12 or 20 in 12 roof. The plan view of each roof still shows the hips at 45 degrees. Figures 2-30 and 2-31 in Chapter 2 show the proper way of cutting rafters.

Cutting Line 6, the Cheek Cut: For the double cheek cut hip you have transferred line (6) to the other side of the rafter. Set the saw at 45 degrees and cut along each line. The ridge line and the shortening line are cut away.

For the single cheek cut hip, also transfer line (6) to the other side, but cut on only one side for one hip and on the other side for the other hip. This gives you an opposite pair.

Notice that this hip grows by one-half the thickness of the hip. The double cheek cut hip does not. If our ridge had been only 1'' material or 3/4'' thick, we would not have been able to draw the ridge line at the corner of the material because the hip shortening would have been less than the growth of the cheek cut.

Remember that 45-degree single cheek cuts grow by half the thickness of the material (Figure 4-56).

Cutting Line 7, the Plancher Cut: Set the saw to 0 degrees and make the plancher cut along this line.

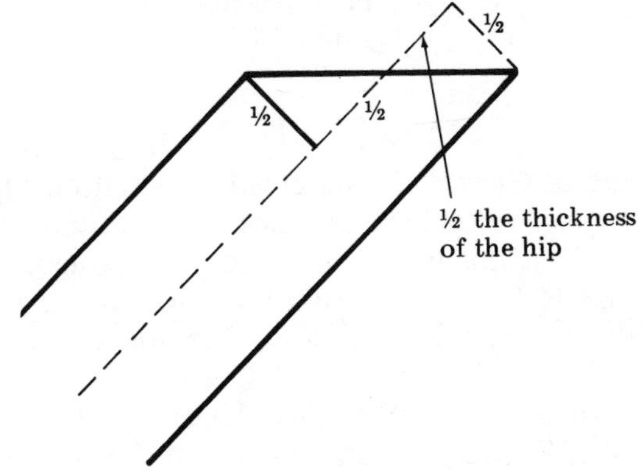

½ the thickness of the hip
Figure 4-56

Cutting Lines 2 and 9a-b, the Birdsmouth: If you are dropping the rafter, cut the birdsmouth carefully along lines (2) and (9ab) and finish the cuts with a hand saw.

Cutting Line 8, the Tail Cut: Since line (8) is a cheek cut, you'll have to reset the saw to 45 degrees. Check to see if line (8) has been transferred to the other side. Then cut along each line at 45 degrees. Line (3) will be cut away.

Cutting Line 9e, the Backing: To cut backing, set the saw to the same angle the hip makes to the plate line. Remember the hip angle is different from the common angle. Both have the same rise. But the hip run is 16.97'' and the common run is only 12''. What's the angle for the hip? See Figure 4-57.

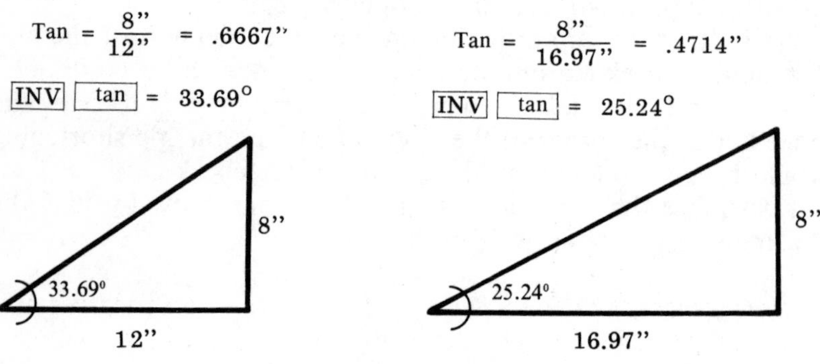

Different angles, same roof
Figure 4-57

Near the end of Chapter 2, under "Two Relationships of Right Triangles," I covered the key sequence on the calculator for figuring common rafter unit triangles. Follow the same procedure for the hip, but use 16.97'' instead of 12''. Your answer will be 25.24 degrees. At the end of this chapter is a hip and valley angle chart, which also gives you this angle.

For the backing cut, lay the rafter flat and set the saw just past 25 degrees. Fix a guide board to the hip so the saw makes a nice straight cut. Then follow line (9e) along both sides. The method in Figure 4-58 works best on 2 x 6 material. For our 2 x 4, a table saw is better.

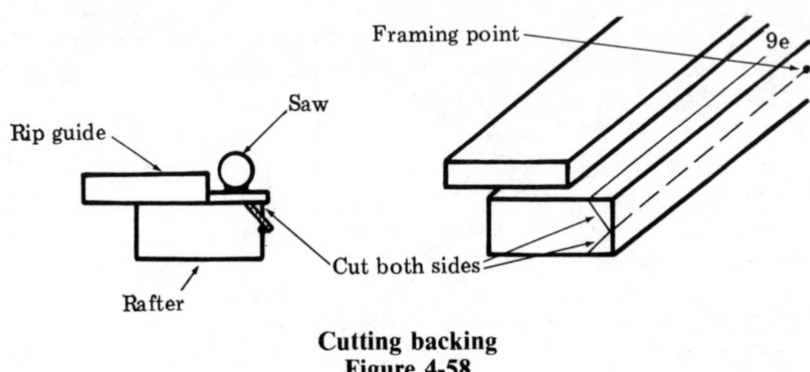

Cutting backing
Figure 4-58

Backing is less important on 2'' material than on larger beams. Figure 4-59 shows the need for backing a larger beam. Part A shows a 12 in 12 or 45-degree roof. The hip, however, is always a lower pitch. This pitch can be found by dividing the unit rise of 12 by 16.97. That gives 0.7071, which is the tangent for this hip. Since rise equals tangent times run, distance (d) in Figure 4-59 can be found by multiplying 0.7071 by 4''. That's 2.828'', or 2$\frac{13}{16}$''.

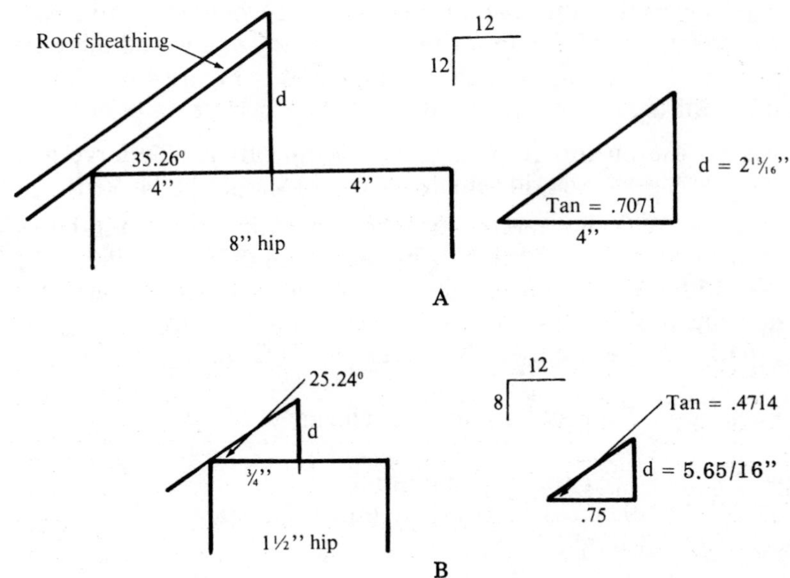

The necessity of backing
Figure 4-59

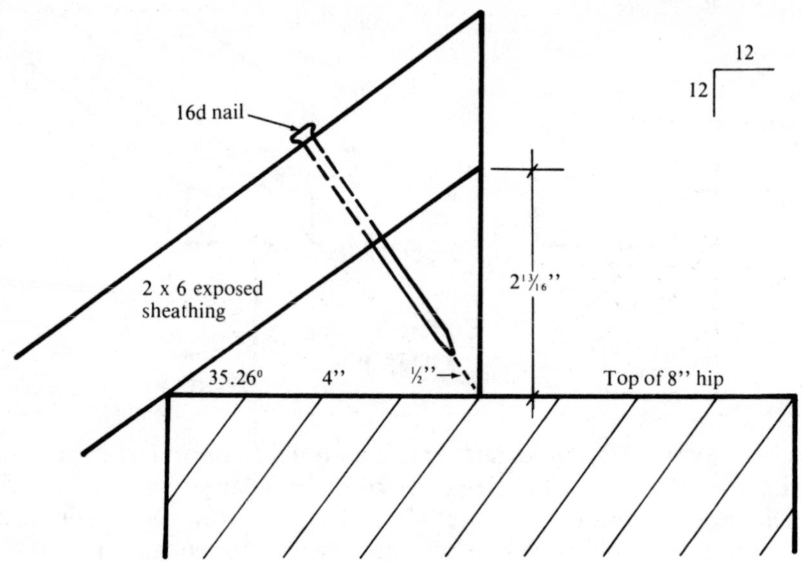

Nailing roof sheathing to a 12 in 12 hip rafter
Figure 4-60

This means that the edge of the sheathing would stick up 2¹³⁄₁₆" above the center of the hip rafter. A 16-penny nail would miss the top of the hip by half an inch, as illustrated in Figure 4-60. An 8d nail in half-inch sheathing would miss by the same amount.

It's no use on this roof to drop the hip rafter. The large gap above the hip would still remain. This hip should be backed.

Figure 4-59 B shows our model roof hip rafter. This hip has a tangent of 0.4714. Multiply this by the run of 0.75 to get 0.3535", or 5.6/16ths. That's our backing dimension. (This is only another arrangement of the formula in the section on dropping the hip using math. In practice there is no need to back this hip.)

Backing = Tangent of Hip x ½ Thickness of Hip

The Double Cheek Cut Birdsmouth

There's one other cut to consider on the hip rafter; the double cheek cut birdsmouth.

Suppose that a building requires 8" x 16" rough-sawn exposed rafters. The outside sheathing is to be 5/8" fir plywood siding trimmed in 2" x 6" rough-sawn lumber.

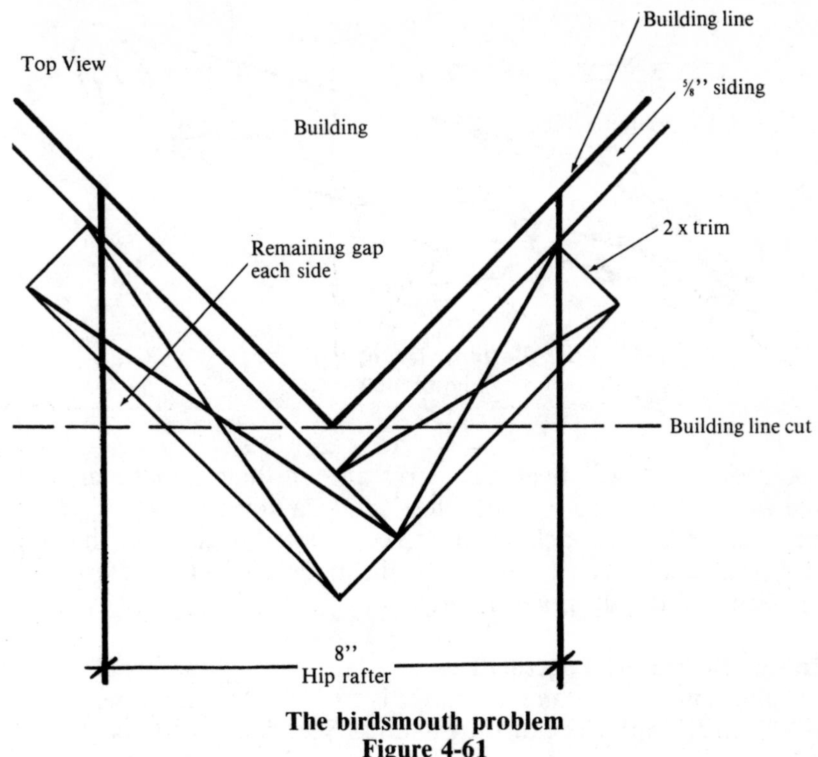

The birdsmouth problem
Figure 4-61

Even with the siding and trim in place, a gap remains between the lower corner of the building line and the trim. See Figure 4-61.

To avoid this, make a double cheek cut birdsmouth. This can also be done on your hip rafter model.

Figures 4-49 and 4-50 show how to draw a construction line. It is actually the set-forward line (9c) for the double cheek cut birdsmouth. Notice that half the thickness of the material is used in three places: lines (6), (8), and (9c). For 1½'' thick material you move forward 3/4''. But for an 8'' rough-sawn beam you would move forward a full 4''.

Look at Figure 4-62. First square the building line across the bottom of the rafter and put a dot in the center. Draw a diagonal line (a) on the bottom of the rafter from the dot in the center, forward to the edge where the set-forward line hits the bottom of the rafter. Transfer the set-forward mark to the far side of the rafter and draw the other line (b). These lines are *not* at 45 degrees.

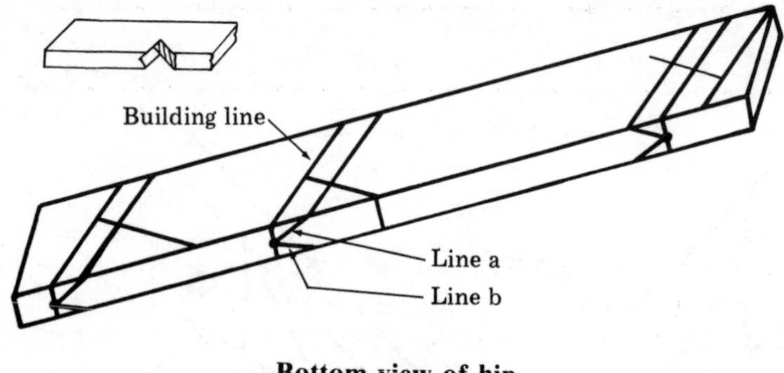

Bottom view of hip
Figure 4-62

Cut on these two cheek cut marks and on the original seat-cut line (4). You'll need a hand chisel and hammer to remove the material. Make the backing cuts to complete this hip. Use this as one of the double cheek cut hips on the model to see the difference between it and a dropped rafter.

Putting the Model Together

Since the run of the common rafter is 18", the framing point will be 18" in from the end of the building. See Figure 4-63. Back up half the thickness of the common rafter so that the center line of the rafter will be in line with the framing point. Make a mark at 17¼" on each rafter plate.

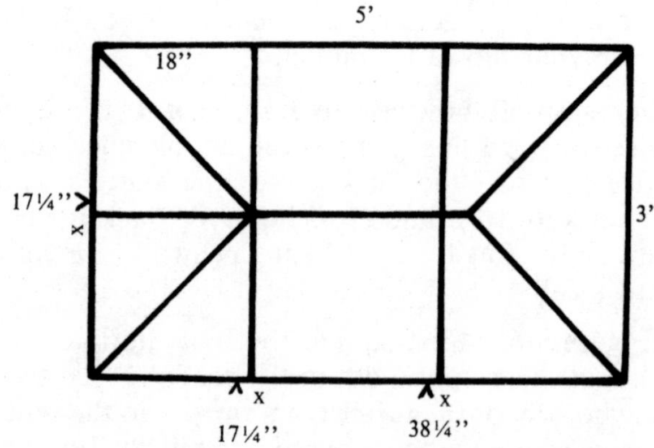

Laying out the rafter plate and ridge
Figure 4-63

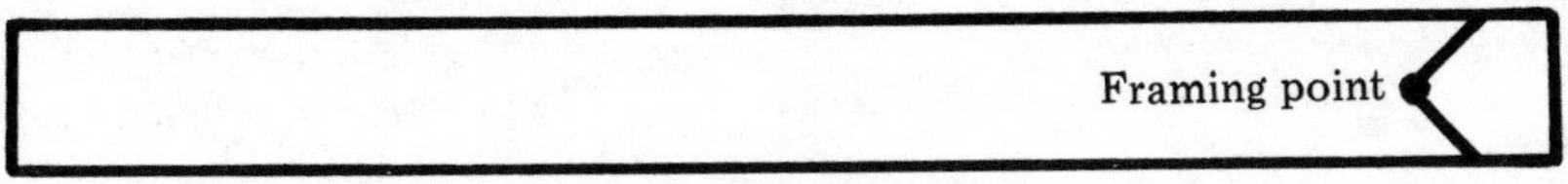

Framing point

The ridge
Figure 4-64

Also mark 38¼'' and place an "X" to the right of those marks. Mark the ridge on top at 21'' and place an "X" to the right of this mark also. On the double cheek cut side mark 17¼'' for the end common rafters.

Lay the ridge in the center of the frame and lay the four opposing common rafters over it. Nail the rafters at the seat cut on the line and over each "X". Lift the ridge in place and nail it off. Add the common rafter on the end.

On the ridge, draw two 45-degree lines from the single cheek cut framing point to help you line up these rafters, as in Figure 4-64. Now add the single cheek cut hips. Notice that a corner of the ridge sticks out and must be removed. The amount of extra material to be removed depends on whether you backed or dropped your hip rafter. For each, the amount is different.

Congratulations on your completed model.

Figure 4-65 is a reference chart for hip and valley roof angles and their respective secants.

(1) Use this secant with the actual run, which is the apparent run x 2.

(2) Use this secant with the apparent run, which is the run of the common rafter. Also see Figure 4-29.

Rise	Degrees	(1) Actual Run Secant	(2) Apparent Run Sec. x$\sqrt{2}$
24	54.74°	1.7320	2.4495
23	53.58°	1.6843	2.3819
22	52.35°	1.6372	2.3154
21	51.06°	1.5910	2.2500
20	49.68°	1.5456	2.1858
19	48.23°	1.5012	2.1230
18	46.69	1.4577°	2.0616
17	45.05°	1.4154	2.0017
16	43.31°	1.3744	1.9436
15	41.47°	1.3346	1.8875
14	39.52°	1.2964	1.8333
13	37.45°	1.2597	1.7815
12	35.26°	1.2247	1.7320
11	32.95°	1.1917	1.6853
10	30.51°	1.1607	1.6415
9	27.94°	1.1319	1.6008
8	25.24°	1.1055	1.5635
7	22.42°	1.0817	1.5298
6	19.47°	1.0607	1.5000
5	16.42°	1.0425	1.4743
4	13.26°	1.0274	1.4530
3	10.02°	1.0155	1.4361
2	6.72°	1.0069	1.4240
1	3.37°	1.0017	1.4167

16.97''   0''

Hip and valley rafter roof angle chart
Figure 4-65

5

Valley Rafters Are Easy

If you've mastered the information in Chapter 4, you'll have no trouble with valley rafters. Hips and valleys are very similar, so most of the same principles apply.

Our first two roof models had rectangular top plates and therefore had only outside corners. A roof with only outside corners could have either a simple gable roof or the more complex hip roof.

Part 1: An Equal Span Addition

If the owner makes an addition to one of the sides, two interior corners are formed and each must have a valley rafter. See Figure 5-1.

Valley rafters support the weight of the valley jack rafters and the surrounding roof. That's why valley stock is usually one or two sizes bigger than the common rafters, or is doubled — two pieces of stock are nailed together side by side to make the valley stronger.

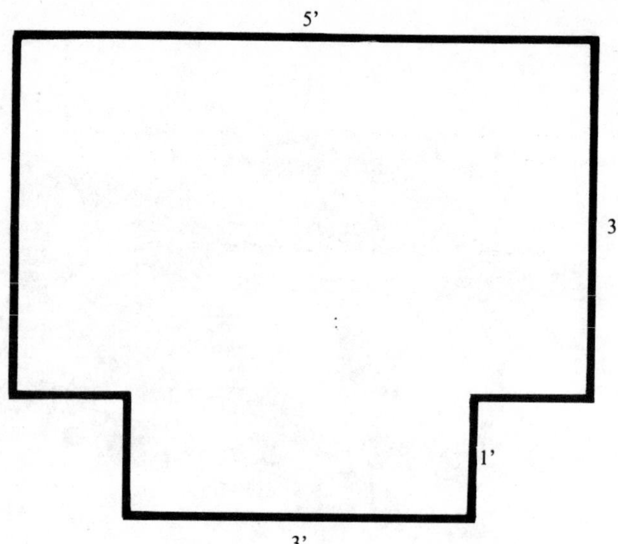

Plan view of an equal span addition
Figure 5-1

The Addition Rafter Plate

Remember the roof you put on the model in Chapter 4? Well, the owner has decided he needs some more living space. He wants to add a room at one side of the house, and you have to frame the roof for the addition.

Here's how to start: First, orient the 3' addition plate to conform with Figure 5-1. Then remove two common rafters at the middle of the appropriate side of the model, leaving the hip rafters and other commons in place.

Then, lay out marks on the plate, one foot in from each side, so that the addition will be centered, as shown in Figure 5-2. Cut a piece of 2 x 6 stock exactly 3' long.

Next, cut two end pieces to the correct length so that the rafter plate of the addition extends exactly 12'' beyond the main structure. Nail the pieces together with three 16-penny nails at each end. With your framing square, set the pieces at right angles and nail a gusset triangle on each corner. See Figure 5-3.

Now turn the piece over, so the gussets are on the bottom. Use two 16-penny duplex nails to fasten the addition rafter plate to the model.

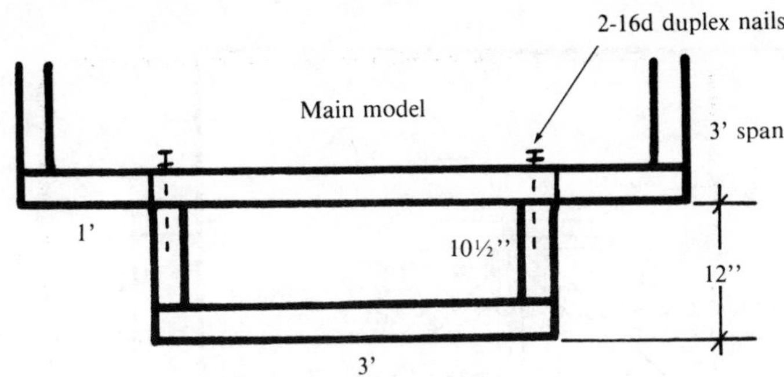

The addition rafter plate
Figure 5-2

Measure between the framing points on the main ridge and mark the center framing point, as indicated in Figure 5-4. Move 3/4'' to one side and draw a line and an ''X'' through the framing point. This marks the layout for the new ridge of the addition.

The 3' Addition Ridge

When the span of an addition is equal to the span of the main building, and the unit rise is the same for each roof, the two ridges will be at the same height. That's the case with the addition we're building.

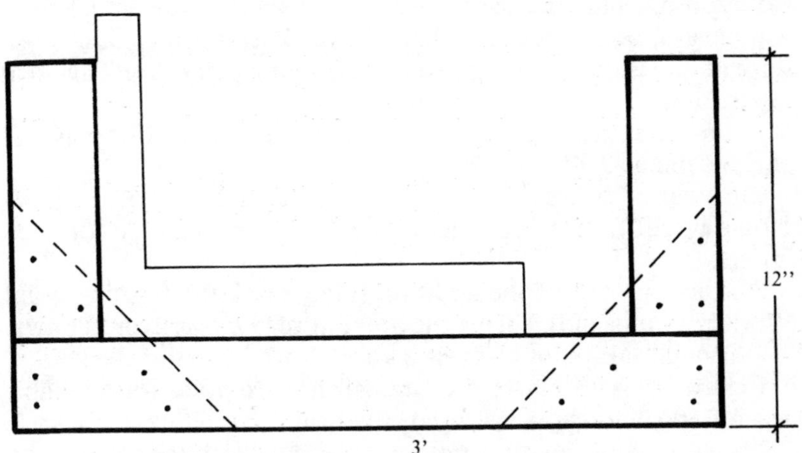

Adding the rafter plate
Figure 5-3

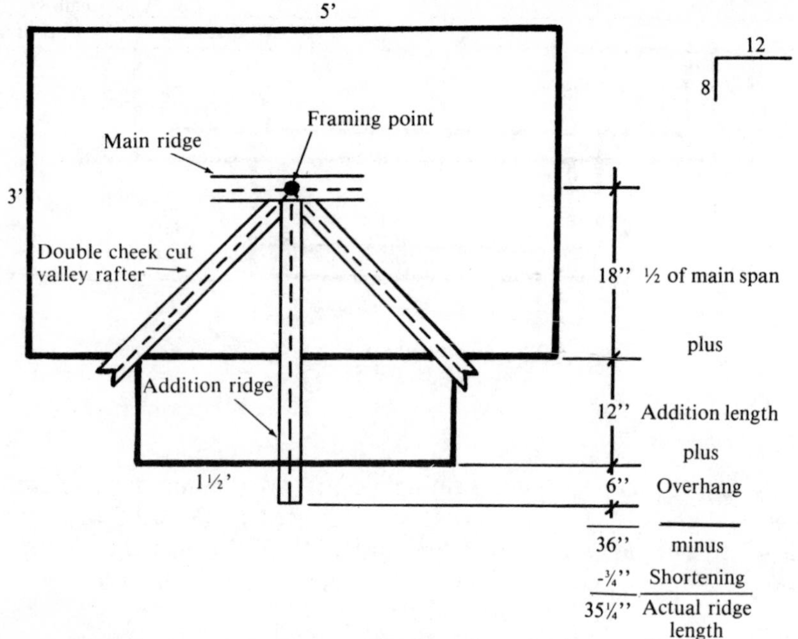

Ridge and valley framing detail
Figure 5-4

The addition ridge (in this case a gable ridge) will have a mathematical length of one-half the main span, plus the length of the addition, plus the overhang. But notice that the actual ridge will have to be shortened by half the thickness of the main ridge, since its end is that far away from the framing point. Look again at Figure 5-4.

In this case, the length of the addition ridge will be 18" plus 12" plus 6" minus 3/4", or 35¼".

Cut your addition ridge to this length from 2 x 6 stock. Ridges should be cut from stock at least one size bigger than the common rafters.

On the top edge of the addition ridge, mark the framing point 6" in from one end. This is the amount of the overhang. Draw a line through this point and place an "X" to the left of the line, as in Figure 5-5. This marks a framing reference point for the gable end rafters. You should also label it "full span addition ridge."

The shortening for the common rafters will be the same as on the main roof. So you can reuse the two common rafters you removed for this gable roof addition.

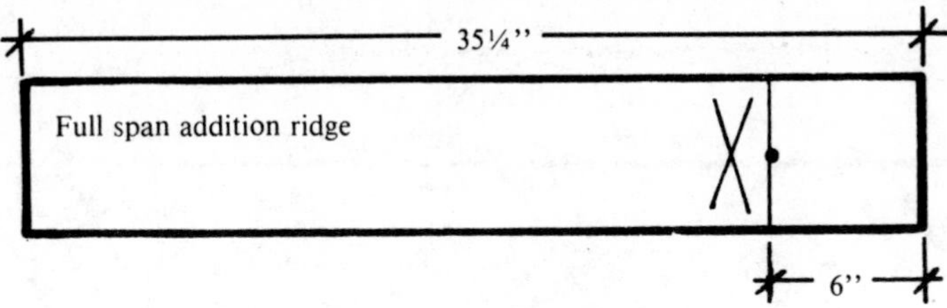

The addition ridge
Figure 5-5

Now we'll erect the new ridge. Use the same procedure as you did for the original ridge. Start with the pair of common rafters you just removed from the model. Nail them to the plate at the addition gable end. Make sure the outside edges of these rafters are flush with the outside of the addition building line.

Drive an 8-penny nail part way into the end of the ridge and bend it over as shown in Figure 5-6. Lay this hook over the main ridge. It holds the new ridge in place on one end while you nail the head of each rafter to the layout "X" on the addition ridge.

Now nail the addition ridge into the main ridge at the layout mark using two 16-penny duplex nails driven from the far side of the main ridge.

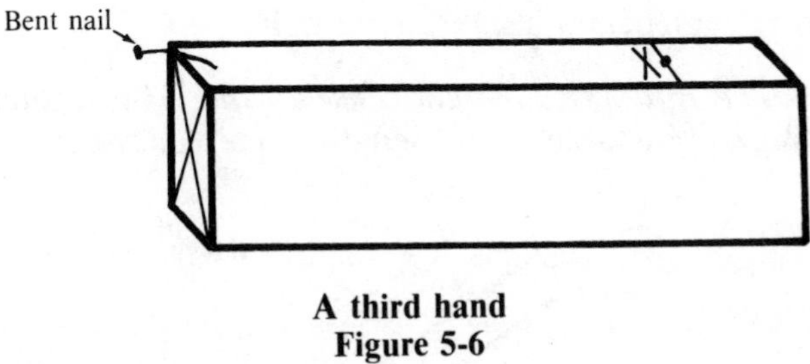

A third hand
Figure 5-6

The Valley Rafter

Figure 5-7 shows the layout lines of a valley rafter for this roof addition. The numbers in the following paragraphs refer to the line numbers in Figure 5-7.

Notice some differences between the valley rafter and the hip rafter you marked for this same model. Here, the cheek cut lines are set back with respect to the building and fascia lines. You'll

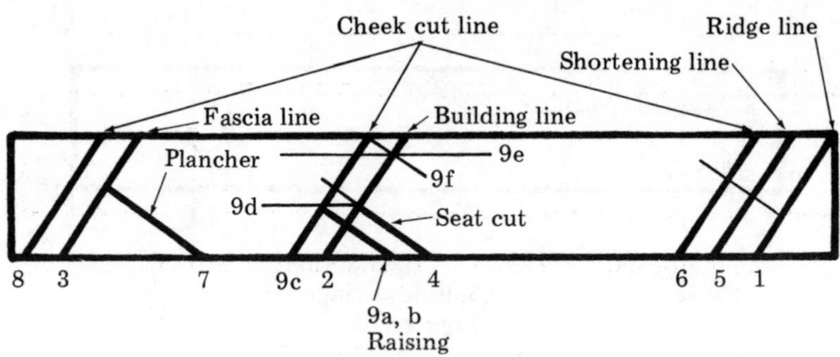

The lines of the valley rafter
Figure 5-7

remember that on the hip rafter they are set forward. For the valley rafter, the seat cut lines (4 and 9a,b) are also reversed. And the shortening is different.

1) The Ridge Line: The model valley rafters will be cut from 2 x 6 material. Select a piece at least 44'' long, crown it, and lay the crown away from you. Since this member runs at 45 degrees to the building line, you set the framing square gauge to 16.97'' on the body. And because all of this roof is 8 in 12 pitch, set 8'' on the tongue.

Draw the ridge line from the corner of the rafter, as in Figure 5-8. This is your beginning reference point.

2) and 3) The Building Line and Fascia Line: The second line of the framing square table reads ''length hip or valley per foot run.''

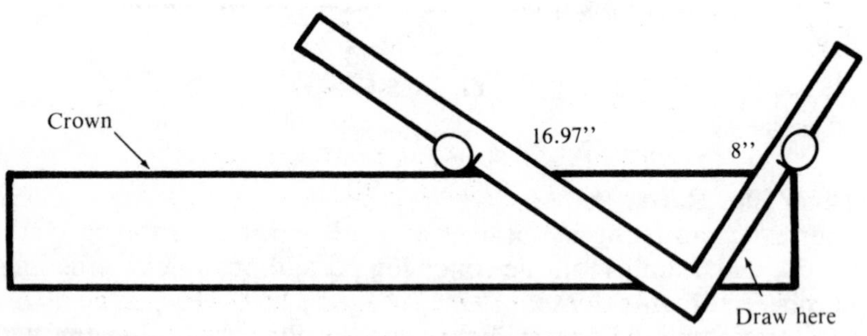

Drawing the ridge line
Figure 5-8

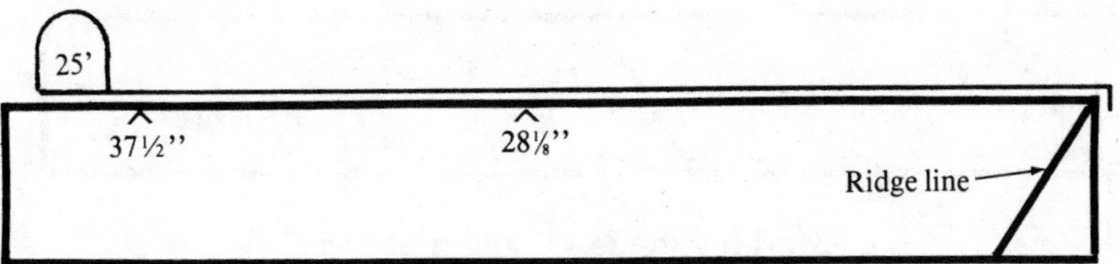

The mathematical lengths of the building and fascia lines
Figure 5-9

The second section of the *Rafter Length Manual* reads "hip or valley rafter lengths." Each of these titles indicates that the unit lengths in the tables are 16.97". The model valley rafter and hip rafter will have the same apparent run, 1½' in this case.

Since the run is the same, the mathematical lengths you calculated for the hips will apply to the valley rafters. Calculated lengths for the hip were 28⅛" to the building line and 37½" to the fascia. Mark these two lengths along the top edge of the rafter. See Figure 5-9.

Draw lines for the building line and fascia line as shown in Figure 5-10, and label them as indicated.

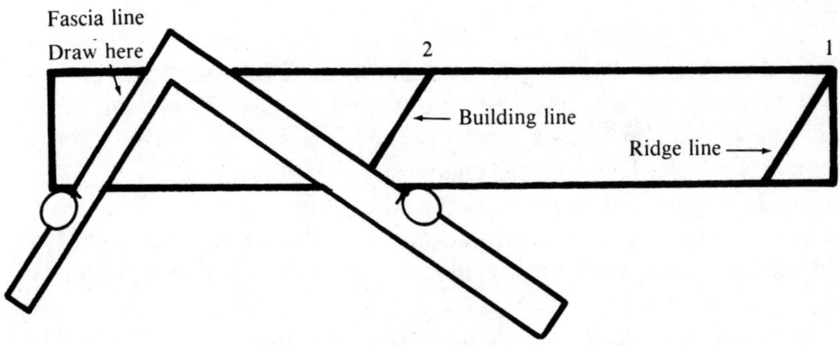

Drawing the building and fascia lines
Figure 5-10

4) and 7) The Seat Cut and Plancher Cut: The HAP and the plancher cut dimensions are the same as on the hip. From the top of the rafter, measure down 2¾" on the building line and then 2" on the fascia line. See Figure 5-11.

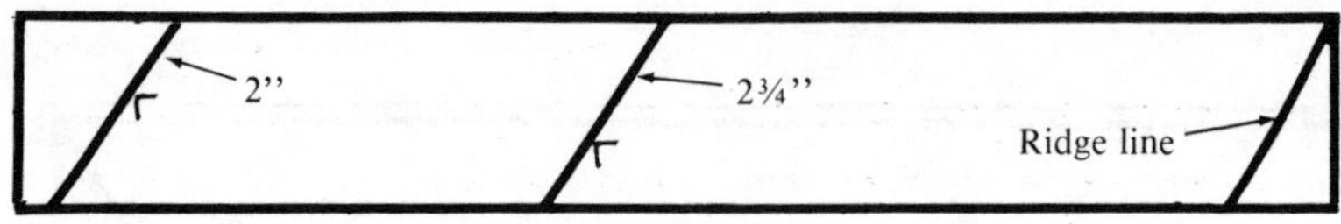

Marking the HAP and plancher
Figure 5-11

Draw the seat cut line and the plancher cut line, as in Figure 5-12. Then label these lines to avoid a mistake and to make checking easier.

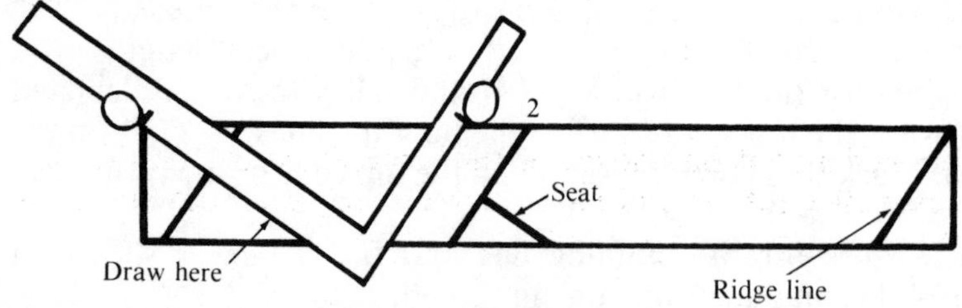

Drawing the plancher and seat cut lines
Figure 5-12

5) and 6) Valley Shortening and Setback: The valley rafter that you're marking will join the point where two ridges of equal thickness intersect. Look back to Figure 5-4. You know by now that it will require a double cheek cut.

Study Figure 5-13. How much shortening is needed for each valley rafter? You can see that where a valley rafter intersects two ridges, the valley must be shortened by half the 45-degree thickness of the largest ridge.

In our model, both ridges are equal and each is 1½'' thick. The shortening will be 0.75'' times the square root of 2. That's 1.06'' or 1$\frac{1}{16}$''.

If the material you're using is a bit thicker, increase the shortening just a little.

The cheek cut will be half the thickness of the valley rafter for both illustrations in Figure 5-13. But the actual cut depends on the configuration and which valley rafter you're cutting. You have to cut a right-side rafter and a left-side rafter if the thickness of the two ridges is different.

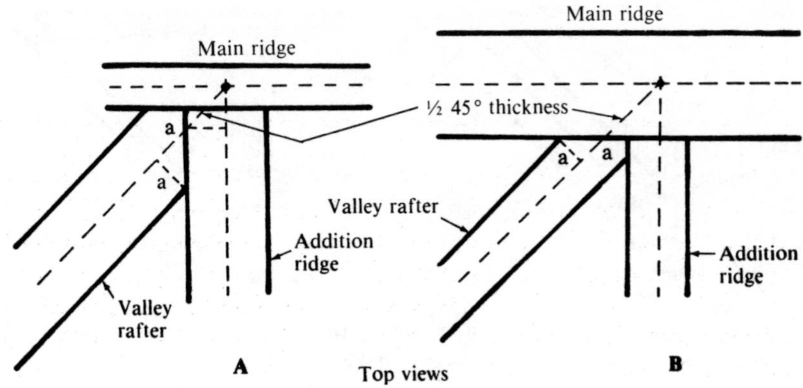

Valley shortening and setback at the intersection of two 90° ridges
Figure 5-13

Our model roof has ridges of the same thickness, so the right and left valley rafters will have double cheek cuts that are equal. Set the saw at 45 degrees for the cuts.

Draw a line perpendicular to the ridge line and then measure off the shortening and the setback. See Figure 5-14.

Draw the lines for the shortening and cheek cuts. Then square across the top and bottom of the rafter at the cheek cut line so you can transfer the cheek cut line to the other side.

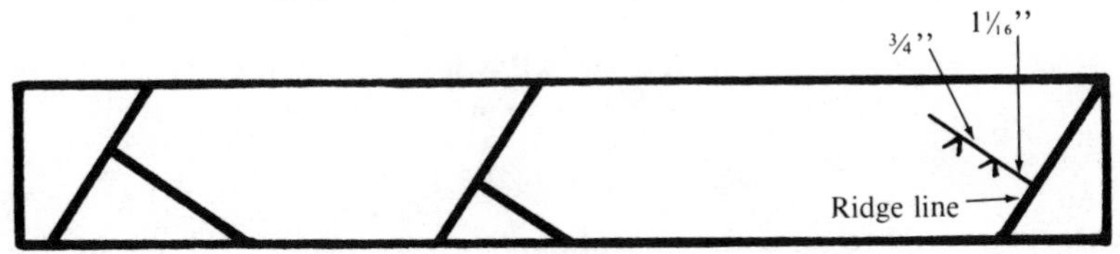

Marking shortening and setback
Figure 5-14

8) The Tail Cheek Cut: The valley rafter tail must be cut for an inside 45-degree fit of the fascia board. Otherwise the fascia won't look right. Look at Figure 5-15. Make the tail cheek cut line short of the fascia line by half the thickness of the rafter on each side.

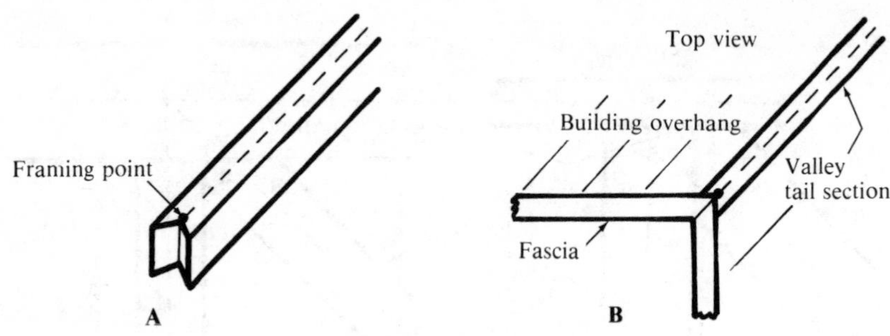

The valley rafter tail cut
Figure 5-15

Extend the plancher line on the valley rafter and measure back along this line to 3/4'' beyond the fascia line. See Figure 5-16. Make a mark and then draw the tail cheek cut line through this point. Do the same on the other side.

To make this cut, first set your saw at 45 degrees. Next, adjust the depth of cut to half the thickness of the rafter. Secure the rafter and cut carefully, always moving the saw forward.

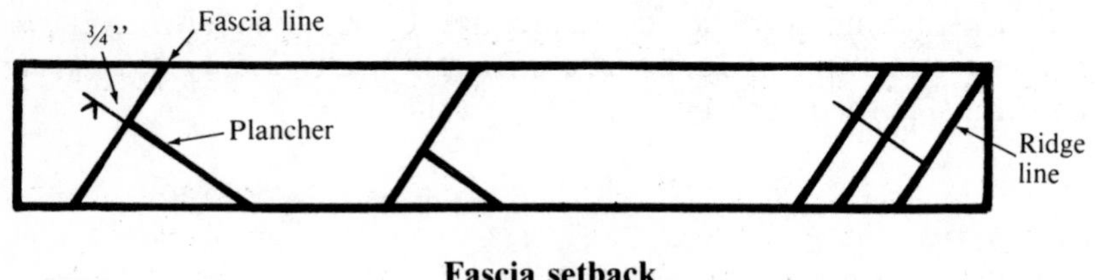

Fascia setback
Figure 5-16

9) Raising the Rafter: Remember from Chapter 4 that the hip rafter had to be either backed or dropped so that it would fit flush with the plane of the roof. The valley rafter is just the opposite. When it has been cut on the seat cut line, the valley rafter already has the edges of the rafter below the level of the roof, as illustrated in Figure 5-17 A.

Usually you'll want to raise up the jack rafters that end in a valley by the amount of the backing on an equivalent hip rafter. On an 8 in 12 roof, that would be 5½ sixteenths inches.

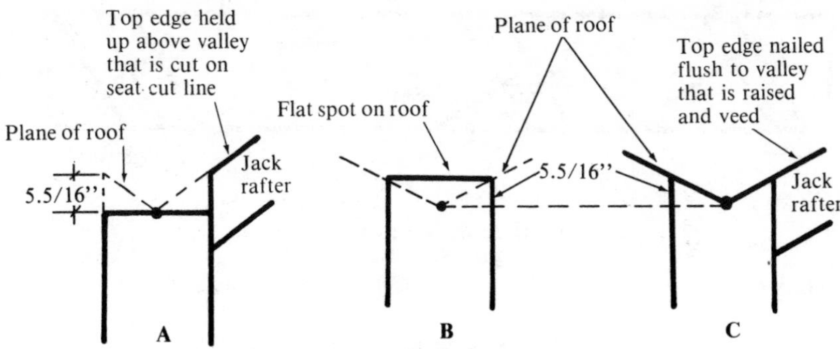

Valley rafters and the framing point
Figure 5-17

This type of construction is good enough if you don't have an exposed beam ceiling with the valley rafters showing. If the jack rafters were held above the valley in an exposed beam ceiling, there would be gaps in the ceiling all along the valley rafter. To avoid this, raise the valley rafter the amount of the backing. That brings the corners of the valley to the proper level. See Figure 5-17 B. But then you must also "vee" the rafter to provide clearance for the sheathing, as shown in Figure 5-17 C.

Unlike a hip that you can either back or drop, a valley rafter must either be left alone or it must be both raised and veed.

9c) Adding the Setback: Raise the valley just the opposite of the way you dropped the hip. Rather than a set-forward mark, you will mark a setback from the building line. Measure back on a line perpendicular to the building line for a distance equivalent to half the thickness of the valley rafter. See Figure 5-18.

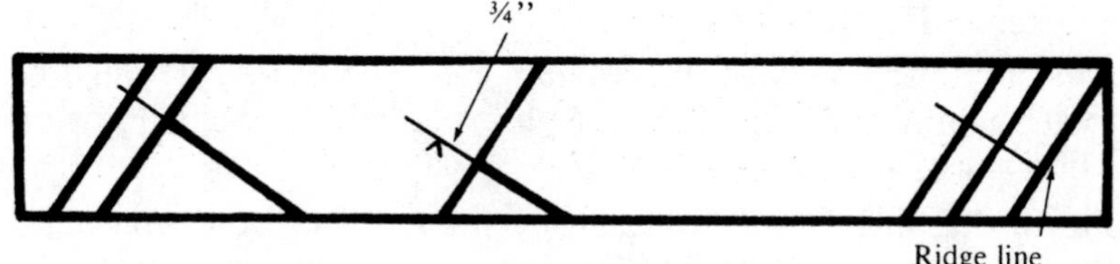

Marking off the setback
Figure 5-18

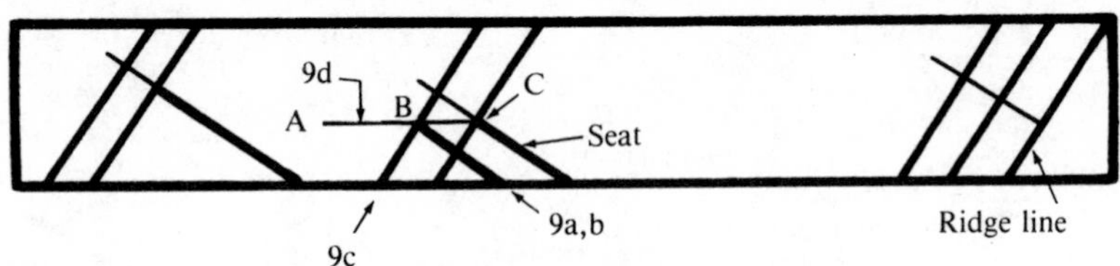

Drawing the cheek cut and raising
Figure 5-19

Draw in the cheek cut line and then construct line (9d) in Figure 5-19. (This line is also labeled "ABC"). Line (9d) begins at the corner of the building line and seat cut line and extends back toward the setback line. (This is just the opposite construction from the hip.) Draw line (9d) parallel to the edge of the valley rafter with a combination square.

From the intersection at (B), draw in the raising seat cut line. This line will appear below the seat cut line. The raising seat cut line is line (9a,b) in Figures 5-7 and 5-19.

Now look at Figure 5-20. After squaring across the top and bottom of the rafter and transferring line (9c) to the far side, draw the two lines on the bottom from the framing point to the edge of each cheek cut line.

There are two ways to cut the birdsmouth on this rafter. The first and most common method is to ignore veeing and raising. Simply cut the birdsmouth with the saw at zero degrees along the seat cut line and the cheek cut line. This leaves a triangular void at the inside corner of the rafter plate when the rafter is in place. With 2 x 6 stock, this gap is insignificant. But thicker material requires a different cut.

To avoid the gap on larger material, set the saw at 45 degrees and adjust the depth to half the thickness of the material. On one side cut along line (9c), with the blade pointing toward the framing point on the bottom of the rafter. See the lines on the bottom of the rafter in Figure 5-20.

Cut only one side with the circular saw. The other side must be cut with a hand saw. Guide it along line (9c) and the angled line on the bottom of the rafter.

Finish the birdsmouth by resetting the saw to zero degrees and cutting along the seat cut line, or, if you're raising and veeing the valley rafter, along line (9a,b). Remember, only the shaded portion is cut away.

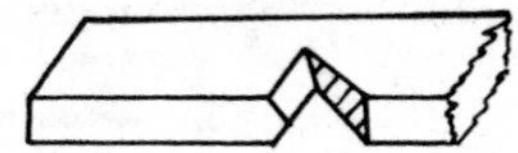

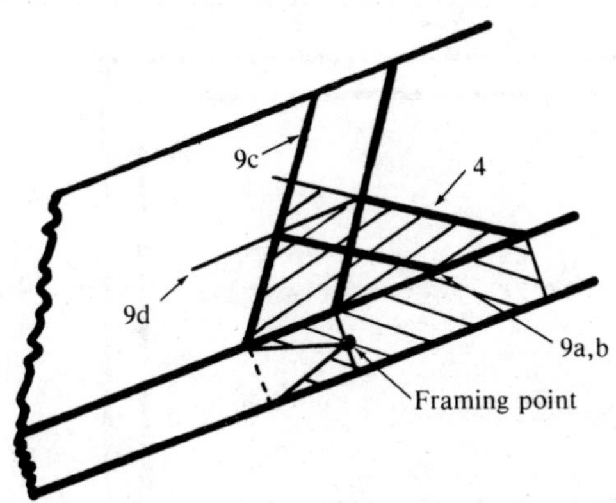

Preparing for cutting
Figure 5-20

Veeing the Rafter

For veeing, first set your saw at the same angle as for backing. On an 8 in 12 roof that angle is 25.24 degrees. Now adjust the depth to half the thickness of the rafter. Be careful to keep the blade on the top of the rafter. Don't cut off any of the edge. That would shorten the effective length of the HAP and plancher.

On your model rafters, making this cut on a table saw will be much easier. Make one valley rafter plain, with only the construction cut. Make the other raised and veed with the proper beveled cheek cut in the birdsmouth. Frame the two different valley rafters on the model.

Congratulations! That completes your full span addition.

Now let's look at how a narrow span addition changes cutting the valley rafter.

Part 2: A Narrow Span Addition

For this addition, the span is 2', or 1' narrower than the main span of the building. See Figure 5-21. You again remove two common rafters, this time from the middle of the opposite side of your model. The next step is to build a plate for the addition.

Cut a piece of 2 x 6 stock exactly 2' long. Again, make the addition 1' wide and nail gussets to the bottom as you did before. Lay out marks 1½' from each end of the main rafter plate and measure

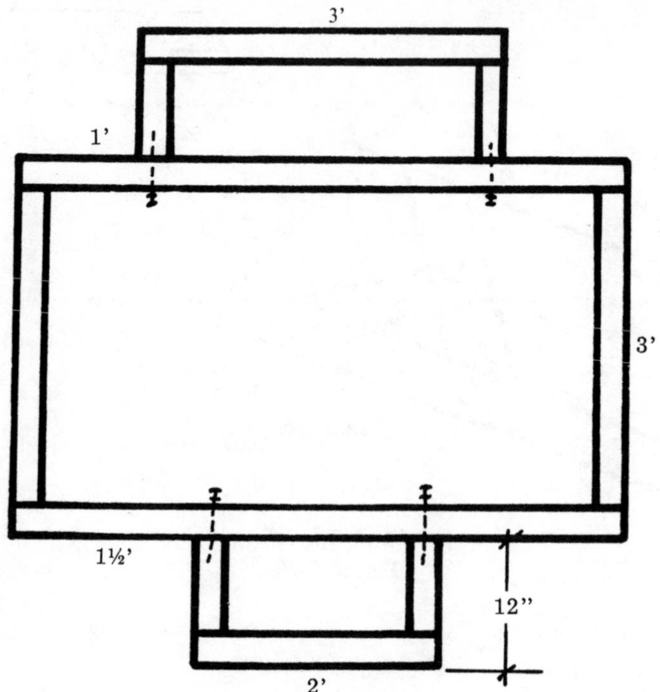

Narrow addition plate
Figure 5-21

between them to make sure you have exactly 2'. Nail the 2' addition rafter plate to the model, as you did the 3' addition.

There are two ways to frame the roof on a narrow span addition. The more popular method is called *blind valley construction* and is described in Chapter 7. In this chapter we'll cover the *supporting valley rafter* method.

The Supporting Valley Rafter

In this type of addition there are two valley rafters. Look at Figure 5-22. One rafter is a supporting valley because it runs to the ridge of the main part of the building and supports the other valley rafter. The other valley rafter is much shorter and is called the *shortened valley rafter*.

Even on the shorter span addition, the supporting valley rafter has the same run as the hip rafter. That makes it nearly identical in several respects to the double cheek cut valley rafter we just cut.

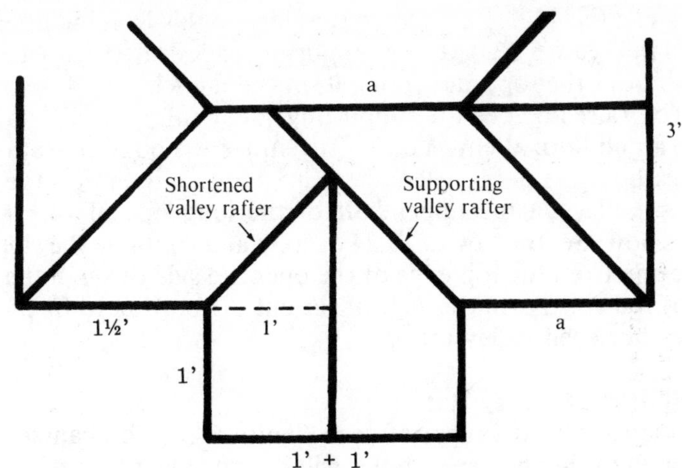

The supporting and shortened valley rafter
Figure 5-22

Lay out another valley rafter as in the last section, but don't transfer line (6) to the far side. This rafter has only a single cheek cut to fit against the ridge.

Notice in Figure 5-22 that length (a) at the ridge line and length (a) at the building line are equal. Since we laid out our addition rafter plate 18'' in from one side, we can also lay out framing point (d) 18'' over from the double cheek cut hip framing point. See Figure 5-23.

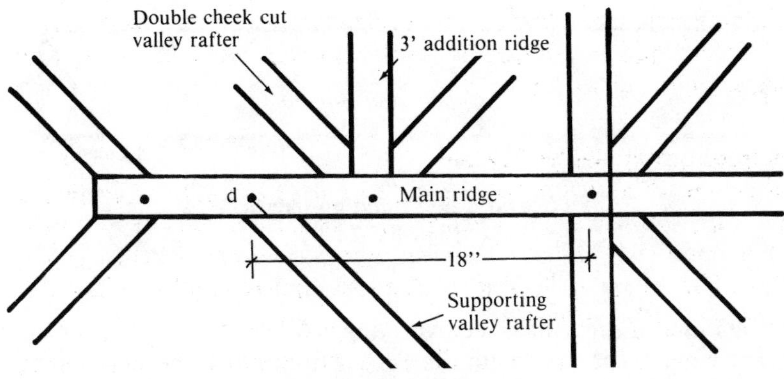

Ridge framing points
Figure 5-23

Draw in the 45-degree line on the ridge at this framing point, as shown in Figure 5-23. Cut the supporting valley rafter with a single cheek cut at the top, and frame it on the model.

If you raise and vee the supporting valley rafter, you'll have to make an additional cut. You must remove the quarter section of veeing above the point where the ridge addition intersects the supporting valley on the upper half of the long side of the rafter. Here's how: Set the saw to 25.24 degrees and continue the slope of the veeing from the top edge of the opposite side through the bottom of the vee. Form the edge at the cut-off portion to fit the end of the shortened valley rafter.

The Shortened Valley Rafter

Find the shortened valley rafter in Figure 5-22. You can see that the length of the shortened valley will be equivalent to the diagonal of a square that is one-half of the addition span. This means that the shortened valley rafter will have a mathematical length to the building line of the unit length times one foot. The unit length is 18.76", or 18¾".

The line numbers in the following section again refer back to the rafter layout lines in Figure 5-7.

1) and 2) The Ridge Line and Building Line: Crown a piece of 2 x 6 material and lay the crown away from you. Place a ridge line at one end as in Figure 5-24. Make sure your framing square stair gauges are on 16.97" and 8". Then make the building line mark at 18¾", as in Figure 5-24, and draw the building line as in Figure 5-25.

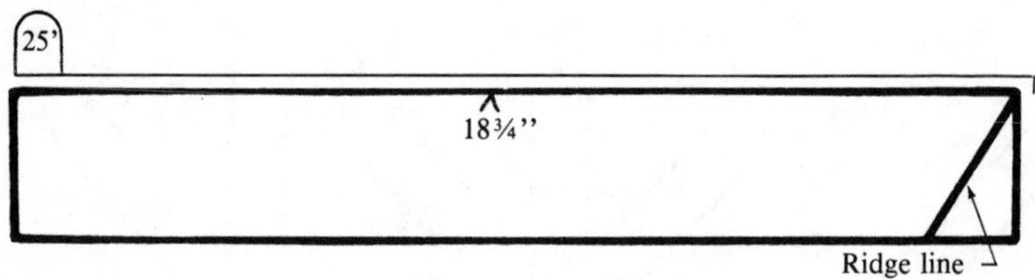

Marking the building length
Figure 5-24

3) The Fascia Line: Since the overhang is 6" on our model, the apparent run to the fascia line is 18". Multiplying the unit length times 18" gives us 28.14", or 28⅛". Mark this point on the rafter, as in Figure 5-25.

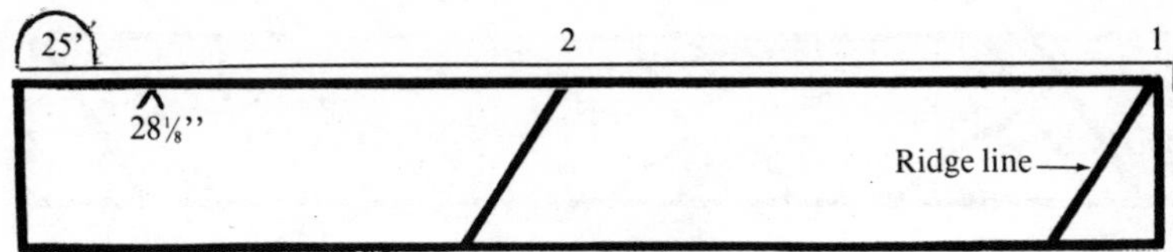

Marking the fascia length
Figure 5-25

4), 7), 8) and 9) The Seat Cut, Plancher and Cheek Cuts: After marking off the fascia line, proceed with the plancher, HAP, and cheek cut lines, as in the other valley rafters. Decide if you will raise and vee the rafter and what type of birdsmouth you'll cut.

5) The Shortening: Look at Figure 5-26. The shortened valley rafter doesn't quite make it to the framing point. It must be shortened half the thickness of the supporting valley rafter, in this case 3/4".

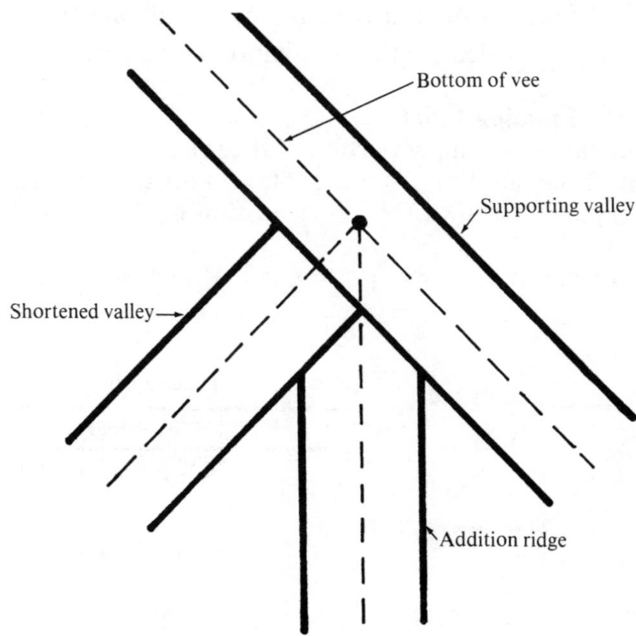

Shortening of the shortened valley rafter
Figure 5-26

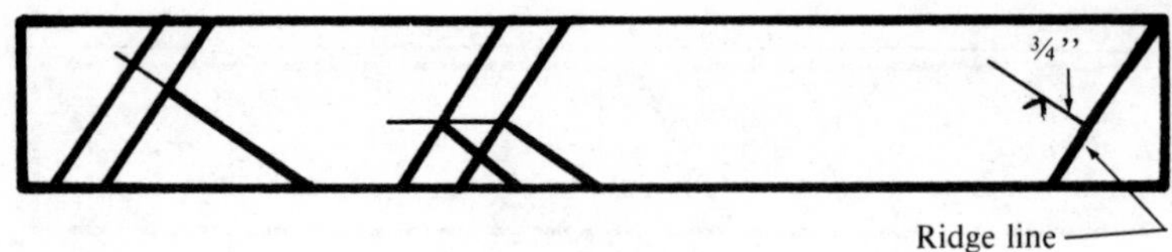

Shortening the rafter
Figure 5-27

Construct a perpendicular from the ridge line, as in Figure 5-27, and measure back 3/4''. Through this point draw and label the shortening line. Cut this rafter at the shortening with the saw at zero degrees, that is, with the saw table perpendicular to the blade.

Line 6 is not needed for this rafter.

Framing the Shortened Valley Rafter

There are two ways to find the framing point where the shortened valley rafter meets the supporting valley rafter. You could square over from the center of the 3' addition ridge and then square up along the 2 x 4 brace in the center of the main building frame.

A better way is to calculate the framing point exactly.

Calculating the Framing Point

First, find the run of the upper portion of the rafter. If we subtract the total run of the addition from the total run of the main roof, we have the apparent run of the upper portion of the rafter. See Figure 5-28 A.

The apparent run of the upper portion is 1½' minus 1' or 1/2'.

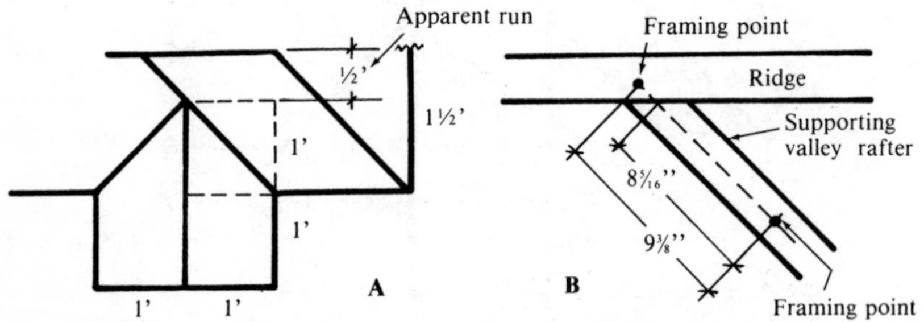

Mathematical location of the framing point
Figure 5-28

Using the framing square method: Apparent Run times Unit Length. That's 0.5' times 18.76, or 9.38", which is 9⅜".

Using the secant method: Apparent Run times Square Root of 2 times the Secant. That's 6" times 1.414 times 1.1055 or 9.38" which is 9⅜".

You could also use the Secant times Square Root of 2, from the last column in Figure 4-65. That's 1.5635 times 6" equals 9.38" again.

Look at Figure 5-28 B. Since the ridge shortening was 1¹⁄₁₆", we must subtract this from 9⅜" to find the framing point correctly. Measure 8⁵⁄₁₆" along the center of the supporting valley rafter and mark the framing point.

Cutting the Small Common Rafters

The total run of the narrow span addition is only one foot. The common rafters we removed from the model to make room for the narrow addition will be much too long. We need to cut one pair of commons for the gable end. They will help support the ridge on the narrow addition.

Change the gauges on your framing square from 16.97" to 12" and switch gears mentally back to 8" and 12" and away from 8" and 16.97".

The unit length for an 8" in 12" roof is 14.42" by either the framing square method or the square root method. Calculate the small common rafter length in Figure 5-29.

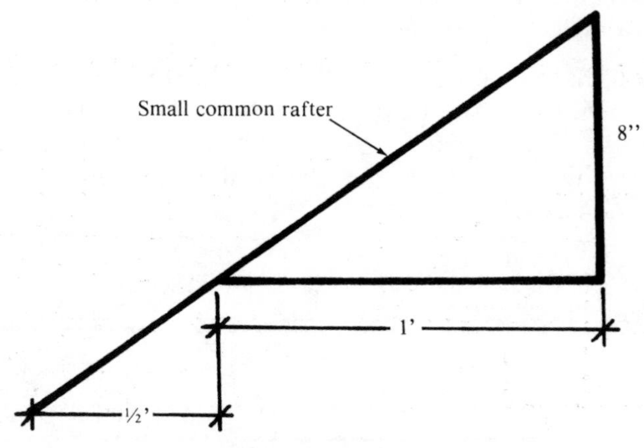

The run of the small common rafter
Figure 5-29

To the building line, the run of common rafter is one foot: 14.42 times 1' is 14.42", or 14⁷⁄₁₆".

To the fascia line, the run of common rafter is 1'6": 14.42 times 1.5' is 21.63", or 21⅝".

Lay out a common rafter that has a building line at 14⁷⁄₁₆" and a fascia line at 21⅝". See Figure 5-30. Use the same HAP and plancher cut as before. Since the 2' span addition ridge is a 2 x 6, use a shortening of 3/4". Cut out two of these rafters and frame them at the gable end as you did for the 3' span addition.

The Narrow Span Addition Ridge

The length of this ridge is calculated in a way similar to the length of the 3' span addition ridge. See Figure 5-31.

The length from (a) to (b) is the length of the overhang. The length from (b) to (c) is the length of the addition. The length from (c) to (d) is always half the span of the addition. Why? Because the two short sides on a 45-degree right triangle are always equal in length.

Ridge Shortening and Cheek Cuts

Cut a piece of 2 x 6 lumber to 30" long and mark a framing point 6" in from one end. Draw a layout line through the framing point and make an "X," as in Figure 5-32. This is where the gable end common rafters will frame.

Look back to Figure 5-26. Notice that the ridge ends short of the framing point by half the 45-degree thickness of the supporting valley rafter. This distance is 1¹⁄₁₆". Therefore, draw line (5), the shortening line, 1¹⁄₁₆" from the edge of the ridge. See Figure 5-32.

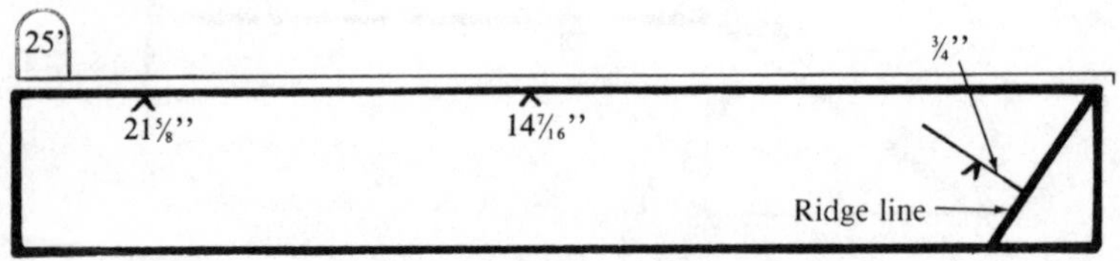

Laying out the small common rafter
Figure 5-30

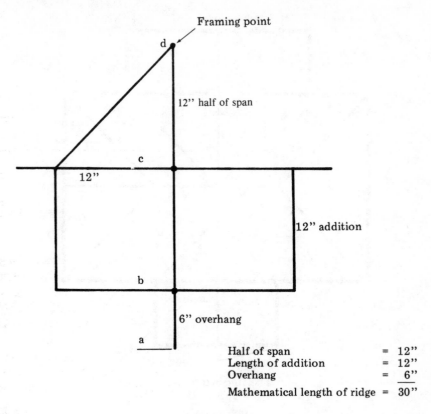

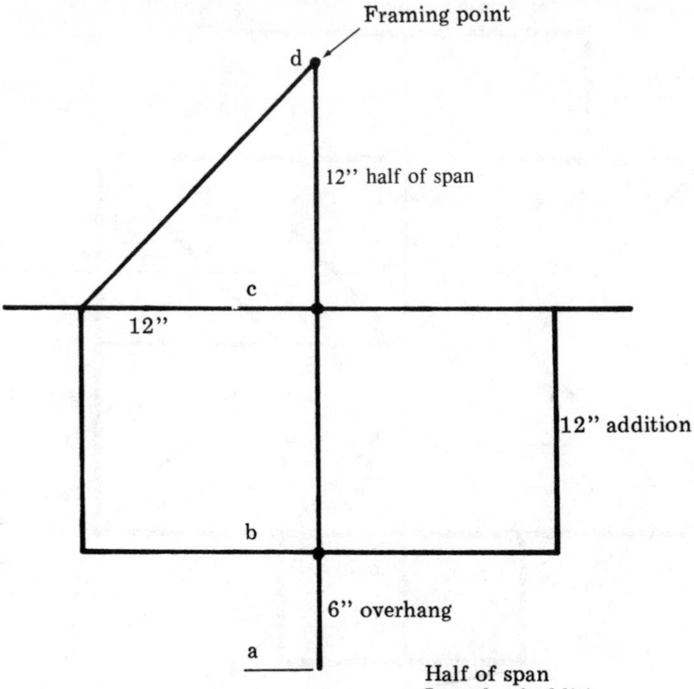

Framing point

d

12" half of span

c

12"

12" addition

b

6" overhang

a

Half of span	=	12"
Length of addition	=	12"
Overhang	=	6"
Mathematical length of ridge	=	30"

Ridge length
Figure 5-31

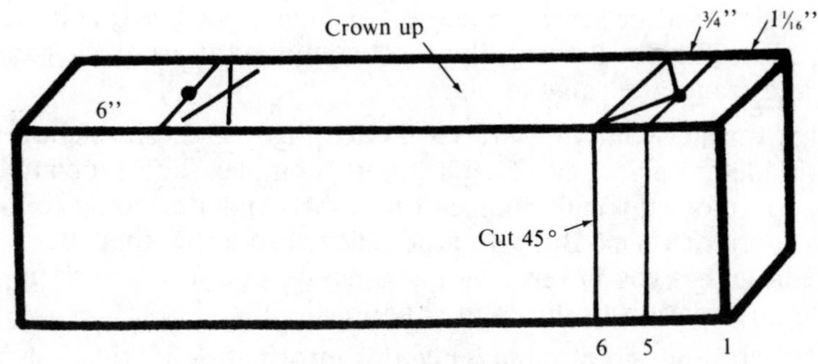

Laying out the ridge
Figure 5-32

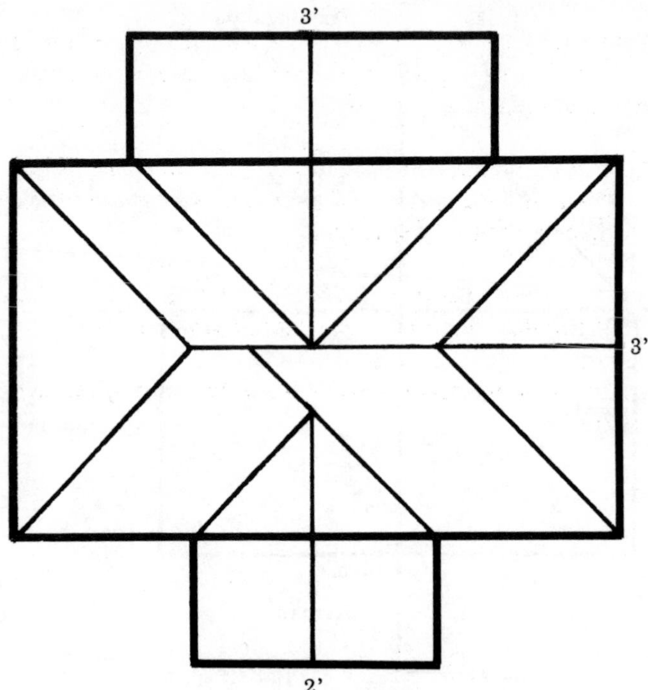

The finished model
Figure 5-33

Since this is a double cheek cut ridge, lay out line (6) another 3/4'' from line (5). Square across the top and transfer this line to the far side. Set the saw at 45 degrees and cut along line (6) on each side. Slip this ridge between the short common rafters at the gable end and frame the ridge in place.

That finishes your valley rafter model. Figure 5-33 shows how it should look in plan view. That's a pretty complex roof. Of course, there are more difficult roof cutting jobs. And I'm going to explain more of them. But you'll be relieved to know that they involve little that you haven't learned already. You just have to apply the information to different situations.

But before we go on, and while this information is still fresh in your mind, let's try some problems that test your understanding of this chapter. The answers are at the back of the book.

Problems

1) *A narrow span addition is to have a span of 18' and a length of 10'. It will have a gable end with a 2' overhang. What is the mathematical length of the ridge?*

2) *A narrow span addition is to have a span of 10' and a length of 18'. It will have a gable end with a 2' overhang. What is the mathematical length of the ridge?*

3) *A 14' span addition is being added to a 63' x 22' building. What is the actual run of the shortened valley rafter?*

4) *In the above problem, what's the actual run of the supporting valley rafter?*

6

Jack Rafters Are Easy Too

You can think of jack rafters as interrupted common rafters. As such their calculations are based on a unit run of 12''.

There are four types of jack rafters:

Jack Rafter	Runs
1. Hip jack	From plate to hip
2. Valley jack	From ridge to valley
3. Hip-to-valley cripple jack	From valley to hip
4. Valley-to-valley cripple jack	From shortened valley to supporting valley

Types of jack rafters
Table 6-1

The first two jacks touch either the plate or the ridge. The last two are cripple jacks and touch neither the plate nor the ridge.

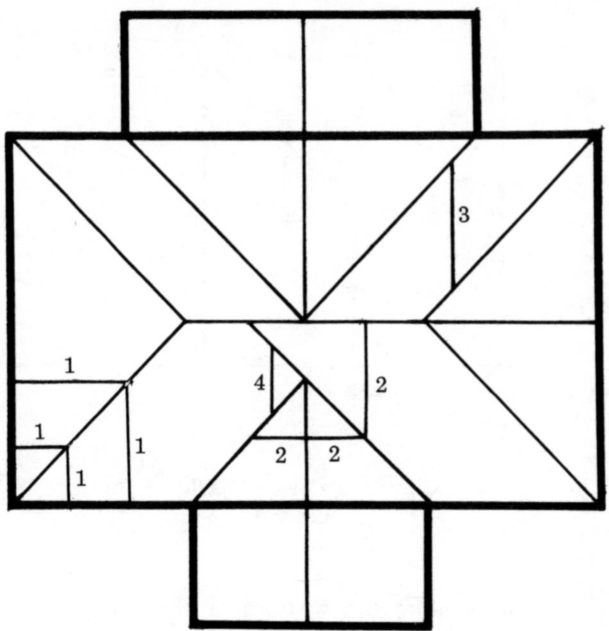

Types of jack rafters from Table 6-1
Figure 6-2

Figure 6-2 shows each type of jack rafter used in a typical plan. The rafter numbers in Figure 6-2 correspond with the numbers in Table 6-1.

Hip Jack Rafters (1)

How are we going to figure the length of these jack rafters?

Here's the key to unlocking the puzzle of the length of the run of jack rafters. Remember that in any figure with 45-degree angles, the two short sides are always equal.

Look at Figure 6-3. Since the hip or valley is at a 45-degree angle, rafter (a) equals rafter (a1) and both are 2' long. Also, (b) equals (b1) and both are 4' long. The run of rafter (a) is 2' and the run of rafter (b) is 4'. The run of this type of jack rafter is the same as the distance from the corner when measured along the plate.

To find that run, all you do is measure along the plate from the corner of the building to the jack rafter center line.

On an 8 in 12 roof, the unit length for a common rafter is 14.42''. Therefore, rafter (a) will have a mathematical length of 14.42 times 2', which is 28.84'', or 28⅞''. Rafter (b) in Figure 6-3

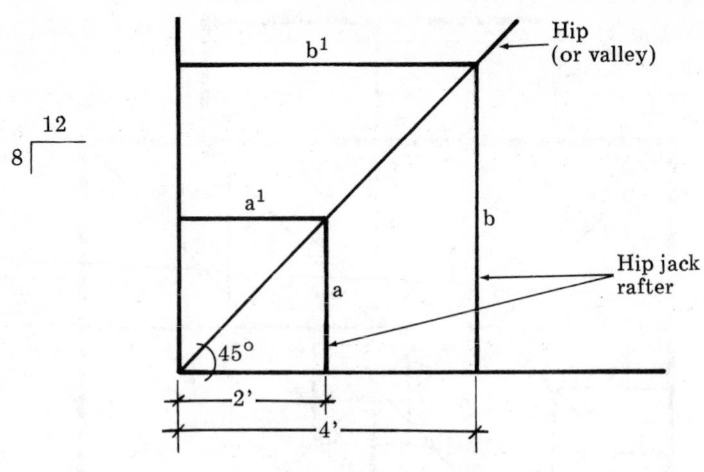

Finding the run of jacks
Figure 6-3

will be twice as long as rafter (a) because the run of rafter (b) is twice the run of rafter (a).

Since the hip jack rafters are evenly spaced, each following hip jack rafter is longer than the one before by the length of the first hip jack rafter. Look at Figure 6-4. For jacks 2' on center, the first rafter is 28⅞'', as we just calculated. For jacks 16'' on center, the

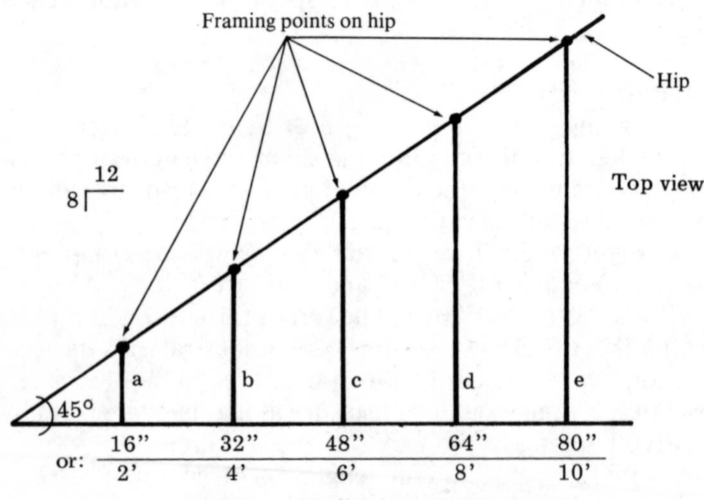

Common difference in hip jacks
Figure 6-4

first rafter is 14.42 times 16'', divided by 12 because the answer has to be in feet.

Sixteen divided by 12 is 1.3333. Multiplying that by 14.42 gives us 19.23'', or 19¼'' for jack (a) in Figure 6-4. The length for jacks (b) through (e) would be found by adding multiples of jack (a), as shown in Table 6-5.

Hip Jack Rafter	a	b	c	d
16'' centers	14.42 x 16/12'' = (19¼'')	2 x a	3 x a	4 x a
2' centers	14.42 x 2 = (28⅞'')	2 x a	3 x a	4 x a

The multiple length of jacks
Table 6-5

Notice that the *common difference* (or length of the first jack) is found by multiplying the unit length by the rafter spacing in feet.

The Model Hip Jack's Common Difference

On our model, shown in Figure 6-6, let's begin our first hip jack 8'' in from the corner. If the hip is 8'' in from the corner, the run of the hip will be 8'', or 0.6667' (which is 8'' divided by 12'').

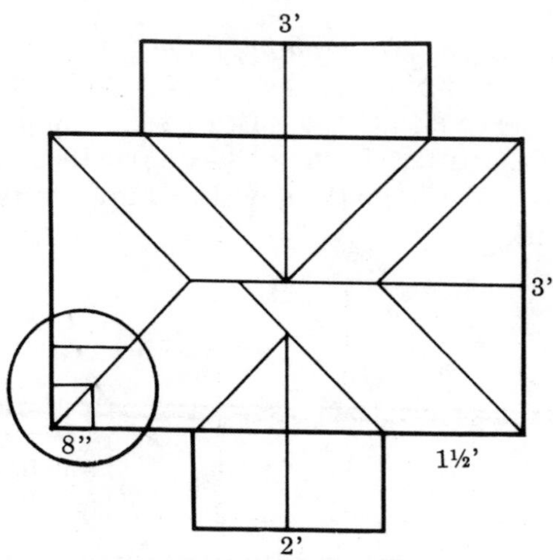

The model showing the hip jacks
Figure 6-6

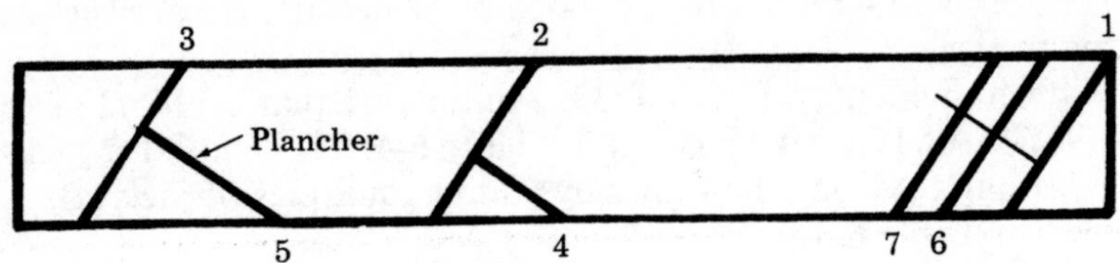

The layout lines for a hip jack rafter
Figure 6-7

The length of this hip jack will be the Unit Length times the Run. That's 14.42 times 0.6667 feet, or 9.61'' (9⅝'').

This measurement, 9⅝'', is both the mathematical length of the first hip jack and the common difference. *But don't start cutting the rafters yet!* We haven't considered the overhang. The 9⅝'' is measured from the building line to the hip. These rafters have to extend all the way to the fascia. The overhang (which is the same as on the common rafters) must be added to the length of each hip jack.

Figure 6-7 shows the layout lines for a hip jack rafter. Again, I'll refer to each of these lines by number. The distance from line (1) to line (2) is the common difference that is to be added to each succeeding hip jack rafter.

Lines (1) and (2), The Hip Framing Point Line and the Building Line: Crown a piece of 2 x 4 and lay the crown away from you. Set the framing square to 8'' and 12''. Draw a hip framing point line on the corner of the rafter. Then measure along the top edge, as in Figure 6-8, and make a mark at 9⅝''. Draw the building line through this mark.

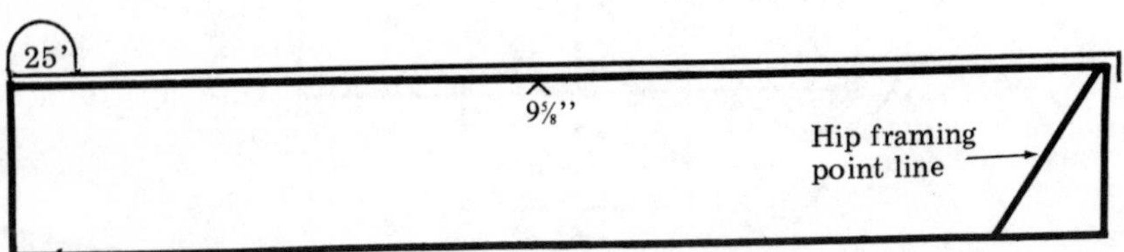

Hip framing point and building lines (Lines 1 and 2)
Figure 6-8

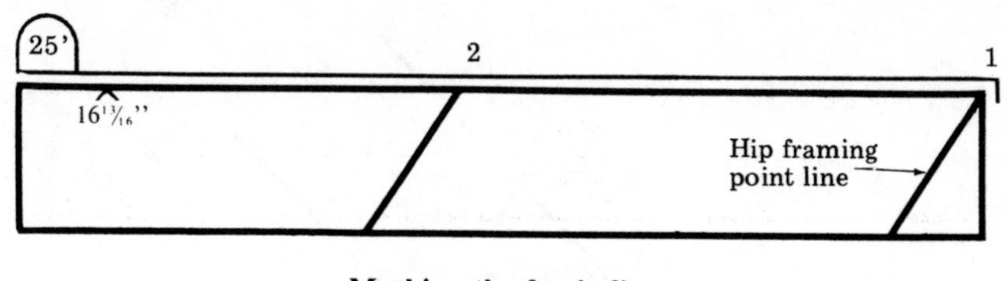

Marking the fascia line
Figure 6-9

Line (3), The Fascia Line: The overhang has a run of 6" and the rafter has a run of 8". Adding 6" plus 8" gives us 14" of total run to the fascia line. To change 14" to feet, divide by 12. Fourteen divided by 12 equals 1.167', and that, multiplied by 14.42, is 16.82", or 16¹³⁄₁₆". This is the overall length of the first hip jack rafter from the framing point on the hip rafter to the fascia line.

Mark 16¹³⁄₁₆" on the rafter, as in Figure 6-9, and draw the fascia line through this point. Notice that the difference between the two measurements is 7³⁄₁₆". If you look back to Figure 2-17 in Chapter 2, you'll see that the difference there was also 7³⁄₁₆", which should be the same if the overhang is to be a consistent 6".

Lines (4) and (5), The HAP and the Plancher: The HAP and the plancher must remain the same throughout the roof. Measure and draw in lines (4) and (5) at 2¾" and 2". Remember to measure only from the top edge of the rafter.

Lines (6) and (7), Hip Jack Shortening and Setback: The hip jack rafter is held away from the framing point by half the 45-degree thickness of the hip. See Figure 6-10. On the model 2 x 4 hip, this is 1¹⁄₁₆". Cheek cuts are always half the thickness of the material you're working on. That's ¾" for this jack rafter.

Draw a line perpendicular to the hip framing point line as in Figure 6-11, and measure back for the shortening and setback.

Lay out two of these rafters. On one of them transfer line (7) to the other side. Cut a rafter on line (7) on each side, with the saw set on 45 degrees as for any cheek cut. That gives you an opposite pair. Move the saw back to zero degrees and make the other cuts.

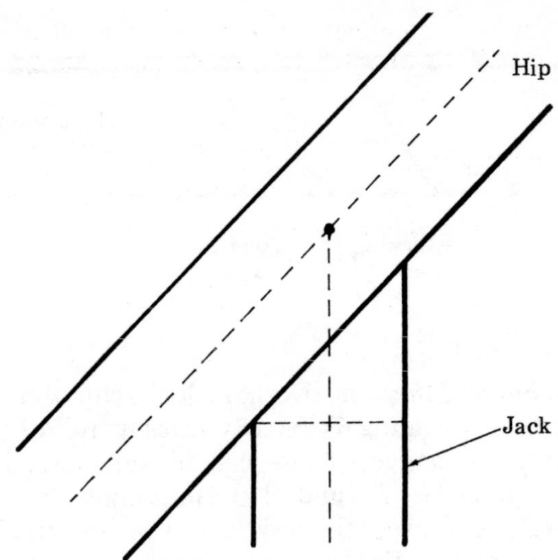

Shortening and setback
Figure 6-10

Next, make marks at 7¼" from each corner of the rafter plate. Then draw layout lines through these marks. Place an "X" at each mark on the side away from the corner. Figure 6-6 shows the center line mark at 8". Since the rafter is 1½" thick, the 7¼" mark makes it come out just right.

Frame the two hip jack rafters on the rafter plate and then nail the plumb cuts against the hip rafter.

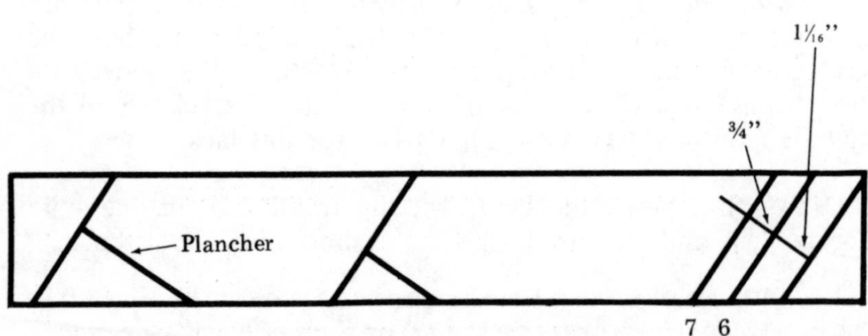

Shortening and setback lines (Lines 6 and 7)
Figure 6-11

Measure-ment	1st hip jack	Common difference	2nd hip jack	Common difference	3rd hip jack
Building line	9⅝"	+ 9⅝ =	19¼"	+ 9⅝ =	28⅞"
Fascia line	16$\frac{13}{16}$"	+ 9⅝ =	26$\frac{7}{16}$"	+ 9⅝ =	36$\frac{1}{16}$"

Adding the common difference
Table 6-12

The Second Hip Jack Rafter

The second hip jack rafter will be centered at 16" from the corner along the single cheek cut hip end of the building. Measure over 15¼" and make a mark. Then place a line and an "X" on the rafter plate to locate the position of the seat cut.

The common difference is 9⅝". Add this difference to the overall length of the first hip jack rafter to find the length of the second.

Adding the common difference to the first jack gives us the length of the second. Adding the common difference to that gives us the length of the third, and so on as far as necessary. See Table 6-12.

Lay out the second hip jack rafter using the new dimensions for the building line and the fascia line. See Figure 6-13. The other cuts and dimensions are the same as in the first hip jack rafter. Cut out a second hip jack rafter and frame it on the layout mark and against the single cheek cut hip.

When the First Space is Unequal

When the first space is unequal, as in Figure 6-14, then the length of the first rafter won't be the same as the common difference. If the corner space is *less* than the common difference, the first rafter will be *smaller* than the common difference.

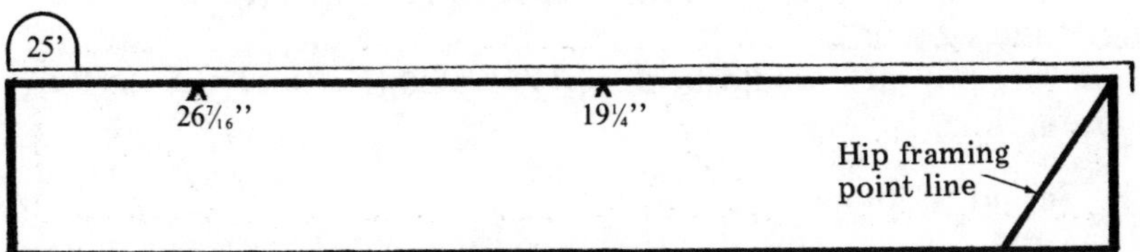

Second hip jack layout
Figure 6-13

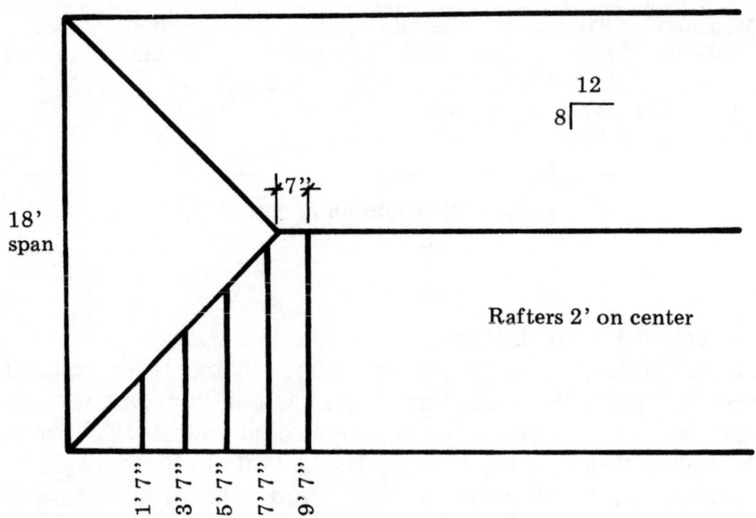

An unequal first space
Figure 6-14

Here's how to find the rafter lengths for the roof in Figure 6-14. Since the span is 18', the total run to the framing point is 9'. The distance to the first common rafter is 9'7". The plans show the rafters at 2' apart. There are four spaces before the corner hip jack. Therefore, the corner hip jack will have to be 8' from the first common rafter.

Subtracting 8' from 9'7", we find that the corner hip jack has to be 1'7" (or 19") from the corner. That's also the run of this corner hip jack.

The mathematical length of this hip jack is 14.42 times 19" divided by 12, or 22.83", which is 22$^{13}/_{16}$". The common difference is found by the common difference formula:

Common Difference = Unit Length x Rafter Spacing (in feet).

In our example: Common Difference equals 14.42 (for an 8 in 12 roof) times 2', or 28⅞".

The second jack would be 22$^{13}/_{16}$" plus 28⅞", or 51$^{11}/_{16}$" to the building line.

The Secant Method

Using the secant method: Rafter Length equals the Secant times the Total Run (in feet). Note that the square root of 2 is not used this time in the secant method because here the apparent run and the actual run are equal.

On your calculator, punch [8] [÷] [1] [2] [=] which gives you 0.6667. Next, find the angle by pushing [INV] [tan]. To find the secant, push [cos] [1/x]. The display will show 1.2, which is the secant. Multiply that by the run of 19" and you have 22¹³⁄₁₆".

To find the common difference, multiply 1.2 by 24" to get 28.8", or 28⅞". So, you see, either method works.

If you've mastered this, you can figure and cut hip jack rafters for any size regular hip roof.

Valley Jack Rafters (2)

You could think of hip jack rafters as the bottom part of a common rafter extending down to the plate. If that's true, then the valley jack rafter is the upper half of the common rafter extending up to the ridge.

The mathematical calculations are identical but the shortening and setback are a little different. See Figure 6-15.

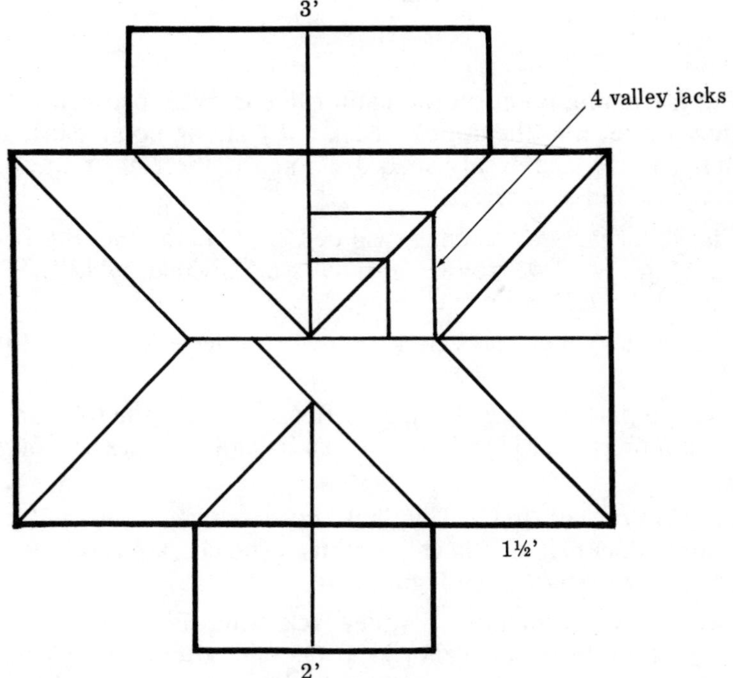

The model showing the valley jacks
Figure 6-15

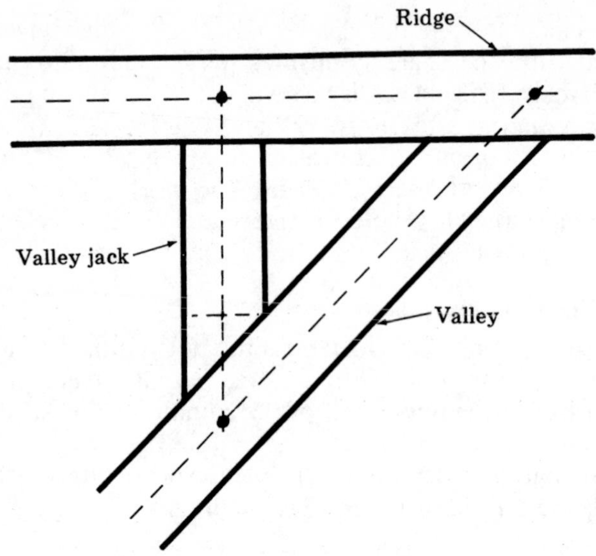

Valley jack detail
Figure 6-16

Put a framing point on the main ridge midway between the 3' addition ridge and the double cheek cut framing point. Since the framing points are evenly spaced 6'' apart, the run of the first valley jack will be 6''.

The valley jack rafter length will be Unit Length times the Total Run. Multiply 14.42'' by 0.5'. (That's 6'' divided by 12''.) The rafter will be 7.21'', or 7$\frac{3}{16}$'' long.

The second valley jack rafter will be twice the length of the first, or 14$\frac{7}{16}$'' long.

The top part of the valley jack rafter is just like the top part of the common rafter. They both shorten by half the thickness of the ridge.

The bottom of the valley jack rafter shortens by half the 45-degree thickness of the valley rafter. The cheek cut is half the thickness of the jack. See Figure 6-16.

Lay out two of each length valley jack rafter. Figure 6-17 shows the layout marks for the valley jack. On one rafter of each length, transfer the cheek cut line to the other side and cut each pair from opposite sides. Frame the four valley jack rafters as indicated in Figure 6-15.

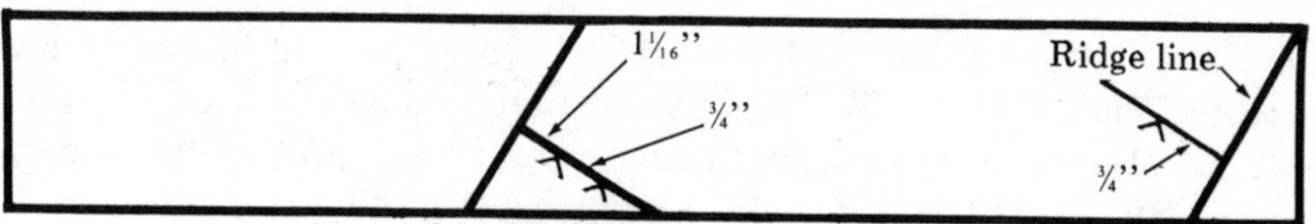

Layout marks of the valley jack
Figure 6-17

One of the valley rafters should be a veed and raised rafter. The other is a construction-cut valley rafter. Notice that the jacks on one side come exactly to the top edge of the vee. If the jacks are placed against the other valley, they must be held up the distance of the veeing.

Hip-to-Valley Cripple Jacks (3)
Look at the lines (a) and (a1) in Figure 6-18. Since the hip cuts through at a 45-degree angle, line (a) is the same length as rafter (a1). Because the hip and the valley are parallel to each other, any hip-to-valley cripple jack rafter will have the same run as the plate line at (a).

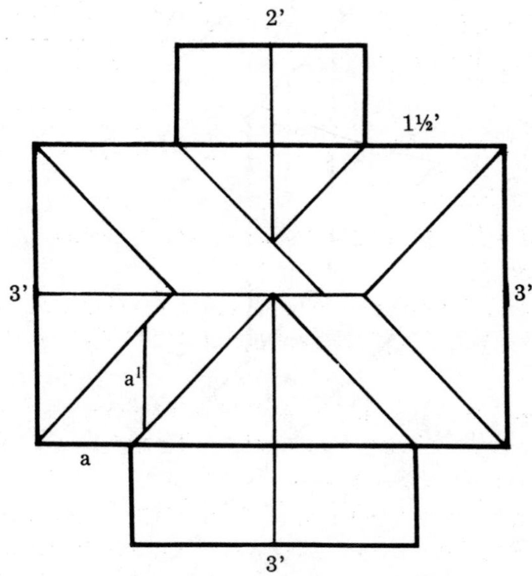

The model showing the hip-to-valley cripple jack
Figure 6-18

On our model, the plate at (a) is 1' long. So the hip-to-valley cripple jack rafter will also have a mathematical run of 1'. The length of the hip-to-valley jack rafter will be the Unit Length times Total Run, or 14.42 times 1'. The answer is 14 7/16".

The shortening at each end is half the 45-degree thickness of the respective hip or valley rafter. See Figure 6-19. In our model, the hip and valley are the same thickness, and the shortening for each is 1 1/16". Since the hip-to-valley cripple jack will be made from 2 x 4 stock, each cheek cut will be set back half the thickness of the jack, or 3/4". On this rafter, the cheek cuts are made on opposite sides.

Transfer the valley cheek cut line to the other side of the rafter. Label the lines as in Figure 6-20. Then set the saw to 45 degrees and make each cheek cut.

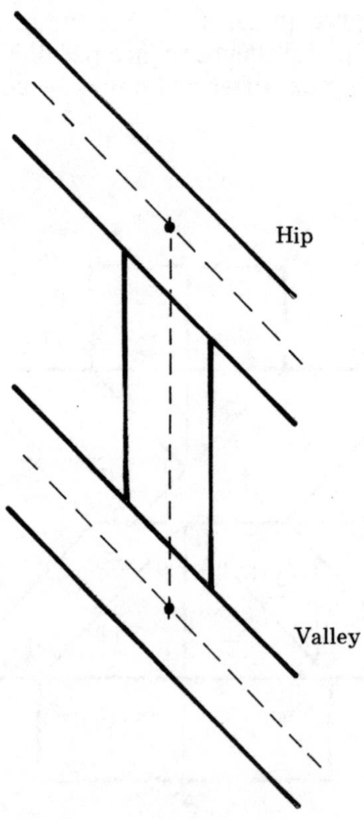

Setback and shortening
Figure 6-19

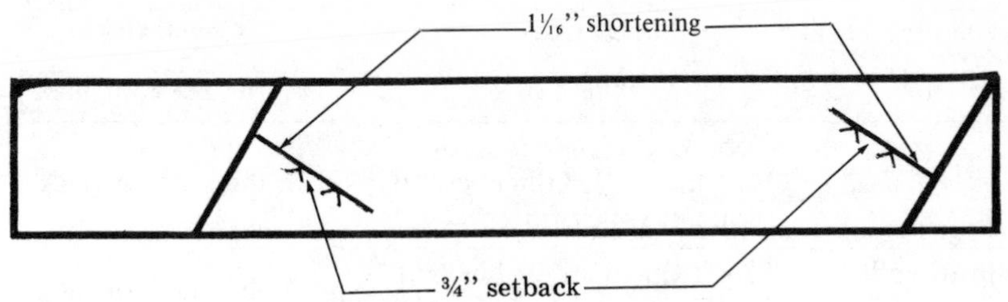

Layout lines for hip-to-valley cripple jacks
Figure 6-20

Valley-to-Valley Cripple Jacks (4)
The run of the valley-to-valley cripple jack is always twice the run of the valley jack it meets. If the valley jack is 9", then the valley-to-valley cripple jack is 18".

For the model, use a framing point distance of 10" for the valley-to-valley cripple jack, as shown in Figure 6-21. The shortening and cheek cut follow the same rule as the hip-to-valley cripple jack. But here the cheek cuts are both made from the same side.

Table 6-22 summarizes shortening and cheek cut rules for jack rafters. Notice that all the cheek cuts are half the thickness of the rafter itself. The valley jack rafter shortens by half the thickness of the ridge, not a 45-degree thickness. All of the shortenings that intersect with hips or valleys (which are at 45 degrees) will have a shortening of half the 45-degree thickness of that member.

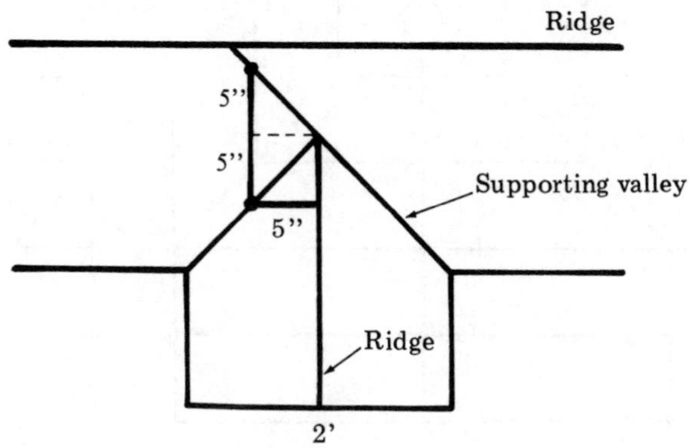

Calculating run of valley-to-valley cripple jack
Figure 6-21

Type of Jack	Shortening	Cheek Cut
Hip jack	½ 45° thickness of hip	½ thickness of jack
Valley jack	Top: ½ thickness of ridge Bottom: ½ 45° thickness of (double) valley rafter	½ thickness of jack
Hip-to-valley cripple jack	½ 45° thickness of hip and ½ 45° thickness of (double) valley	½ thickness of jack
Valley-to-valley cripple jack	½ 45° thickness of supporting valley and ½ 45° thickness of shortened valley	½ thickness of jack

Jack shortenings and cheek cuts
Table 6-22

Problem

Here's a problem that tests your ability to calculate jack rafter lengths. Answers are at the back of the book.

Find the mathematical length of the jack rafters in Figure 6-23 and give the name of each. You must determine the run of each rafter. Use the chart on the next page for your answers.

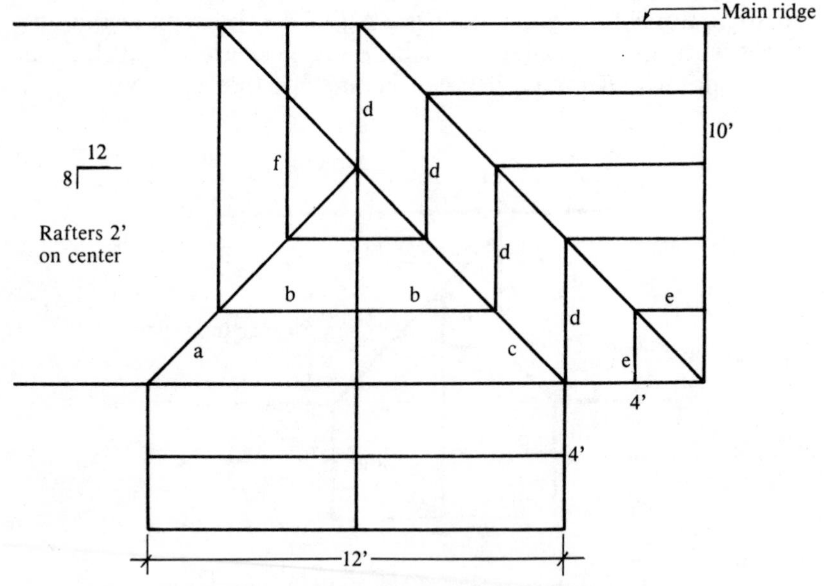

Calculate the length of the jack rafters
Figure 6-23

Rafter name	Apparent run	Actual run	Mathematical length
a			
b			
c			
d			
e			
f			

Chart for Figure 6-23

7

From Dutch to California

The Dutch

A Dutch is a section of hip roof in which the hip rafters don't go all the way to the framing point. Notice in Figure 7-1 that the hip rafters stop against a horizontal cross member (a). That horizontal member forms the base of a vertical wall called a *gable end*.

Let's frame a Dutch gable on our model. Start by removing the double cheek cut hip rafters and the center common rafters from the end of your model shown in Figure 7-2.

Let's say the plans call for the last common rafter at 10¼'' from the outside edge of the rafter plate, as in Figure 7-2 B. Place a framing point on each side of the plate at 10¼'', then mark 9½'' (to allow for half the thickness of the 2 x 4) and make a layout mark as indicated. Nail a common rafter to the layout mark on the 2' span addition side.

The tail cut of the supporting valley rafter and the common rafter should just meet, as shown at (a) in Figure 7-3.

The Cheek Cut Common Rafter

For the side with the 3' addition, you will need to make a special common rafter with a proper tail cut. Lay out a common rafter as you did in Chapter 2. Crown a piece of 2 x 4 and lay the crown

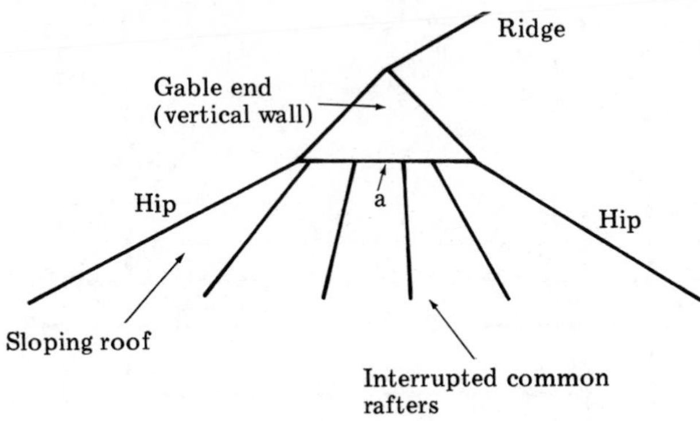

The Dutch
Figure 7-1

away from you. Check the stair gauges on your framing square to be sure they are on 8'' and 12'' because this rafter is for an 8 in 12 roof.

Lay out the common rafter exactly as you did before, but don't bother with the fascia line and plancher line. Cut the birdsmouth and shortening.

Look at Figure 7-3. Edge (b) will be the inside edge of this common rafter. But part (c) must be added so that the tail of the common rafter will fit snugly against the valley rafter to form a neat overhang.

Line (d) on top of the rafter in Figure 7-4 is not a 45-degree angle. To cut this line correctly, set the saw to 45 degrees, lay the rafter flat, and cut along this building line. Cut away the shaded part in Figure 7-4. Frame this to the rafter plate.

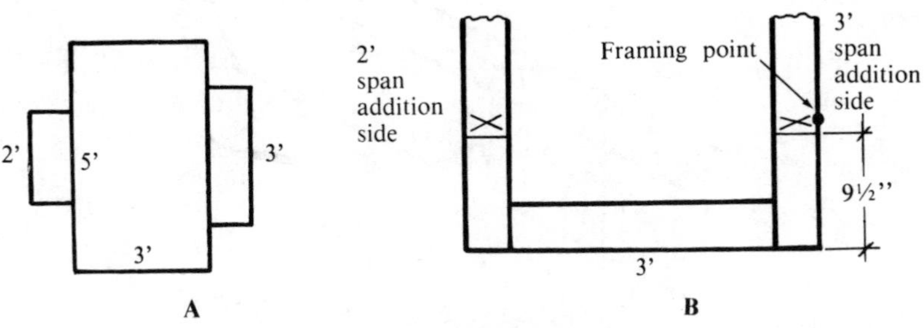

Locating the framing point
Figure 7-2

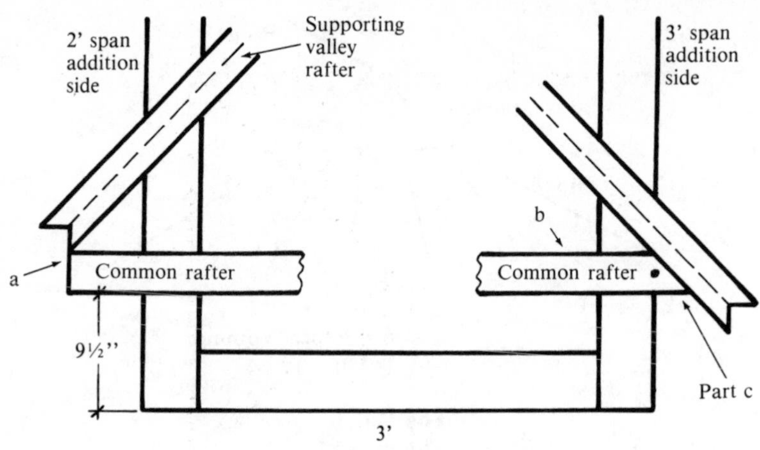

Tail positions
Figure 7-3

The New Ridge

The framing point for the model ridge was at 18'', one-half the 3' span. Now the ridge framing point would extend another 7¾''. (18'' minus 10¼'' equals 7¾''.) See Figure 7-5. The first framing point was extended by half the thickness of the common rafter. Since that 3/4'' amount is also needed at the new framing point, an extra 7¾'' will fit just fine.

There are two ways to extend the ridge board. You can either tear the model apart at this point to put in the correct length ridge, or add a 7¾'' piece of 2 x 6 to the ridge.

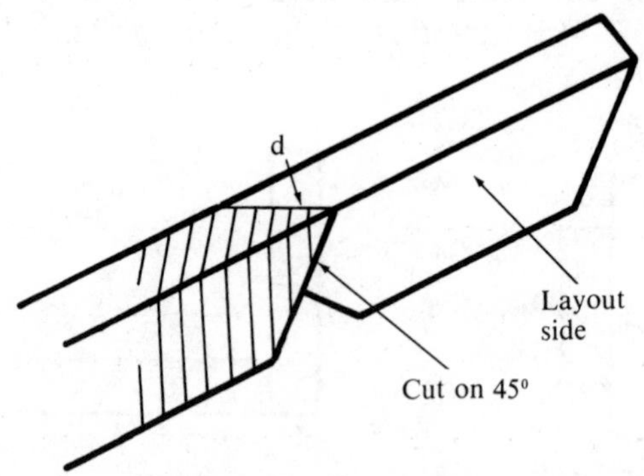

The common rafter tail cut
Figure 7-4

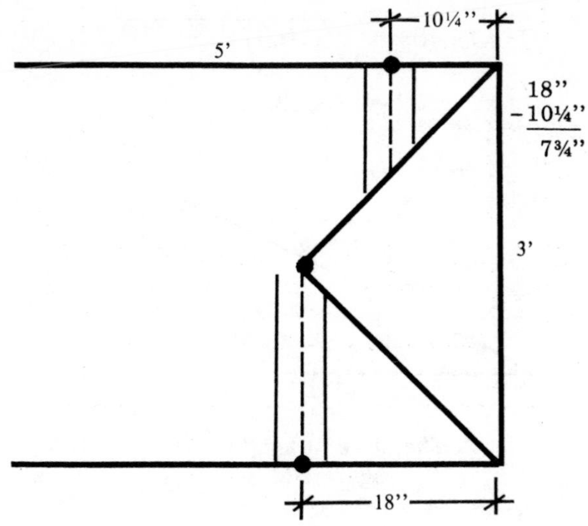

Calculating ridge length
Figure 7-5

Notice that adding a piece to the ridge doesn't work very well on your model. The same is true on a full-size roof. The small piece makes it difficult to keep the ridge both straight and level. When we frame the Tudor roof in the next section, we'll need to add 12" more to the other end of the ridge. So let's plan ahead by replacing our 26$\frac{9}{16}$" ridge with a 46$\frac{5}{16}$" piece of lumber.

The Interrupted Hip

Now we must calculate the length of the interrupted hip rafters. We can do it three ways: with a framing square table, with a secant calculation, or using a ratio. Here's how to do each of them:

The Framing Square Method:

Hip Mathematical Length = Unit Length x Apparent Run

This method requires that the unit length of the hip be multiplied by the common rafter run, which is the apparent run of the hip.

The unit length of an 8 in 12 hip rafter is found by using the Pythagorean Theorem, as in Figure 7-6. You could use your calculator, of course, but the second line of the rafter table on your framing square shows the number 18.76 under 8".

$$\text{Unit length} = \sqrt{(16.97)^2 + (8)^2}$$

$$= 18.76''$$

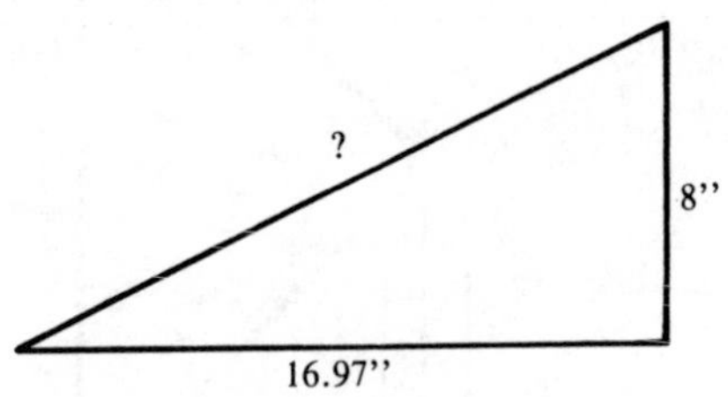

The hip unit length
Figure 7-6

The common rafter run on this Dutch is the distance to the framing point, which was marked at 10¼''. What part of a foot is 10¼''? Divide 10¼'' by 12 to get 0.8541'.

Hip Mathematical Length = 18.76'' x .8541 = 16.024''

Let's call it an even 16'' from the framing point to the building line.

The Secant Method:

Hip Mathematical Length = Secant x Actual Run

Hip Mathematical Length = Secant x Common Rafter Run x $\sqrt{2}$

To use this method, you have to find the secant of the hip and multiply it by the actual run of the hip.

Find the secant on your calculator by first finding the tangent. You know that the tangent is 16.97 divided into the rise. Here's how you do it:

Press [8] [÷] [1] [6] [·] [9] [7] [=] and 0.4714 appears in the calculator display. Press [INV] [tan] to find the angle, 25.24 degrees. Then push [cos] [1/x] and 1.1055 appears. That's the secant of the hip rafter. It's different from the secant of the common rafter because you divided by 16.97 rather than 12.

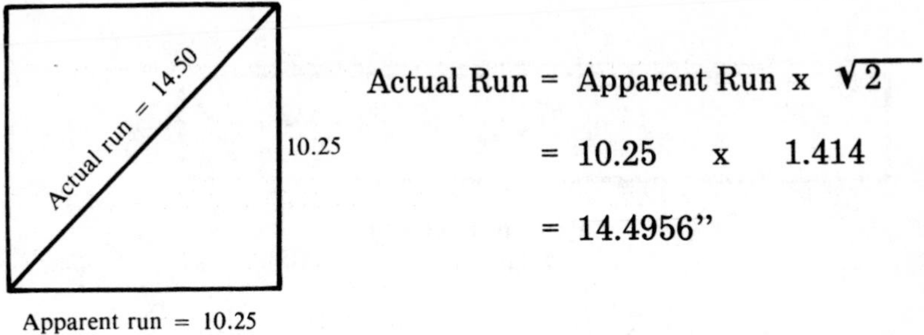

Calculating actual run of the hip
Figure 7-7

The framing point was marked at 10.25", but this was only the apparent run. Figure 7-7 shows how to calculate the actual run of the hip rafter.

$$\begin{aligned} \text{Math Length} &= \text{Secant x Actual Run} \\ &= 1.1055 \times 14.4956'' \\ &= 16.02'' = 16'' \end{aligned}$$

The Ratio Method: Look back to Figure 7-5. Notice that changes in the distance to the original framing point at 18" produce proportionate changes in the length of the hip rafter. When the horizontal distance changes from 18" to 10¼", the length of the hip rafter has to change proportionately. We'll use this proportionate change to find the length of the new Dutch hip rafter.

The hip rafter was 28⅛" long when the horizontal distance was 18". That distance is now 10¼". What's the length of the new hip rafter?

Set up the ratio as two fractions and solve for the answer by cross multiplying:

$$\frac{18}{10.25} : \frac{28.125''}{?}$$

$$18 \quad ? = 10.25 \times 28.125''$$

$$18 \quad ? = 288.28''$$

$$? = \frac{288.28''}{18''}$$

$$? = 16.01'' = 16''$$

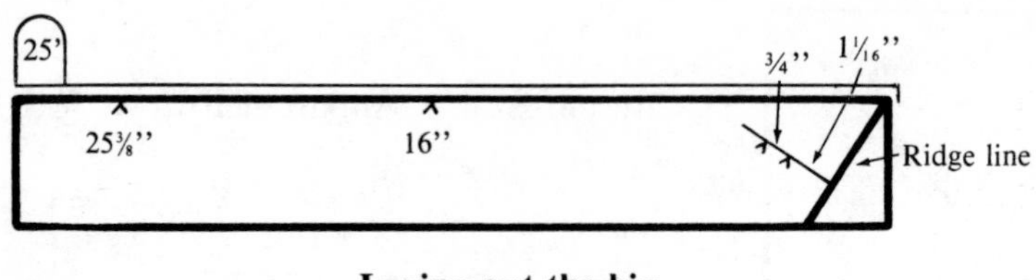

Laying out the hip
Figure 7-8

It's 16'' from the building line framing point to the framing point part way up the common rafter at the base of the gable end.

Let's cut and frame this hip rafter. Crown a piece of 2 x 4 and reset the framing square to 16.97'' and 8''.

Lay out a ridge line, a shortening line and a setback line for a double cheek cut hip. Mark off the building line length and the fascia length on the rafter, as in Figure 7-8.

The original hip was 28⅛'' to the building line and 37½'' to the fascia line. To find the overhang, subtract 28⅛'' from 37½''. Your answer is 9⅜''. Add the 16'' to 9⅜'' to get 25⅜'' to the fascia line for this interrupted hip.

Lay out the hip rafter as usual and make a double cheek cut on the upper end.

The Scab Ledger

Look at Figure 7-9. The double cheek cut hip will rest against the common rafter and against the end grain of the *scab ledger.* What is the length of this ledger?

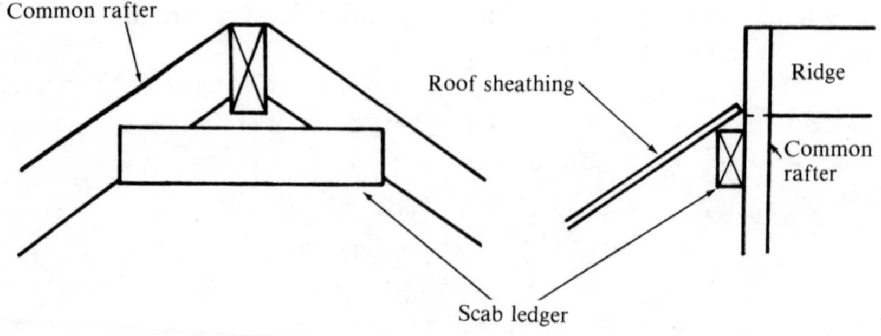

The ledger
Figure 7-9

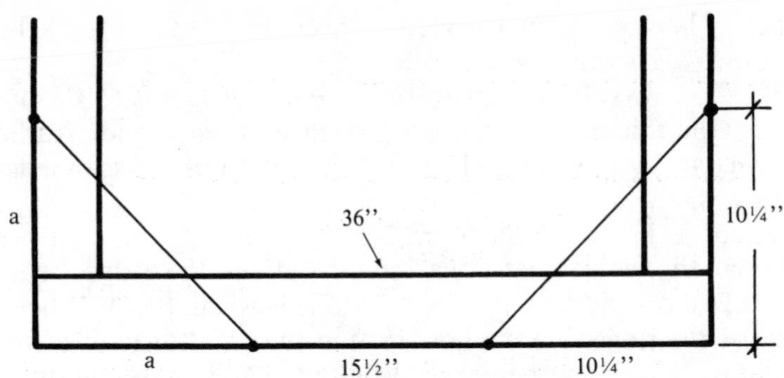

Calculating ledger length
Figure 7-10

The span of the building as shown in Figure 7-10 is 36". We know that the hip rafter runs at 45 degrees from the corner of the building to the scab ledger. The framing point at the scab end will be 10¼" from the corner of the rafter plate. Thus both sides labeled (a) in Figure 7-10 will be 10¼". If that's true, then the distance between the framing points at the ends of the scab ledger will be 15½". (Subtract 10¼" twice from 36" to get 15½".)

You can think of the scab ledger as a ridge brought forward, with each end receiving a cheek cut of its respective hip rafter.

Now look at Figure 7-11. The joining line (abc) requires the ridge to grow by half the thickness of the common rafter on each end. Since our interrupted hips are cut the same way and the ridge

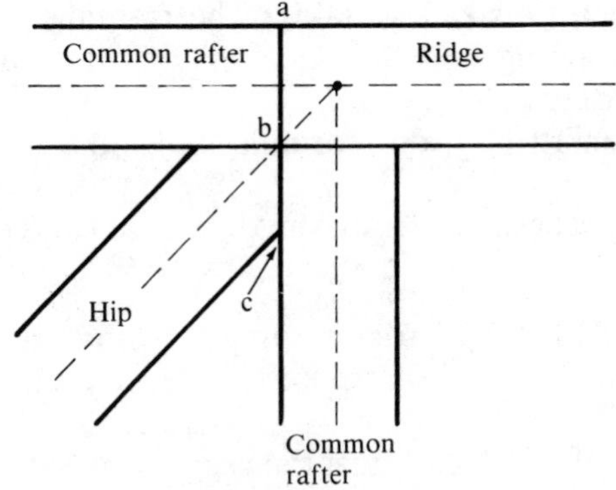

The double cheek cut hip to ridge
Figure 7-11

has merely been brought forward to become a ledger, the ledger will require a growth at each end.

Here's how we solve this problem: The scab ledger is 15½'' long. Add half the thickness of a common rafter at each end. Adding 3/4'' twice to 15½'' gives us 17''. So, the length of the scab ledger is 17 inches.

Now cut a 17''-long piece of 2 x 4 and mark the center so you know the point that falls directly under the ridge board. Frame this short piece so the top of the ledger will be 6³⁄₁₆'' below the top of the ridge. Here's how to find the 6³⁄₁₆'' distance: Since the main roof has a mathematical height of 12'' and the run to the center of the ledger is 8¾'' (shown in Figure 7-12), the mathematical height of the ledger must be 8.75 parts of a foot times the 8'' unit rise. A quick calculation shows that 8.75 divided by 12, multiplied by 8, equals 5.83''. Subtract the 5.83'' from the 12'' height of the main roof, and you have 6.166, or 6³⁄₁₆'' below the main ridge. Now frame in the hips.

The Interrupted Commons

Hip jack rafters and interrupted commons will fill the area between the fascia and the Dutch gable end. Look back to Figure 7-1. The Dutch gable end, remember, is the vertical wall. You know how to calculate the length of the hip jacks from the last chapter. Now we'll look at the interrupted common rafters.

All of the tail cuts on these rafters will be the same, since the overhang will be the same. Now, what's the new building line length and shortening?

Figure 7-12 clearly shows that the interrupted common rafter will fit against a surface 8'' in from the building line. We can use this figure directly instead of taking the framing point figure of 10¼'' and subtracting half the thickness of the common and the width of the ledger.

Stated mathematically, our formula looks like this:

Mathematical Length = Unit Length x Run (in feet)
= 14.42'' x .6667' = 9.61 = 9⅝''

If we had used the 8¾'' for the run, as in the previous section, then we would have had to shorten the rafters by half the thickness of the ledger.

Take a piece of 2 x 4 and turn the crown up. Set the gauges on your framing square to 12'' and 8'' and lay out the marks as indicated in Figure 7-13. The 16¹³⁄₁₆'' is calculated the same way as the jack rafter overhang. Add 7³⁄₁₆'' to the 9⅝'' of building line.

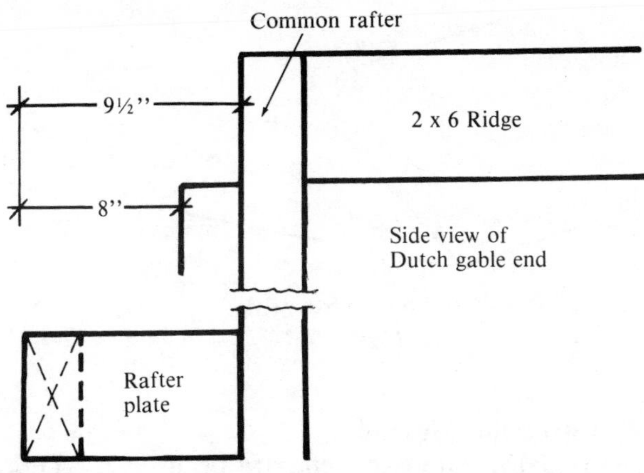

The run of the interrupted common
Figure 7-12

Cut out the rafters and make a framing point mark on the plate at 12" and 24". Make a layout mark 3/4" back from each framing point, and frame the commons to the model.

With a straightedge at least 4' long, check across the tail plumb cuts from hip to hip to see if they line up for the fascia board.

Congratulations on your Dutch gable.

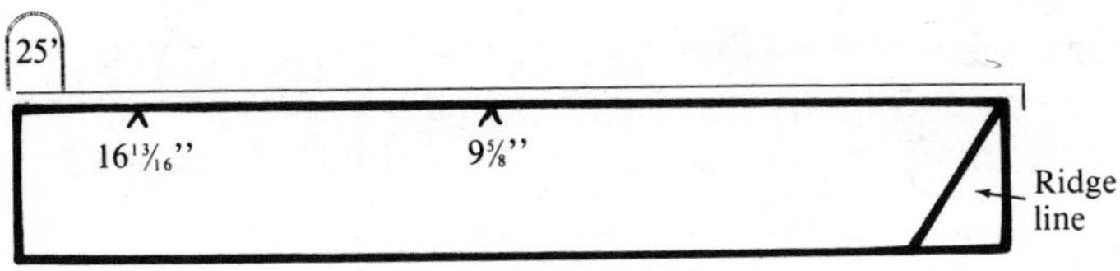

The interrupted common
Figure 7-13

The Tudor Peak

There are two ways to frame a Tudor peak. The first has a soffit at the overhang. The second allows the rafter tails to remain exposed.

You'll remember that the Dutch gable began at a slant and ended in a vertical wall at the peak. The Tudor peak is just the opposite. It begins with a vertical wall and moves to a hip roof at the peak. See Figure 7-14. As defined earlier in this chapter, the gable end is the triangular vertical wall above the rafter plate.

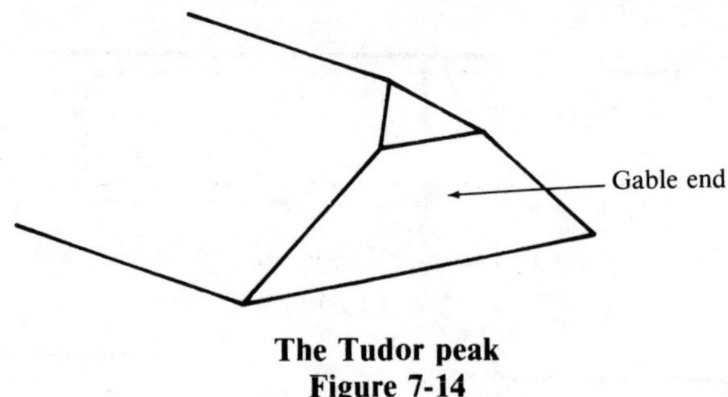

The Tudor peak
Figure 7-14

The First Construction Method

Look at Figure 7-15. The gable-end frame in every Tudor peak roof has two interrupted common rafters with a Tudor plate between them. Let's say that the plans show the run of this plate as being one foot long.

The Interrupted Common Rafters

First, let's calculate the length of these interrupted common rafters. The common rafter overhang on our 8 in 12 model roof will continue to be 7³⁄₁₆". We'll add this amount to the building line to find the fascia line measurement. The mathematical length of the rafter will be the unit length times the 1' run. Unit length is 14⅜" (14.42" per foot of run on an 8 in 12 roof). Adding 7³⁄₁₆" to 14⅜", we get 21⁹⁄₁₆".

Crown a piece of 2 x 4 and lay the crown away from you. Set the stair gauges on your framing square to 12" and 8". Lay out the plumb cut, which in this case will be the cut line. Next, lay out the

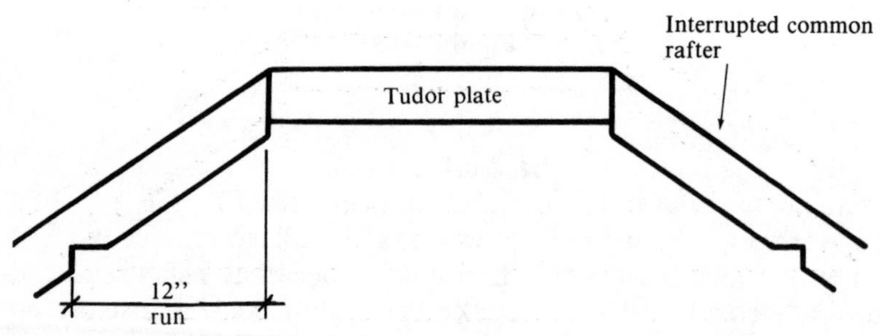

The Tudor plate
Figure 7-15

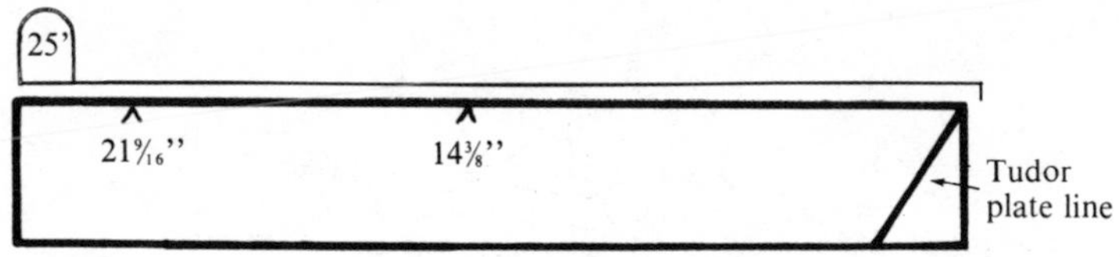

First Tudor interrupted commons
Figure 7-16

building line and fascia line as indicated in Figure 7-16. Finish the layout and cut out two rafters. There is no shortening at the top.

The First Tudor Plate

Now look at Figure 7-17. The run of the full common rafter is 1½'. The run of the interrupted common rafter is 1'. Their difference is half a foot. Since there's a right triangle on each side of the ridge, there will be half a foot on either side, or 1' overall. As a rule then:

The length of the Tudor plate equals two times the difference between the run of the full common rafter and the interrupted common rafter.

Since the framing points are at the joint, there's no need to shorten the interrupted commons, nor is growth required for the gable plate. Cut a piece of 2 x 4 to 12" and nail the three pieces together.

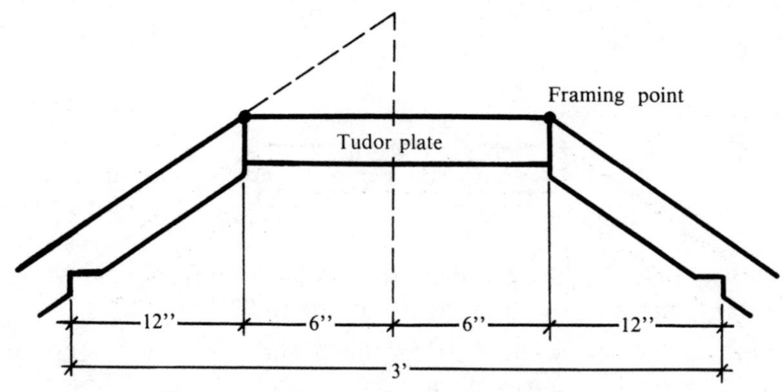

Length of the Tudor plate
Figure 7-17

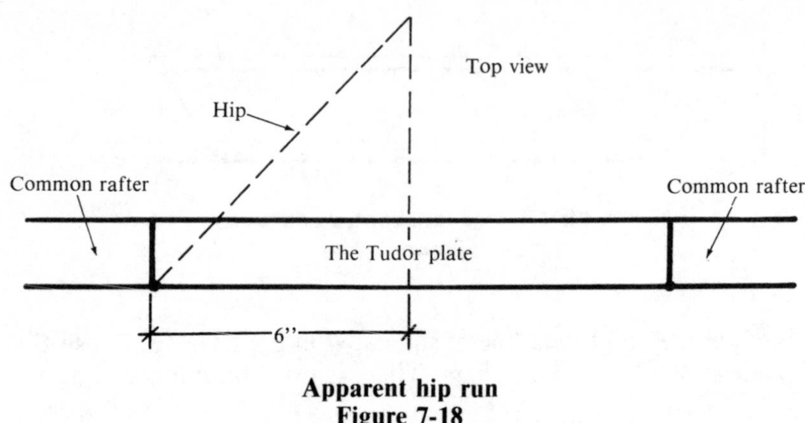

Apparent hip run
Figure 7-18

Now let's prepare the model to accept the Tudor peak framing. At the single cheek cut hip end, remove the hip jack rafters, the hip-valley cripple jack rafters and the single cheek cut hips. Frame the Tudor gable end along this 3' side, keeping the outside face of the common rafters flush with this end of the 3' by 5' frame.

The First Tudor Hip Rafters

The apparent run of the Tudor hip is the same as half the length of the Tudor plate, or 6''. See Figure 7-18.

There are two ways of calculating the hip length. Both give us the same answer.

The Framing Square Method:

(Math) Hip Length = Unit Length x Apparent Run
(Math) Hip Length = 18.76'' x .5 = 9.38'' = $9\frac{3}{8}$''

The Secant Method:

(Math) Hip Length = Secant x Actual Run
= 1.105 x $.5 \times \sqrt{2}$ = .78' = 9.38'' = $9\frac{3}{8}$''
or = 1.5635 x .5 = .7818' = 9.38'' = $9\frac{3}{8}$''

Crown a piece of 2 x 4 and lay the crown away from you. Move the stair gauges of the framing square to 16.97'' and 8''.

For the model, we're cutting single cheek cut hips. But of course, we could just as well cut double cheek cut hips. These first hips won't require the overhang portion. Therefore, you'll need only the $9\frac{3}{8}$'' dimension.

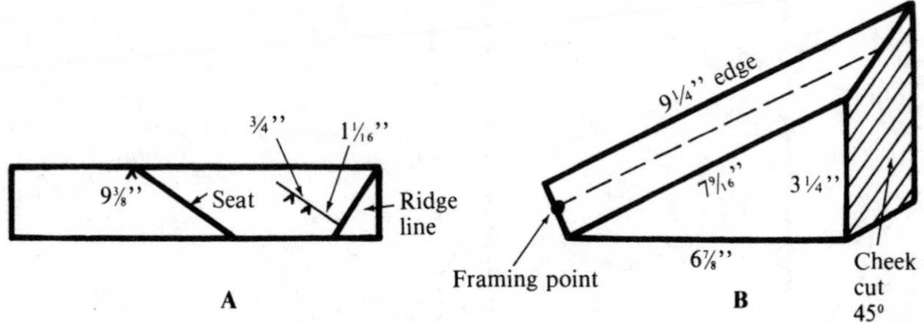

An odd single cheek cut hip
Figure 7-19

We'll be cutting the cheek cut line on 45 degrees. Figure 7-19 shows you how to do this. Transfer the mark from one side of your rafter to the other to make a pair. Draw the seat cut as in Figure 7-19 A. This hip will not be dropped or backed. The result will look like Figure 7-19 B. Notice the framing point location at the bottom in the center of the hip.

In Figure 7-20, the plate framing point is on the front edge at the joint of the Tudor plate. Measure over 1¹⁄₁₆'' (half the 45-degree thickness of the hip) and draw a 45-degree line. This is your rafter layout reference mark.

The Ridge Growth

Refer back to Figure 6-18 in Chapter 6. That was where we calculated the run of the hip-valley cripple jack. Notice how similar that was to calculating the distance from the single cheek cut framing point to the Tudor framing point. The situations are similar because both involve 45-degree angles. Distance (a) in Figure 7-21 is the run of the interrupted common rafter and is the same as distance (a1) between the framing points.

Since these hips will be single cheek cut hips and ridge growth is the same in both cases, we need only add 12'' to the length of the

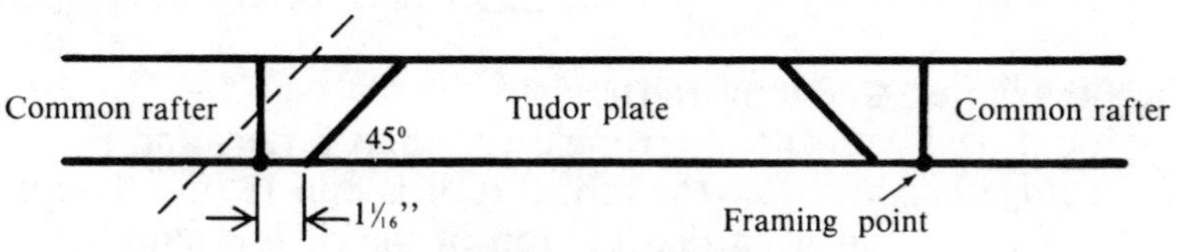

Finding the framing point
Figure 7-20

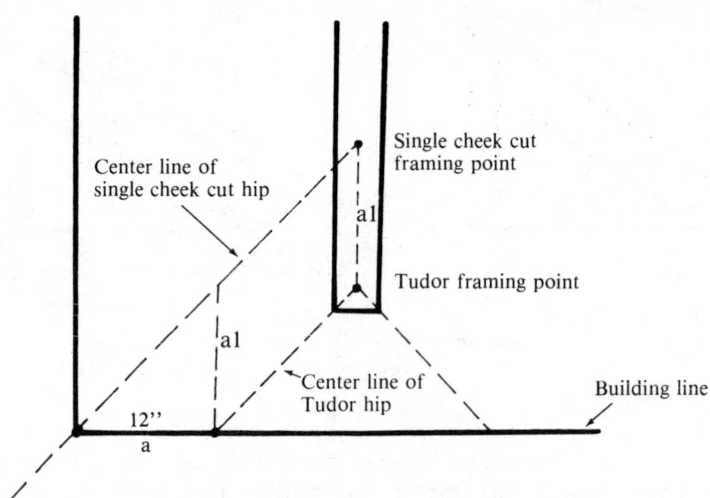

Ridge growth
Figure 7-21

ridge. The original ridge was 26$\frac{9}{16}$''. We added 7¾'' to the other end. Now we'll add 12'' to this end for a total ridge length of 46$\frac{5}{16}$''.

Next add the single cheek cut framing point. That's 1$\frac{1}{16}$'' plus 3/4'' (for half the thickness of the ridge) or a total of 2$\frac{13}{16}$'' in from the end. Draw in two 45-degree lines as in Figure 7-21. These show the hip center line. Frame the two hips.

In practice, these hips should be backed and the ridge edge should be cut off. Also, the tail needs to be cut at 45 degrees to each side of the framing point. You would have to do this from the bottom. *But be careful.* Trying to make these cuts on a piece this size is very dangerous. Don't bother to back one-quarter scale hips. Wait until you're framing a *real* Tudor peak roof.

The Second Construction Method

Now let's look at our second method of framing a Tudor peak. This time we're going to allow the rafter tails to remain exposed.

The Interrupted Common Rafters

First we'll deal with the interrupted common rafters and the apparent ridge height. Since the model roof is 8 in 12 and the total run is 1½', it would seem that the top of the ridge would be 12'' above the rafter plate, as in Figure 7-22 A. In actual practice, however, as you learned in Chapter 3, the HAP (Height Above

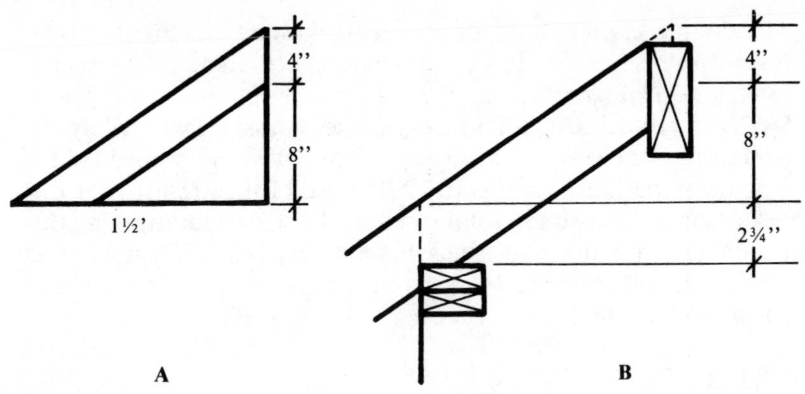

Apparent and actual ridge height
Figure 7-22

Plate) must be added to this total. Adding 2¾" to 12" gives us 14¾", as shown in Figure 7-22 B.

Calculating Loss of Ridge Height

Now look at Figure 7-23. It turns out that 14¾" is not really the case either, because the ridge is flat on top rather than peaked. Both the tangent method and the ratio method show that in an 8 in 12 roof with a 1½" thick ridge, you lose 1/2" in height. So, we'll subtract 1/2" from our ridge height of 14¾". This results in an overall height of 14¼".

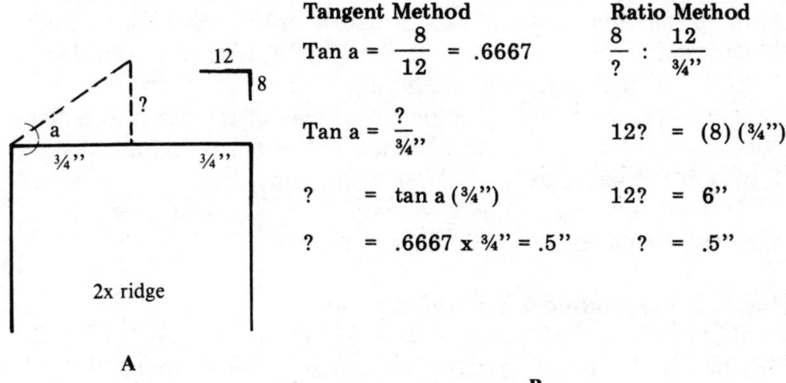

Calculating loss of ridge height
Figure 7-23

Here's an explanation of the calculations in Figure 7-23: Tangent is the rise divided by the run. Since angle (a) in Figure 7-23 A is 33.69 degrees (on an 8 in 12 roof), we know the tangent is 0.6667. The run is 3/4". To find the rise, you simply multiply the tangent times the run. Multiply 0.6667 by 3/4" and you get 1/2".

The ratio method says 8 is to 12 (the established figures for this roof) as some rise is to the total run of 3/4". Cross multiplying this formula and dividing gives the same answer, 1/2". If your model is correct, it will measure 14¼" from the top of the rafter plate to the top of the ridge.

The HAP

For this second construction, the hip is to have a 6" overhang. In order to have an overhang, a rafter must be given a HAP. Look back at Figure 7-17. Our new hip rafter will sit on top of the Tudor plate. Here is the problem. The Tudor plate is fastened to the common rafter, and this common rafter already has a 2¾" HAP in it. If we were to put a HAP on our hip rafter, that would stack a HAP on top of a HAP, and the hip rafter would overshoot the ridge. Yet we need this other HAP.

Now look at Figure 7-24. We have dropped the Tudor plate a given distance to provide for this second required HAP. This is unique to the Tudor peak with exposed rafter tails.

If we drop the plate 2¾", we will provide space for the needed HAP in the Tudor hip rafter.

For our model, let's back the hip to provide a smooth transition from the 8 in 12 common, which runs at 33.69 degrees, to the hip rafter, which runs at 25.24 degrees.

Use a backing of 5/16". Subtracting this from 2¾" leaves 2⁷⁄₁₆". If we drop the Tudor plate 2⁷⁄₁₆" down from the front top corner of the common rafter, as shown in Figure 7-24, we'll still be able to cut 5/16" worth of backing on the hip.

Furthermore, since the framing point was at the joint and the hip goes by at 45 degrees, the common rafter needs to have a setback of half the 45-degree thickness of the hip (1¹⁄₁₆") and must be cut at 45 degrees to frame flush against the hip. The framing points remain stationary and the members are re-cut.

Laying Out the Second Interrupted Common

Lay out another pair of interrupted common rafters as before, but on this pair add the setback to the ridge cut. See Figure 7-25.

Because of the growth of the 45-degree cut, you'll have to set the ridge line at least half an inch in from the corner of the rafter. On the other hips there were both a shortening of 1¹⁄₁₆" and a setback

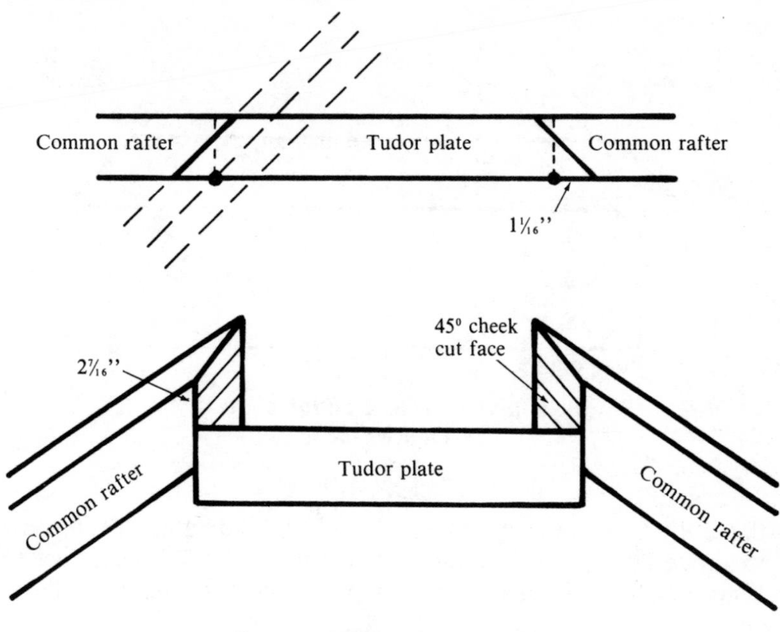

Common rafter framing detail
Figure 7-24

of 3/4". Here we have a setback of only 1¹⁄₁₆", so there won't be quite enough length to finish the cheek cut.

Mark 2⁷⁄₁₆" along the cheek cut line for a framing mark, as shown in Figure 7-24. Set the saw to 45 degrees for this cheek cut. You'll need to cut two opposite rafters.

The Second Tudor Plate

Take a look at Figure 7-26. The plate must grow on each end by half the 45-degree thickness of the hip. Lay out these lines and

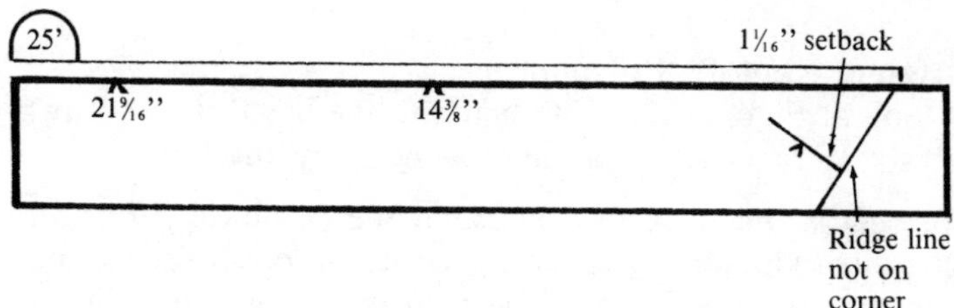

Second Tudor interrupted common rafter
Figure 7-25

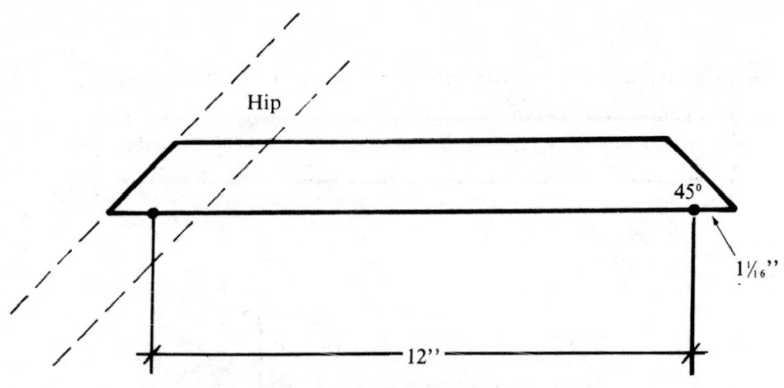

Top view second Tudor plate
Figure 7-26

make both bevel cuts on the same side as indicated. Frame the pieces together. Make sure the plate is dropped 2⁷⁄₁₆", as indicated in Figure 7-24. Remove the first set and frame this one in its place.

The Second Tudor Hip

The apparent run of this hip is 6", and the mathematical length is 9⅜". The overhang is also 6". The layout mark for the fascia line will be 18¾". See Figure 7-27.

Set the gauges on your framing square to 16.97" and 8". Crown a piece of 2 x 4 and lay the crown away from you. Lay out the ridge line, the building line and the fascia line. Now shorten this single cheek cut hip by half the 45-degree thickness of the ridge, and add a setback of half the thickness of the hip. Add the plancher line and a set-forward line for a 45-degree tail cut. This will be a single cheek cut as shown in Figure 7-27 B. Finally, cut the two cheek cuts and the plancher cut.

A Different Kind of Birdsmouth

Let's look at Figure 7-28. The seat cut line hasn't been drawn yet. For this rafter we won't be measuring along the building line.

Line (ef) is the joint face between the common rafter and the Tudor plate. The hip is against the common rafter face and directly over the framing point. Line (ab) is the building line on the rafter, which has been laid out and drawn at point (b). Line (cd) is a 45-degree line on the bottom of the rafter, which is the back of the birdsmouth.

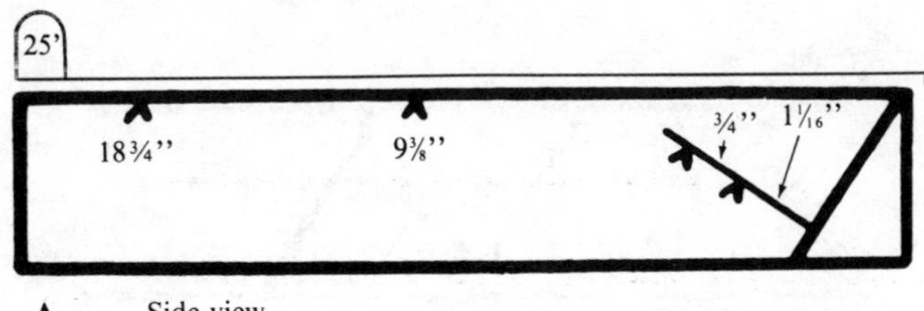

A Side view

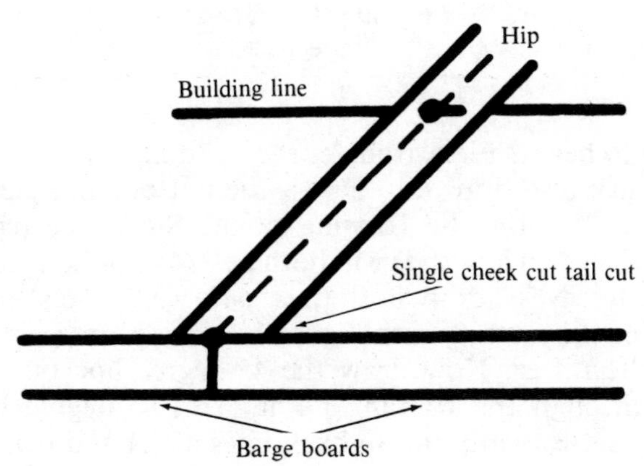

B Top view

The second Tudor hip
Figure 7-27

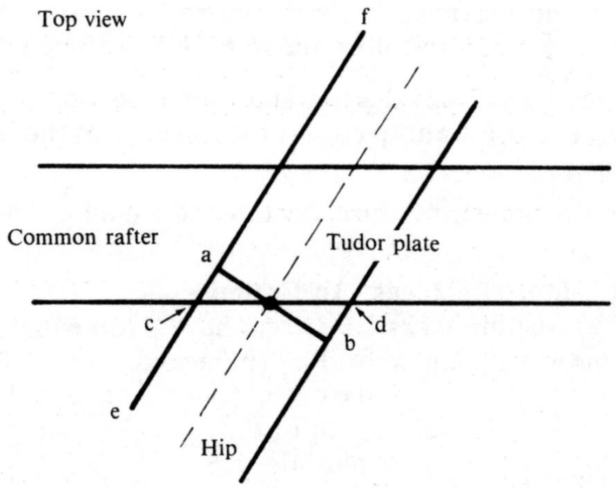

The birdsmouth cut
Figure 7-28

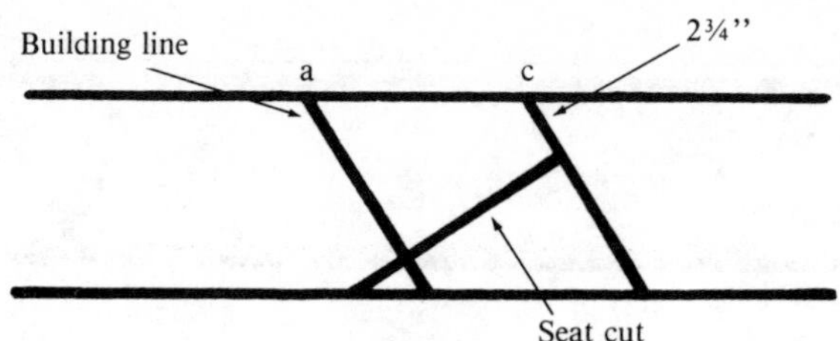

Marking the seat cut
Figure 7-29

Ready to begin? First, transfer the building line to the far side at position (a); also draw (ab) across the bottom and place a dot in the center to show the framing point. Since line (cd) is at 45 degrees, lines (ac) and (bd) will both be 3/4'', or half the thickness of the rafter. Next, draw both the set-forward line at point (d) on one side of the rafter and the setback line at point (c) on the other side of the rafter. Now draw the 45-degree line on the bottom directly through the framing point. This 45-degree line should touch the set-forward and setback lines at (c) and (d).

On line (c), the setback line, measure down 2¾'' if you are going to back or drop the hip, but only 2⁷⁄₁₆'' if you do neither. Now, draw the seat-cut line as shown in Figure 7-29. This location (c) rests against the point on the common rafter where you measured to drop the plate 2⁷⁄₁₆''.

Place the hip upside down in a vise. With a hand saw, cut along the 45-degree line and also follow the setback line along point (c).

Cut the backing at 25.24 degrees and frame the hip to the plate and ridge. Again, you want to cut off the top edge of the ridge for proper framing.

Following the same procedure, lay out and cut an opposite hip.

The Run of Tudor Hip Jacks: An Example

In Figure 7-30, the hip jacks labeled (a) have a run equal to their distance from the framing point. Pair (b) jacks are longer than the (a) jacks by the distance from the outside building line to the framing point. In this example, the run of the (a) jacks equals 2'. The run of the (b) jacks equals 2' plus 4', or 6'.

To cut rafters like these, follow the directions for hip jacks in Chapter 6.

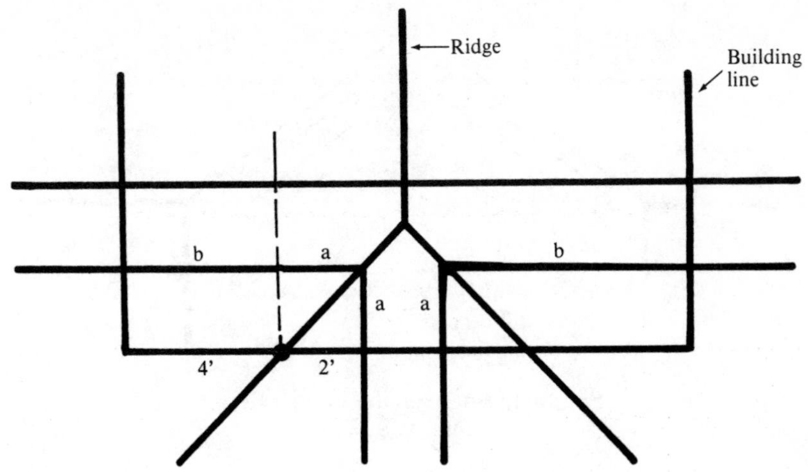

Top view of hip jacks: example
Figure 7-30

The Blind Valley

Figure 7-31 shows a small span addition on a home. You could frame this addition with a supporting valley rafter, a shortened valley rafter and all the jacks. But on a small addition like this it will be easier to build the main roof as a simple gable roof. Then, once the roof sheathing is in place, build what is called a *blind valley*.

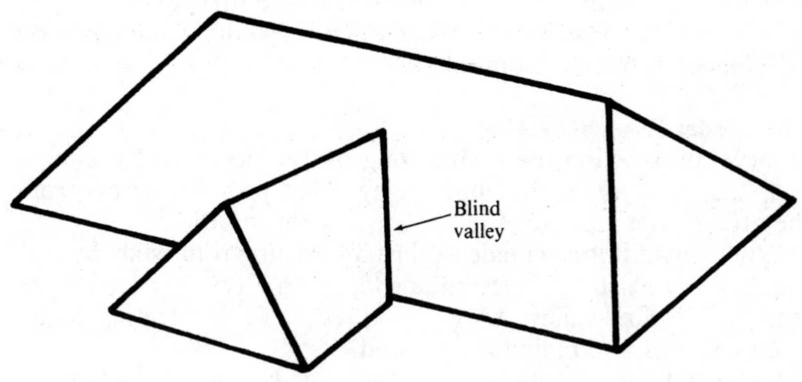

Blind valley construction
Figure 7-31

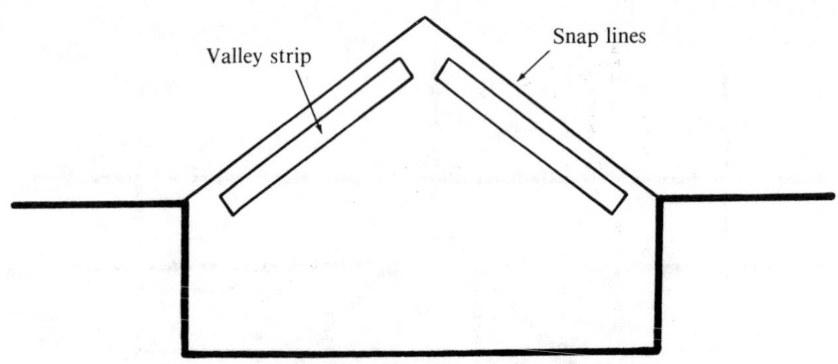

Snapping lines for the blind valley
Figure 7-32

The blind valley consists of two valley strips, which are usually 1 x 6 material. These strips help distribute the weight of the addition roof onto the main roof sheathing. They also provide a nailing surface for the tails of the valley jack rafters. The back part of the ridge can rest on the upper portion of the valley strips. The ridge extends to the end of the overhang. Add common rafters and valley jacks as needed.

Here's a summary of how to frame a blind valley addition. First, snap valley lines on the main roof sheathing, from the rear point of the addition ridge to the inside plate corners. Then nail two valley strips into place a short distance in from the chalk lines, as shown in Figure 7-32. The next step is to cut a length of 2 x 6 material about the right length for the ridge. Raise it into place. Then measure the long-point or downhill side of the first valley jack, lay it out, cut, and frame. The other valley jacks have the common difference subtracted from them.

The Model Roof Sheathing

In new construction, omit sheathing on the portion of the main roof that falls under the blind valley. That provides access from the attic of the main roof to the attic of the addition.

When an addition is made against an existing roof, start by cutting off any existing overhang and opening a crawl space before framing the blind valley. Also, provide room for the heating ducts, electrical runs and piping as required.

For our model, we'll eliminate the crawl hole and calculate the length and width of our one-piece sheathing. On a full-sized roof, these calculations wouldn't be needed.

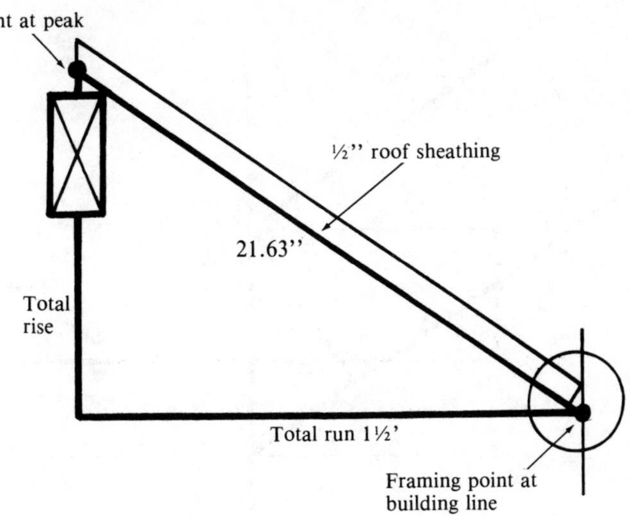

Mathematical width
Figure 7-33

The Sheathing Length and Width

Get the model ready for sheathing by removing all the valley rafters. Frame in either the Dutch-Tudor ridge or the gable ridge and add any necessary common rafters.

The ridge length of our gable roof is 6', or 72''. A barge board will be added on each side, adding 3'' to the 72'' length for a total length of 75''. Adding the barge board will be explained later on in this chapter.

From the framing point at the center of the ridge to the framing point at the building line, the mathematical width is 21.63''. But look at Figure 7-33. The top edge of this 1/2'' sheathing is flush with the outside edge of the building. But the bottom edge of the sheathing is short of the building line. If we keep the top edge right on the building line, there is an effective gain in the length of the sheathing equal to 0.33'', or 5/16''.

In Figure 7-34, parts A and B show the triangle formed at the sheathing end. The calculations used to find the 0.33'' (5/16'') gain are in part C.

This 0.33'' must be subtracted from 21.63'' for a total of 21.30''. Thus, if the top edge of the sheathing is cut square, the width of the sheathing for our model would be 21$\frac{5}{16}$''. If a plumb cut is to be made at the top, 0.33'' must be added back for the long point measurement.

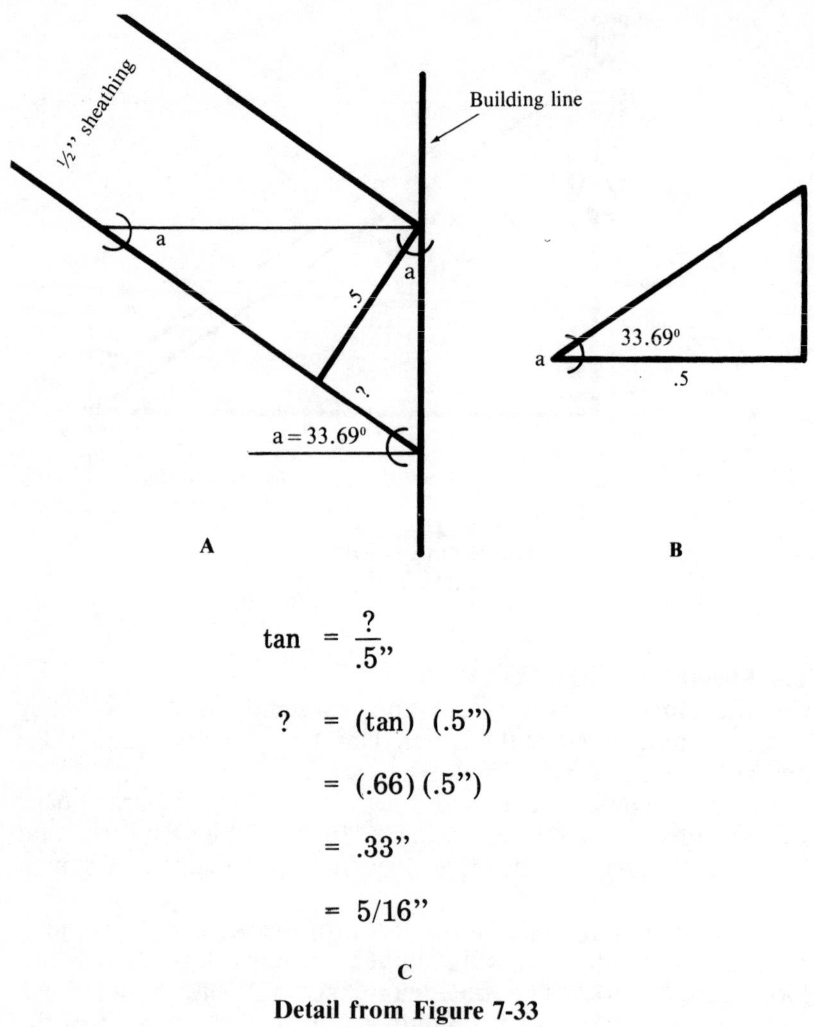

Detail from Figure 7-33
Figure 7-34

Cut the sheathing to the right width and frame it against the building line as in Figure 7-34 A and 7-39.

The Blind Valley Ridge

Most roof framers don't bother to calculate the length of the blind valley ridge. They use material that's plenty long enough to extend beyond the end of the overhang. After the addition is framed, they cut the ridge off at the right length. This saves some effort. But calculating the ridge length isn't hard. Figure 7-35 is an elevation

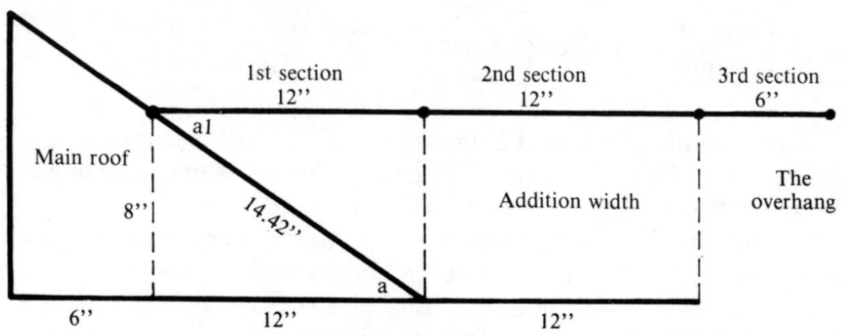

Blind ridge calculation
Figure 7-35

view of the main roof and proposed addition. We'll calculate the length of the ridge in three steps, labeled 1st, 2nd and 3rd sections in Figure 7-35.

1st Section: Angle (a1) in Figure 7-35 is the same as angle (a), which is the common rafter angle of the roof. We know that for an 8 in 12 roof, the angle is 33.69 degrees. The tangent is the rise divided by the run. Eight divided by 12 is 0.6667. Using your calculator, punch [.] [6] [6] [6] [7] [INV] [tan] and 33.69 degrees appears in the display.

It just so happens on this job that the addition ridge height (8 inches) is the same as the main roof unit rise. Therefore, you can see that the first section has a length of 12". It's not always this easy to find the first section length. So let's develop a general formula that finds the first section length on any blind valley ridge. There are four steps to the general formula we'll use:

If: 1) $\text{Tan } a^1 = \dfrac{\text{Addition Unit Rise}}{\text{1st Ridge Section}}$

Then: 2) $\text{1st Ridge Section} = \dfrac{\text{Addition Unit Rise}}{\text{Tan } a^1}$

If 3) $\text{Tan } a^1 = \text{Tan } a$

Then 4) $\text{Tan } a = \dfrac{\text{Main Ridge Unit Rise}}{12''}$

Substituting:
(4 into 2)

$$\text{1st Ridge Section} = \dfrac{\text{Addition Unit Rise}}{\dfrac{\text{Main Ridge Unit Rise}}{12}}$$

$$\text{General Formula} : \text{1st Ridge Section} = \frac{\text{Addition Unit Rise x 12}}{\text{Main Ridge Unit Rise}}$$

This formula gives us the apparent mathematical length of the first section of any blind valley ridge. Applying the numbers in our model, 8'' times 12'' divided by 8'' equals 12''.

2nd and 3rd Sections: Now we want to add the next two sections in Figure 7-35, the length of the addition and the length of the overhang. The addition is 12'' and the overhang is 6'' in our model. Thus the three sections total 30'', the apparent mathematical length of the addition ridge. Now let's find the actual length after shortening.

Ridge Shortening

Remember that the top of the ridge is 1/2'' below the framing point. Refer back to Figure 7-23 to understand why this is so. This 1/2'' drop causes a 3/4'' loss in mathematical length of the ridge. See Figure 7-36 A.

The actual mathematical length of the ridge then, is 12'' plus 12'' plus 6'' minus 3/4''. That gives us 29¼''. Now we're ready to lay out the ridge.

Laying Out the Ridge

Crown a piece of 2 x 6 with a squared end and lay the crown away from you. See Figure 7-37. Mark at 29¼'' and lay out a common

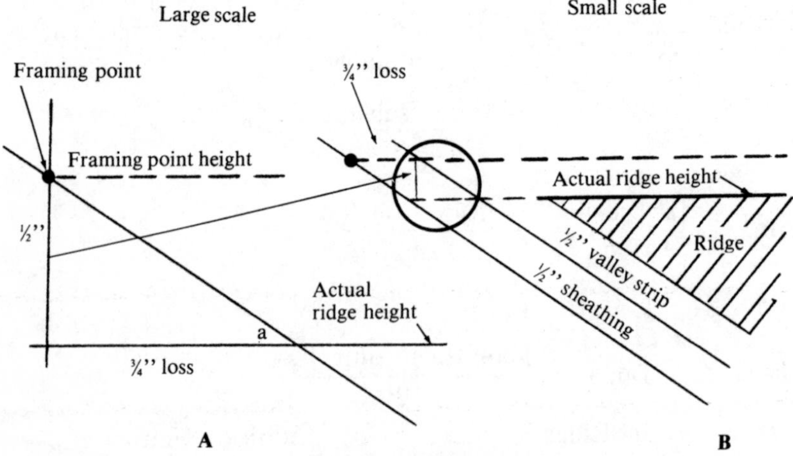

Three losses of ridge length
Figure 7-36

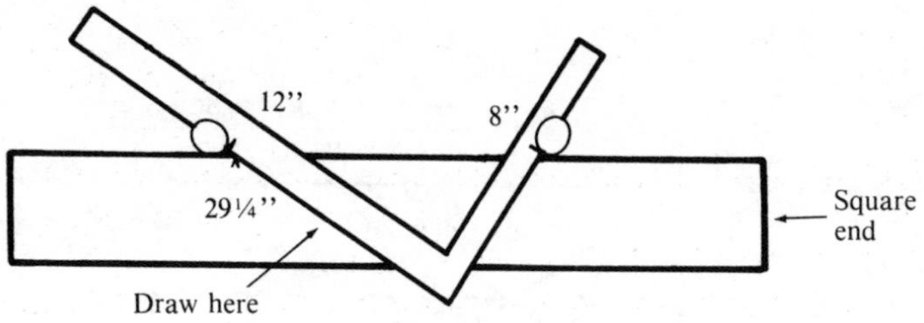

Laying out the ridge board
Figure 7-37

rafter seat cut line through that point. Notice that the seat cut of the addition ridge is the same as the seat cut of the main roof common rafter.

Look back to Figure 7-36 B. The ridge is away from the framing point by the thickness of the roof sheathing and also the thickness of the valley strip. Since both are 1/2'' thick, mark off these distances perpendicular to the seat cut line as indicated in Figure 7-38 and draw the new seat cut line. Finally, cut out the ridge.

Marking the Roof Sheathing

In general practice, the ridge is framed between several pair of common rafters with little attention to where the ridge meets the main roof. Backing is then placed on the underside of the sheathing directly under the ridge end. Next, the valley lines are snapped, then valley boards without plumb or seat cuts are nailed in place.

This practice is acceptable on the job. But for our model, let's calculate these distances so you understand what's involved. Refer to Figure 7-39 as we go through these calculations.

We used a formula to find the length of the first ridge section. The length was 12'', or 1 foot. (The run must be in feet for the work we are doing now.) For every foot of run in an 8 in 12 roof,

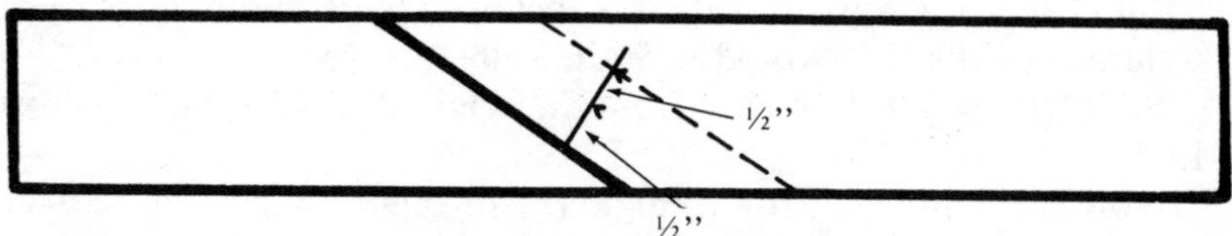

Ridge shortening
Figure 7-38

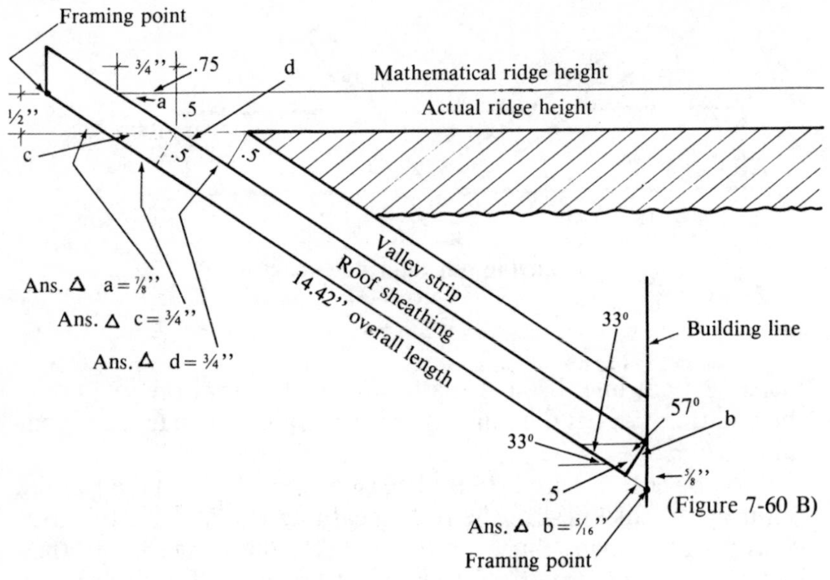

Layout on the sheathing
Figure 7-39

the diagonal or hypotenuse is 14.42''. So we start with the ridge framing point at 14.42'' above the framing point at the plate. Note the ''14.42 overall length'' in Figure 7-39. But this is only the mathematical length. Now we'll subtract the four triangles identified as (a) through (d) in Figure 7-39.

Triangle (a) at the upper left end shows the loss on the sheathing because the ridge is 1/2'' lower than the mathematical height. How do we find the actual point where the top of the ridge meets the sheathing? We find the length of the diagonal (the question mark) in Figure 7-40, which is our triangle (a) from Figure 7-39.

One way is to use the sine. The length of the diagonal is equal to 0.5 inches divided by the sine of 33.69 degrees. Your calculator will tell you that the answer is 0.9013 inches. That's about 7/8''. The square root method would give the same answer.

Subtracting 0.9013 inches from the original 14.42 inches gives us 13.5187''.

Now we'll subtract for triangle (b) in Figure 7-39. Figure 7-41 shows the problem. The triangle has been rotated so it rests on its 1/2'' base. The angle is the same 33.69 degrees and the tangent of that angle gives us the missing part. The tangent of 33.69 degrees is

$$\text{Sin a} = \frac{.5}{?} \qquad ? = \frac{.5}{\text{Sin } 33.69^\circ} = .9013" = 7/8"$$

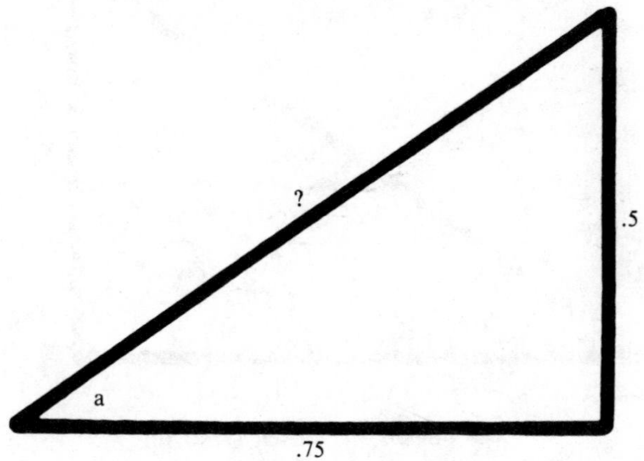

Loss due to lack of ridge height
Figure 7-40

$$\text{Tan b} = \frac{?}{.5} \qquad ? = (\text{Tan } 33.69^\circ)\ (.5) = .3333" = 5/16"$$

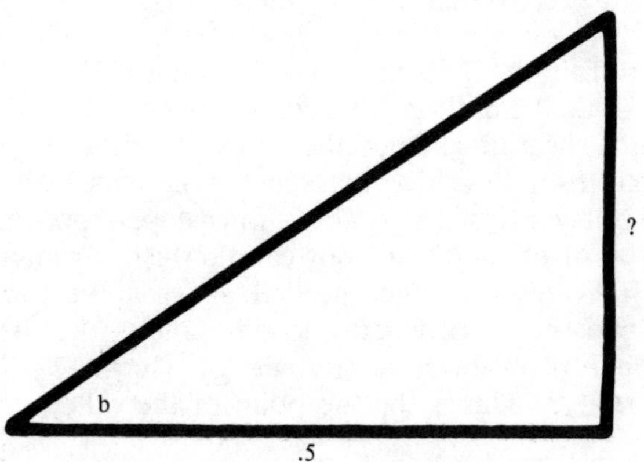

Loss because of corner of sheathing
Figure 7-41

$$\text{Tan c} = \frac{.5}{?} \qquad ? = \frac{.5}{\text{Tan } 33.69^\circ}$$

$$? = .75'' = \tfrac{3}{4}''$$

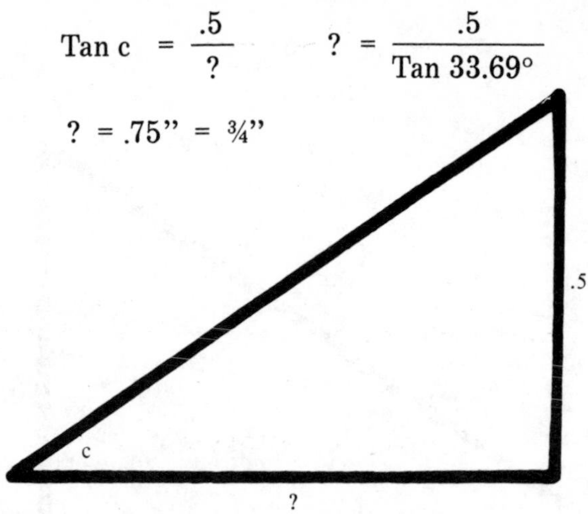

Loss due to ½'' thick sheathing
Figure 7-42

0.6667''. The answer to triangle (b) is 0.5'' times 0.6667, or 0.3333''. Subtracting that from 13.5187'' gives us a new length of 13.1854''.

The mathematical length of 14.42'' is measured at the bottom side of the roof sheathing. Triangle (c) allows for the thickness of the sheathing. The length of the missing side in Figure 7-42 is 0.5'' divided by the tangent of 33.69 degrees. That's 0.5'' divided by 0.6667, or 3/4''.

Subtracting 0.75'' from 13.1854'' leaves us 12.4354'', or about 12 7/16''.

Measure up from the bottom of the sheathing 12 7/16''. Plumb up on each side along the building line corner to locate the valley corner on the roof sheathing. Draw the two valley lines. Figure 7-43 shows the main roof sheathing marked for the blind valley.

Leave the valley strips just slightly downslope from the valley lines so the top of the addition roof can lie flush. Triangle (d) in Figure 7-39 indicates the space needed between the top of the valley strips and the point where the ridge meets the sheathing. This is the same problem as in triangle (c). Here, 12 7/16'' minus 3/4'' equals 11 11/16''. This is the top point of the valley strips.

Length of the Valley Strips

Once the ridge is in place and the valley lines are snapped, most roof framers would rightfully throw a piece of 1 x 6 on the roof,

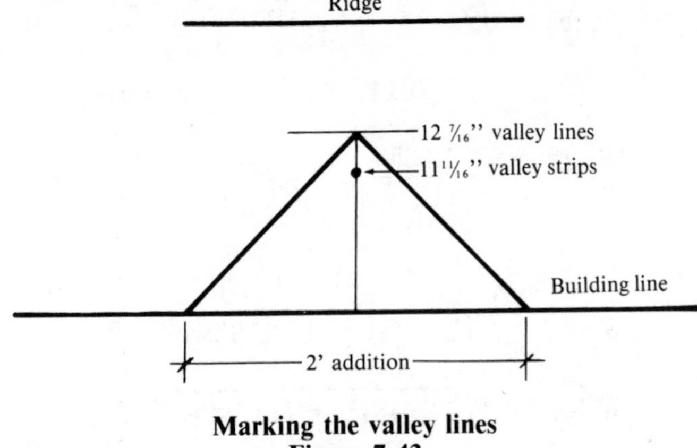

Marking the valley lines
Figure 7-43

hold it below the line, and nail it down. I'll calculate the actual length here so you can refer to these pages if needed.

Figure 7-44 shows the problem. We're looking for the length of (c), the valley strip. But we have to start with (h), the mathematical length. In step one, we find that to be 17.2811 inches. Next, find (a) so we know the length of (b). In step two we set up a ratio: 3/4" is to "x" as the overall height is to the overall length of the base. (The "x" in this calculation stands for (a) in Figure 7-44.) Length (a) in Figure 7-44 turns out to be 0.7237 inches. Subtracting that from 12 inches gives us 11.2763 inches. That's step three, the length of the base of the angle which has the valley strip as the hypotenuse.

In step four, we use the square root of the base and height squared to find the length of the valley strip, 16¼".

The Plumb and Seat Cuts

Set 11 11/16" on the tongue of your framing square. That's the plumb cut side. Set 11¼" on the body for the seat cut at the bottom. Cut the valley strips from 1/2" plywood, 3½" wide, and put them in place. Frame the two small-span common rafters in their place and then frame the ridge.

Height of the Blind Valley Ridge

The height of the blind valley ridge should be 10¼" above the plate. For this 1' run, the rise is 8". Add 2¾" for the HAP and subtract 1/2" for the loss in ridge height and you have 10¼".

① $(h) = \sqrt{(12\tfrac{7}{16})^2 + (12)^2}$

$(h) = 17.2811"$

② $\frac{.75}{x} : \frac{12.4354}{12}$

$x = .7237 = (a)$

③ $(b) = 12 - (a) = 11.2763"$

④ $(c) = \sqrt{(11.2763)^2 + (11.6854)^2}$

$(c) = 16.2389 = 16\tfrac{1}{4}"$

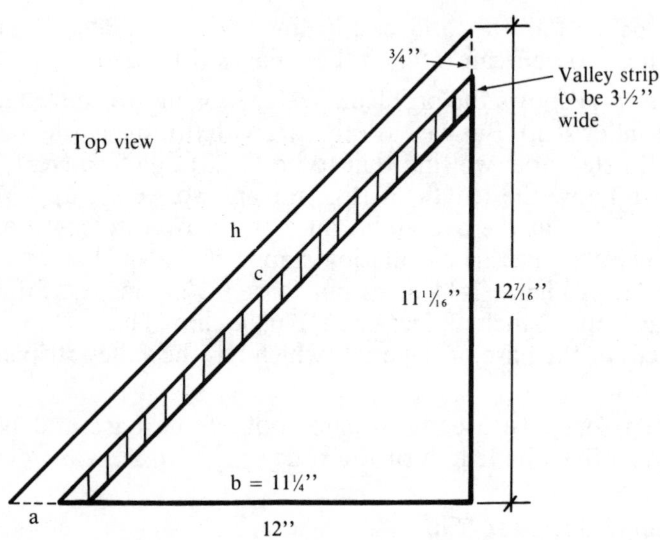

The valley triangles
Figure 7-44

First Blind Valley Jack

Most roof framers would measure the length of the first jack rafter from the top of the ridge to the lower outside corner of the valley strip. But you could also calculate that length. Figure 7-45 shows the problem and the solution.

$$\frac{11.27}{12} \times 14.42 = 13.5503 = 13\tfrac{9}{16}''$$

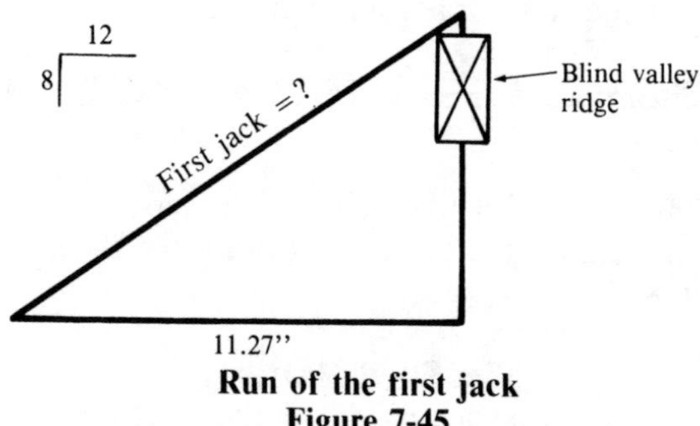

Run of the first jack
Figure 7-45

The run to the edge of the blind valley strip is 11¼''. Since the addition is an 8 in 12 roof, you have the usual calculation for common rafters.

Crown a piece of 2 x 4 and lay the crown away from you. Reset the stair gauges to 8'' and 12''. Mark 13⁹⁄₁₆'' for the mathematical length. Then mark off the ridge shortening. See Figure 7-46. The seat cut at the bottom must be shortened 1/2'' because of the thickness of the model valley strip.

The seat cut moves uphill at the same angle as the main roof common rafter. Therefore, set your saw as close to 33.69 degrees as possible to make this compound seat cut. Make an opposite pair and frame one on each side.

Notice that the rafter tail does not come all the way out to the edge of the valley strip. This is because the outer edge of the valley

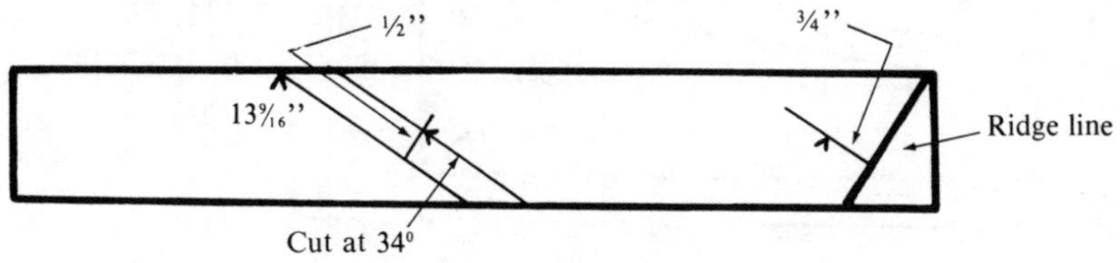

Layout lines of the first jack
Figure 7-46

strip has not been cut to an 8 in 12 slope. When the edge material has been removed, the rafter will fit perfectly.

Finding the Common Difference

If we wanted to frame another model rafter 4" in from this first valley jack rafter, we would calculate the common difference and subtract it from the first jack's mathematical length.

$$\text{Common difference} = \frac{4''}{12''} \times 14.42 = 4.8066'' = 4\tfrac{13}{16}''$$

$$\text{Subtracting:} \quad 13\tfrac{9}{16}'' - 4\tfrac{13}{16}'' = 8\tfrac{3}{4}''$$

The mathematical length of the long point for the second valley jack would be 8¾".

The Shed Rafter

It's fairly easy to frame a shed roof. But there are a few points that aren't so obvious. Let's build a 3 in 12 shed roof rafter to fit on the 3' model rafter plate.

A shed roof is like one side of a gable roof. See Figure 7-47. The building span becomes the total run. The total rise is found in the usual way, by multiplying the total run by the unit rise. In this case, 3 times 3" equals 9" of total rise.

Building the Pony Wall

A short wall is sometimes called a *pony wall*. The model requires a pony wall with an overall height of 9" as shown in Figure 7-47 B. If each plate is 1½" thick, then a 6" piece of 2 x 4 is required at each end. But check the milled thickness of your 2 x 4's. Cut out and frame the pony wall to the model.

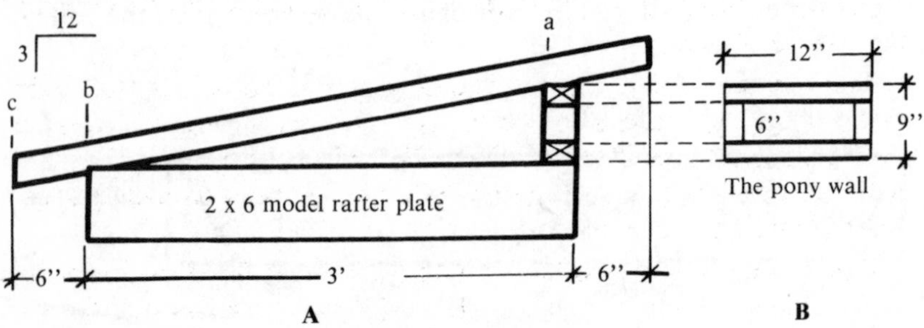

A model shed roof rafter
Figure 7-47

Finding the Rafter Length

The secant for a 3 in 12 roof can be found in a trigonometry table if the tangent is known. In this case the tangent is 3 divided by 12, or 0.25. The inverse tangent is listed as 14 degrees and 2 minutes. To find the secant on the calculator, punch [3] [÷] [1] [2] [=] to find the tangent, 0.25. Then push [INV] [tan] to find the angle, 14.0362. Now, to find the secant, push [cos] [1/x]. 1.0308 appears in the display. This is the secant for a 3 in 12 roof. Figure 1-14 also lists this secant. You'll multiply this figure by the run to find the rafter length, so push [STO] to store the secant to use in the calculations below.

Figure 7-47 shows a 6" overhang on each side of the building line. There are three runs to a shed rafter. They're marked (a), (b), and (c) in Figure 7-47.

Run (a) is the run of the overhang plus the width of the upper plate. This is 6" plus $3\frac{1}{2}$", or $9\frac{1}{2}$".

Run (b) is the total run of the building plus the run of the upper overhang. That's 36" plus 6", or 42" on our model.

Run (c) is the total run of the building plus the run of both overhangs. This is 36" plus 6" plus 6", or 48".

Now we can multiply each of these lengths by the secant, 1.0308, to find the rafter length. Push [RCL] to retrieve the secant from memory. For (a), secant times $9\frac{1}{2}$" equals 9.7924" or $9\frac{13}{16}$". For (b), secant times 42" is 43.2926" or $43\frac{5}{16}$". For (c), secant times 48" is 49.4773, or $49\frac{1}{2}$".

Laying Out the Rafter

Crown a piece of 2 x 4 and lay the crown away from you. Set the stair gauges on 12" and 3". With these settings, the framing square will not fit across a 2 x 4. If we add half the setting to each side, the framing square would be set on 18 and $4\frac{1}{2}$". This works fine for laying out a 2 x 4.

Mark a plumb cut line on the corner and mark off the three rafter length calculations. See Figure 7-48. At (a) and (b) make a $2\frac{3}{4}$" HAP. At a 3 in 12 pitch, a $2\frac{3}{4}$" HAP allows full bearing without allowing part of the seat cut into the upper overhang area. There is no plancher cut at (c).

Cut out the rafter and frame it to the model. Shed rafters are a little different from common rafters, but the same principles apply.

Congratulations. You've mastered cutting another type of roof!

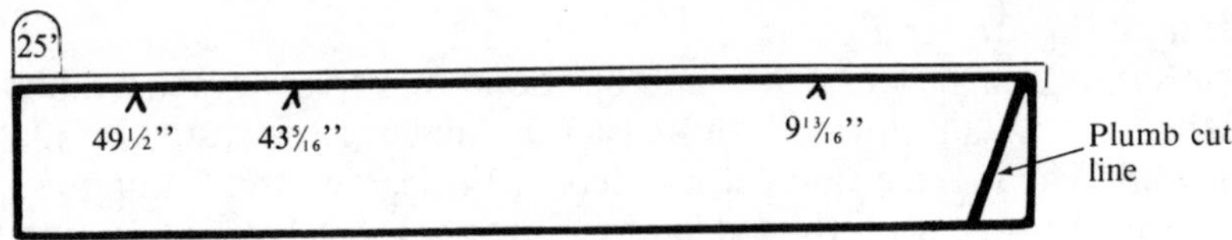

Laying out a shed rafter
Figure 7-48

Gable End Framing

Sometimes it will be your responsibility to cut the studs for the gable end. Figure 7-49 shows a gable end on our model.

Here are some considerations when framing the gable end. The gable end will usually include framing for either an attic vent or a window, which will be centered below the ridge. You don't want a gable stud directly under the ridge. If plywood siding is used, these gable studs must fall on 16'' centers.

Figure 7-50 shows how our model will be framed.

The Longest Gable Stud

We calculated the mathematical height of the model ridge as 14¾''. On our model, the 1½' run at 8 and 12 produced a 12'' rise.

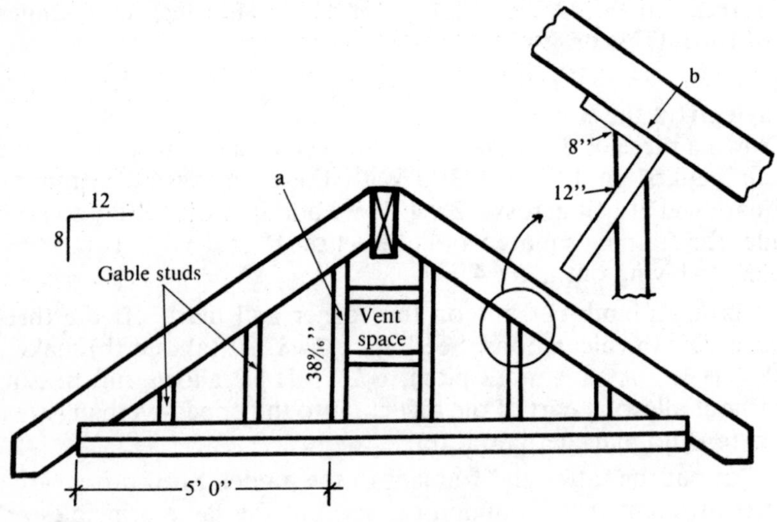

The gable end
Figure 7-49

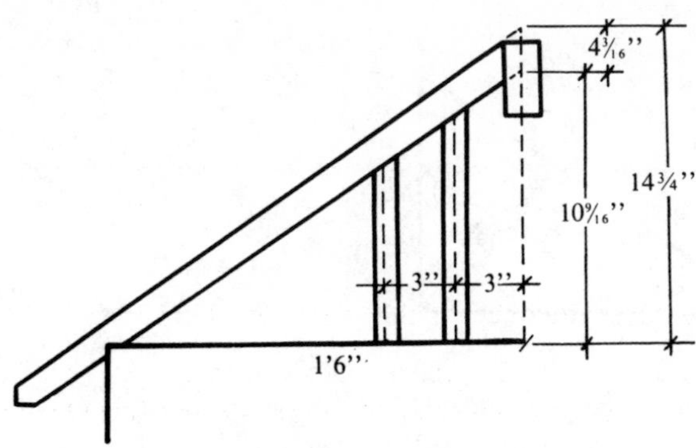

The length of the longest gable stud
Figure 7-50

The HAP adds another 2¾''. If we installed a gable stud centered under the ridge point, its length would be the total rise minus the distance of the plumb cut. What is this distance? We'll figure it out by graphing and by calculation.

On 1/4'' grid graph paper, lay out an 8 in 12 line as in Figure 7-51. Construct a second line parallel to the first line and 3½'' from it. This represents the 2 x 4 rafter. Measure the plumb cut distance. It will be 4 3/16''.

Figure 7-52 A shows an 8 in 12 unit triangle and a dotted line which represents the 3½''-wide rafter. Line (ab) is drawn at 90 degrees to the rafter so that the 90-degree angle of the unit triangle

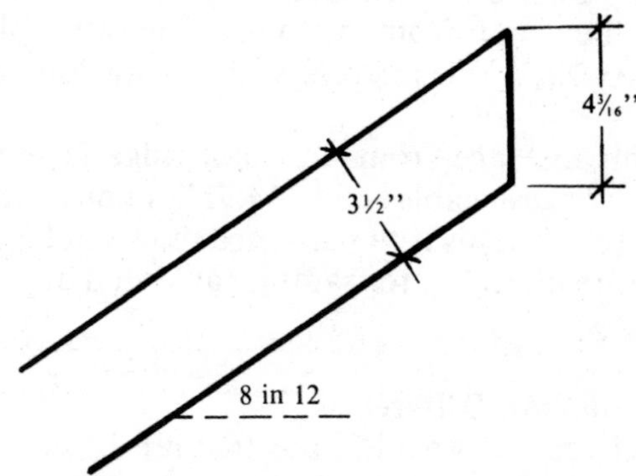

Finding the plumb cut distance
Figure 7-51

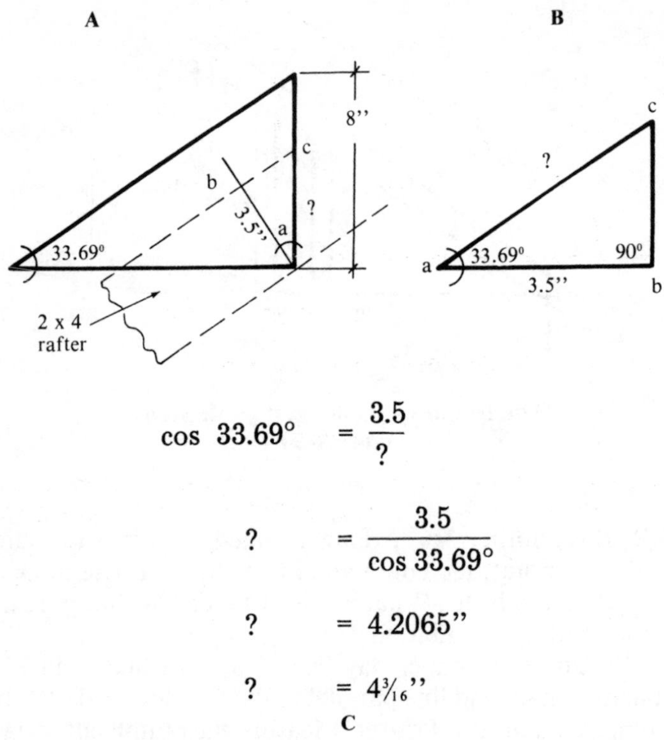

The length of the plumb cut
Figure 7-52

is broken into two complementary angles. Triangle (abc) is shown repositioned in Figure 7-52 B. Use the cosine of the 33.69-degree angle to find the length from (a) to (c). The rafter plumb cut will be 4.2065'', or 4 3/16''. This agrees with our measured distance in Figure 7-51.

Subtract this distance from the total ridge height to find the length of the longest gable stud: 14.75'' minus 4.2065'' equals 10.5435'', or 10 9/16''. This is the distance shown in Figure 7-50. But of course, you're not going to frame a gable stud directly under the ridge.

Finding the Common Difference

For our model you'll cut gable studs that are 3'' on center to each side of the ridge, as shown in Figure 7-50. This means that the gable stud short point will have to be 3 3/4'' from the ridge center line. Figure 7-53 shows this 3 3/4'' projected against an 8 in 12 line.

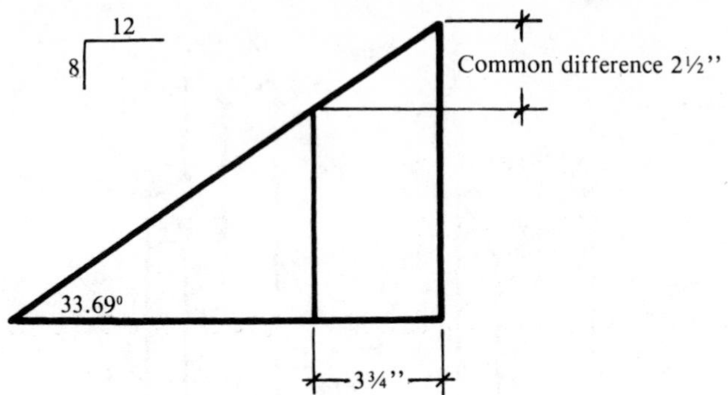

Finding the common difference
Figure 7-53

The Graph Method— Using the graph line just constructed, measure, from the plumb line toward the left, 3½'' and project the line up to the 8 in 12 line. Then project across, as indicated in Figure 7-53, and measure the common difference. The distance will be 2½''.

The Mathematical Solution— For an 8 in 12 roof, every 12'' of unit run produces 8'' of rise. If the gable studs were 12'' on center, their common difference would be 8''. In our problem however, the distance is only 3¾'', or 3.75 twelfths of 8'':

$$\frac{3.75}{12} \times 8 = 2.5''$$

That 2.5'' is the same amount just measured on Figure 7-53. The general formula to find the common difference for gable end studs is:

$$\text{Common Difference} = \frac{\text{On-center Distance (in inches)}}{\text{Unit Run (in inches)}} \times \text{Unit Rise (in inches)}$$

If it's too much trouble to go through the mathematics or make a scale drawing to find the longest gable length, just measure the longest gable stud and use a simple common difference formula to find the next gable stud length.

There's yet another way to figure these gable studs. Assume that gable stud (a) in Figure 7-49 had a run from the short point to the corner of 5', as indicated. Since the drawing shows it as an 8 in 12 roof, there are 5 times 8'', or 40'' of height in the total triangle. The HAP must be added, giving us an additional 2¾'', or 42¾''.

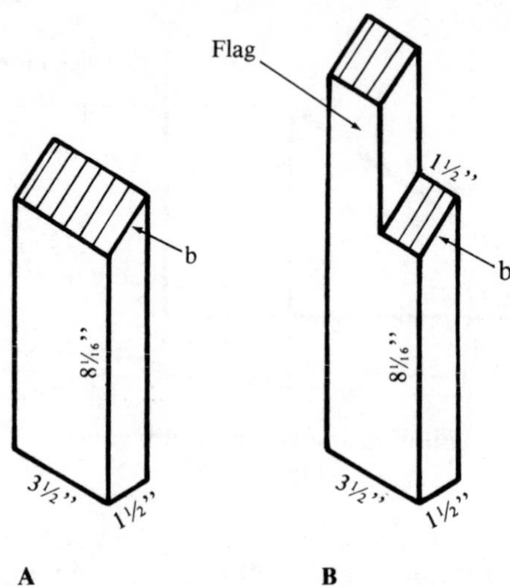

The first gable stud
Figure 7-54

Then subtract the thickness of the rafter plumb cut to find the correct short point distance: 42.75'' minus 4.2065'' equals 38.5435'', or 38⁹⁄₁₆''.

Finding the Second Gable Stud

The longest gable stud measurement was found to be 10.5435'', as marked in Figure 7-50. The first gable stud short point run is 3'' plus 3/4'' from the center line of the roof. That 3¾'' gives us a common difference of 2.5''. From 10.5435'' subtract 2.5'' to get 8.0435'', or 8¹⁄₁₆''. This is the short point distance of the first gable stud. The second stud short point is 3'' away from the first. Using the general formula to find this 3'' on-center distance:

$$\text{Common Difference} = \frac{3''}{12''} \times 8'' = 2''$$

$$8.0435 - 2 = 6.0435 = 6\tfrac{1}{16}''$$

The second gable stud short point has a distance of 6¹⁄₁₆''.

For the first gable stud, measure up 8¹⁄₁₆'' on one edge of a 2 x 4. The line (b) in Figure 7-54 A and Figure 7-49 is the plumb cut angle of the roof (the tongue side of the framing square). For the style in Figure 7-54 A, set the saw to the roof angle, in this case 33.69 degrees. For the style in Figure 7-54 B, set the saw depth to 1½'' at zero degrees. Make the cut against the 1½'' side along line (b), the plumb cut line. Readjust the saw for full cutting depth and remove 1½'' of material so that the flag appears. This style provides stronger framing.

Cut out one gable stud with a 6¹⁄₁₆'' short-point measurement and another one at 8¹⁄₁₆''. From the ridge center line, the short-point layout is at 3¾'' and 6¾''.

Gable Vents and Frieze Vents

Gable ends, whether part of the main roof or a blind valley addition, require venting for air circulation in the attic space. Venting requirements are established in most building codes. Usually you'll need clear vent area at both the gable and frieze vents equal to 0.25% of the enclosed area.

If a building has a blind valley gable roof addition that measures 14' x 24', the addition area is 336 square feet. Multiply 336 by 0.0025 to get 0.84 square feet of vent at both the gable and frieze vent.

A standard 14'' x 18'' gable vent has clear vent area of 1.04 square feet. So this size will work fine. Since the joists are 24'' on center, 22⅜'' by 3½'' frieze vents can be used. Each vent has clear vent area of 0.31 square feet. Dividing 0.84 by 0.31, we find that three vents can do the job. For the sake of symmetry, we'll install four frieze vents.

The Barge Board and the Overhang

On a hip roof, rafter tails extend perpendicular to the building line on all sides. The fascia board is framed to these tails, which are all at the same level.

A gable roof is different. The gable end overhang is usually the same as the overhang along the length of the building. But in the gable end only, the ridge board extends the width of the overhang. The fascia board must turn the corner when it reaches the end of the long edge of the building, and run up the gable to the ridge. This sloping fascia is called a *barge board.* It isn't cut like the common rafter, but there are some similarities.

You can't frame the barge board until the overhang "ladder" is installed for support. So that's where we'll start.

The Overhang Ladder

Generally, the end common rafter is placed directly above the building line and the rafters are spaced 2' on center.

If the ridge were cut for a 2' overhang, as in Figure 7-55, the 2 x 4 "ladder" lookouts would be cut to 4'. The building line common

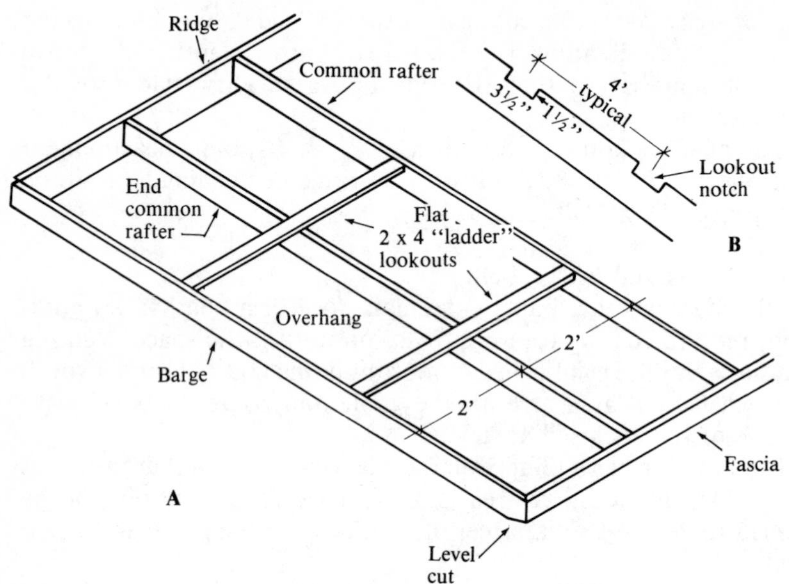

The overhang ladder and barge board
Figure 7-55

rafter should have notches 3½" wide and 1½" deep to receive the lookouts. The notches are spaced 4' apart. See Figure 7-55 B.

On our model, the overhang is 6" and the space to the second rafter is 19½". Look back to Figure 2-35. Add 6" to 19½" and you have 25½". Cut out four lookouts from 2 x 4 material. Each should be 25½" long.

Make two notches centered at 7" and 15¾" on each of two common rafters. See Figure 7-56. Frame these rafters at the building line. Then frame the lookouts in place, nailing them to the notch and back-nailing them to the second rafter.

The Barge Board

The common rafter shortens by half the thickness of the ridge. The barge board does not shorten since it is nailed against the end grain of the ridge at the centerline of the roof. Another difference is that there is no birdsmouth cut on the barge. See Figure 7-57. Also notice the difference in the plancher cut.

The Plancher and Tail Cut

A normal barge board is at least 2 x 6 material. But a 2 x 6 would be too deep to frame on our model. You'll remember that the com-

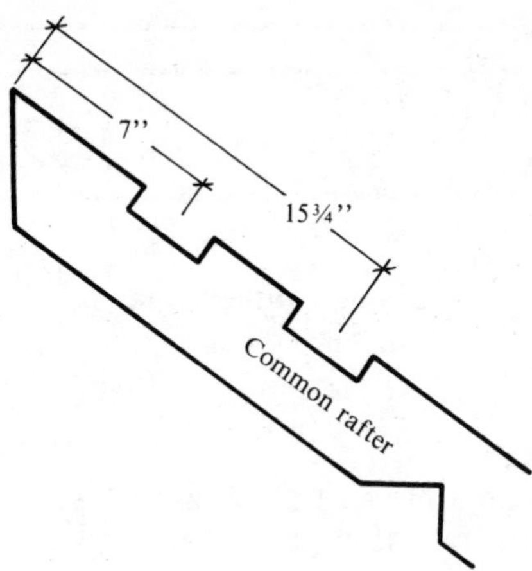

Laying out the model notches
Figure 7-56

mon rafter plancher cut was designed to receive 2 x 4 fascia board. Since the barge miters to the fascia, the plancher cut for the barge board will have to be the depth of the fascia. For the model this distance is 3½".

Crown a piece of 2 x 4 and lay the crown away from you. Set the stair gauges on 12" and 8". Draw the ridge line on the corner and mark the fascia line as in Figure 7-58. This line will be the short point of the 45-degree cut. Add the plancher cut of 3½" on the other side. Lay out and cut two opposite barge boards.

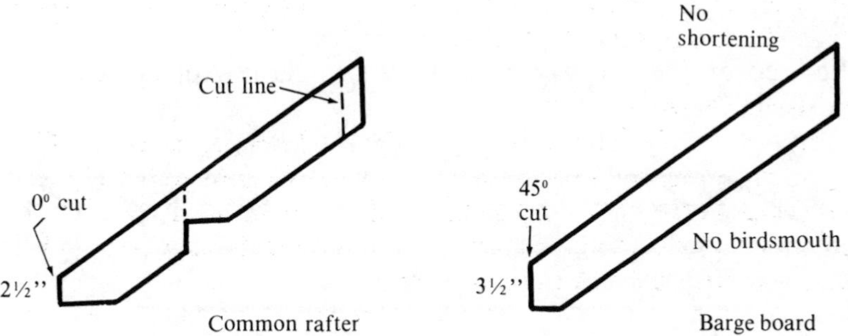

Comparing the common rafter and the barge
Figure 7-57

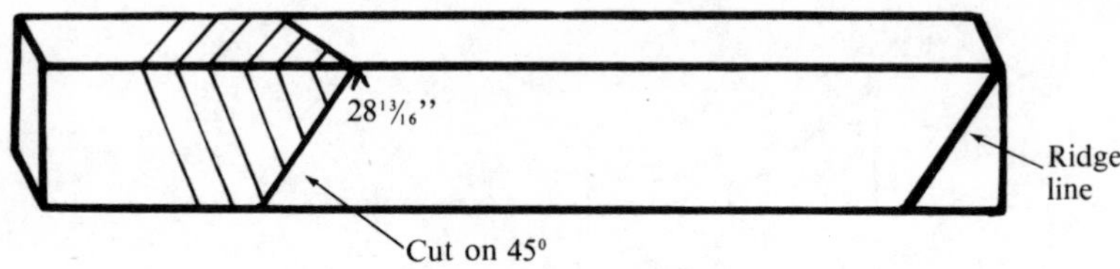

Laying out the barge
Figure 7-58

Cutting the Fascia

Measure off 72" on a piece of 2 x 4, as in Figure 7-59. These are the short points for two 45-degree cuts.

The top of the fascia must be cut to the common rafter angle of 33.69 degrees for an 8 in 12 roof. Figure 7-60 A and line (b) in Figure 7-39 show this sheathing triangle at the building line. The fascia should be framed 5/8" above the common rafter so the edge of the 1/2" sheathing is hidden. Figure 7-60 B shows why we move the fascia up 5/8" instead of just 1/2". To find the length of the hypotenuse in Figure 7-60 B, we divide 0.5" by the cosine of 33.69 degrees to get 5/8".

In Figure 7-60 C, we calculate the amount of loss on the outside of the fascia board due to the 8 in 12 cut. The tangent of 33.69 degrees is 0.6667. Multiplying that by 1.5, we get 1 inch, as indicated in Figure 7-60 A.

Make these cuts and frame the fascia and barge boards to the model. Use a scrap of 1/2" plywood to set the distance for the fascia board. The barge tail will follow the fascia. Using the 1/2" scrap, hold the ridge cut above the ridge by the thickness of the plywood.

We'll cover another way to frame the fascia in Appendix H.

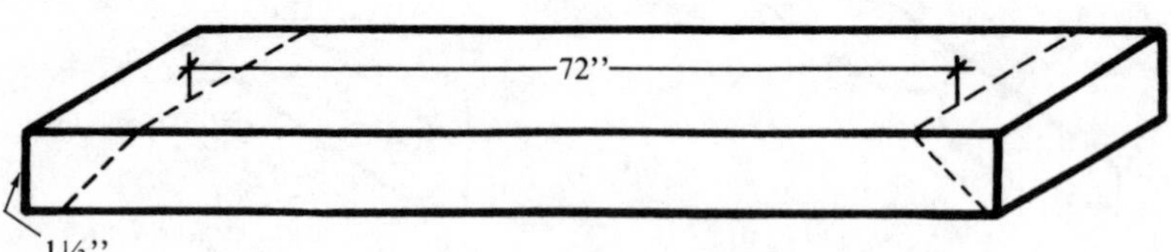

Laying out the fascia cuts
Figure 7-59

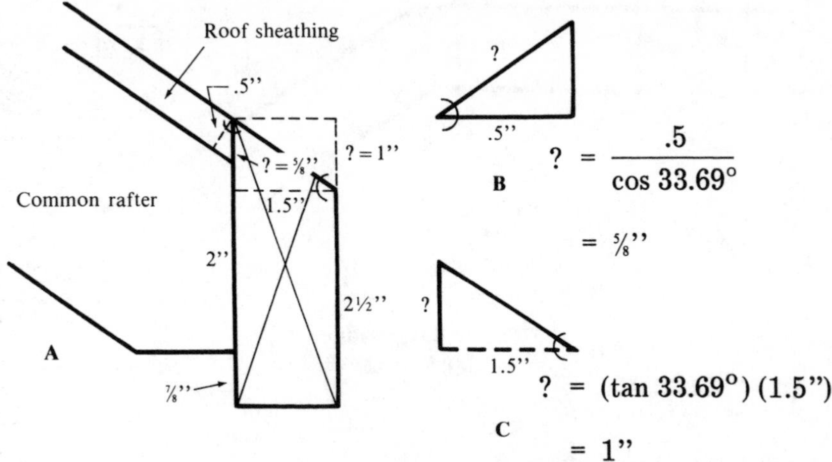

$$? = \frac{.5}{\cos 33.69^\circ}$$

$$= \tfrac{5}{8}''$$

$$? = (\tan 33.69^\circ)(1.5'')$$

$$= 1''$$

The fascia slope and placement
Figure 7-60

The Protruding Ridge

A 2 x 6 ridge board will protrude well below a 2 x 4 barge board at the top of the gable. That will be true on most jobs, because the ridge is generally a size larger than the rafters.

Figure 7-61 shows two ways to trim the ridge. In part A, the lower corner of the ridge has been cut off so 1'' of barge hangs below the shoulder of the chamfer. The ridge could also be square cut as in B. Then the exterior siding is brought up, as shown, to cover the end grain of the ridge.

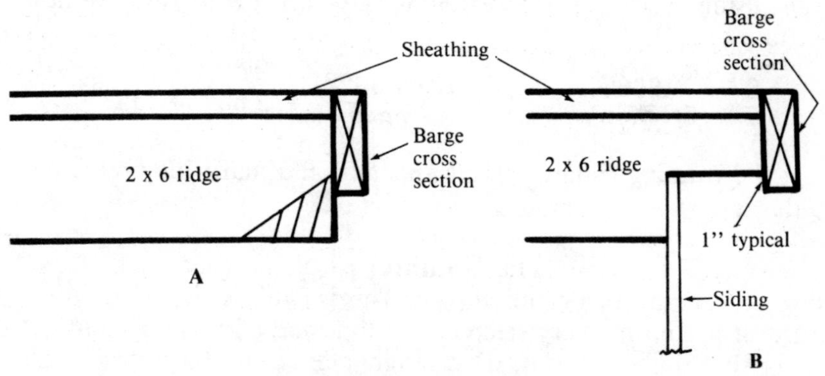

Trimming back the ridge
Figure 7-61

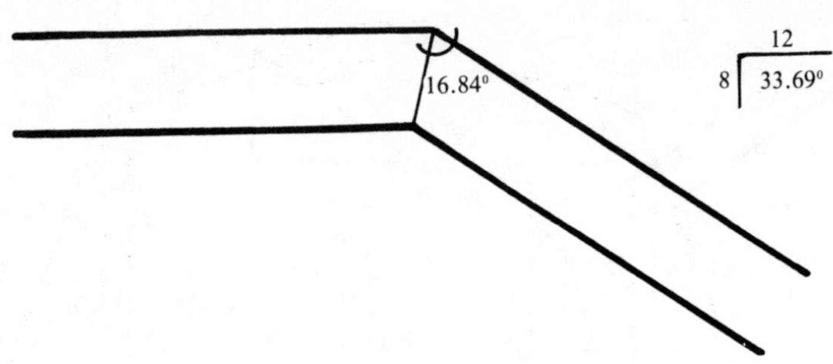

The Tudor barge ridge cut
Figure 7-62

Dutch Barge Tail Cuts

The barge board tail cut for a Dutch gable is the same as that of a valley jack rafter framed to a blind valley. See Figure 7-46. Provide another shortening for the thickness of the roofing material. Also, remember to leave the overhang sheathing loose so the roofer can lift it to shingle underneath.

Tudor Barge Ridge Cut

Figure 7-17 shows the Tudor plate framed against the common rafters with plumb cuts. This type of construction is fine for rough framing. But it isn't acceptable for a barge board joint.

The correct angle of the cut is half the angle of the roof rise. The angle of an 8 in 12 roof is 33.69 degrees. Half of this angle is 16.84 degrees, as in Figure 7-62. The inverse tangent of this is 0.3028. To find the framing square setting, use the 12'', but adjust the 8'' dimension. Set up a proportion as shown below to find the new setting of 3⅝''.

$$\frac{8''}{?} : \frac{.6667}{.3028} \qquad ? = \frac{(8)\ (.3028)}{.6667} = 3.63'' = 3\tfrac{5}{8}''$$

Set the framing square to 12'' and 3⅝'' and mark the plumb cut side.

The Gambrel Roof

Figure 7-63 shows a gambrel roof. This is Dutch colonial architecture. It has both a very steep lower slope and a low-pitched upper slope that closes the roof. It's a familiar design for barn structures.

Look at Figure 7-63. The total run of the lower, steeper slope is identified as (a), the total run of the upper slope is (b). Good

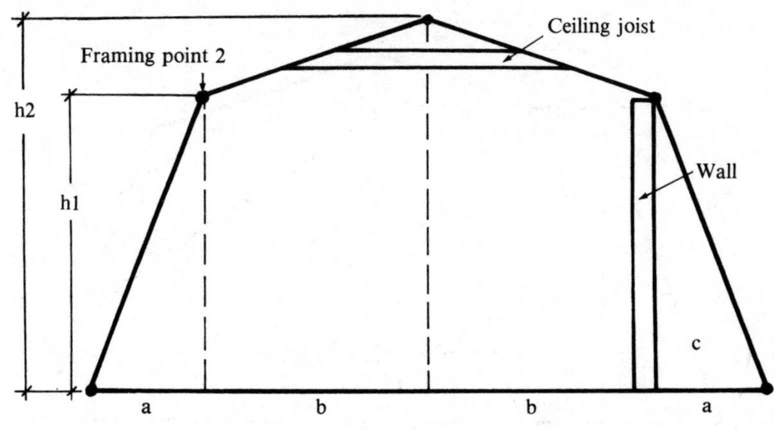

The gambrel roof
Figure 7-63

design requires a ratio of (b) to (a) of about 2:1. But the ratio can vary somewhat with building width and height.

Think of a gambrel roof as two gable roofs stacked one on top of the other. The point marked "framing point 2" in Figure 7-63 acts as the ridge for the lower roof and a plate for the upper roof.

Structural Design

There are two important design considerations in a gambrel roof. The first is the downward thrust of the roof. The second is resistance to wind loads. High winds can exert a strong sideward thrust against a gambrel roof. One common way to provide good support on a gambrel is to erect a bearing wall directly below the lower framing points. This makes the rafter plate on those walls serve double duty. It's both a ridge for the lower slope and the rafter plate for the upper slope.

The wall in Figure 7-63 isolates section (c) of the building, but creates some additional storage space as a closet. Use of this bearing wall provides less habitable interior space. But notice that the room walls will be vertical. Only the closet walls are sloping. The underside of the roof retains both its distinctive shape and strength if reinforced with ceiling joists as shown in Figure 7-63.

If the interior rooms are to have the full architectural effect of the gambrel, then use post and beam construction rather than an interior wall to support the roof.

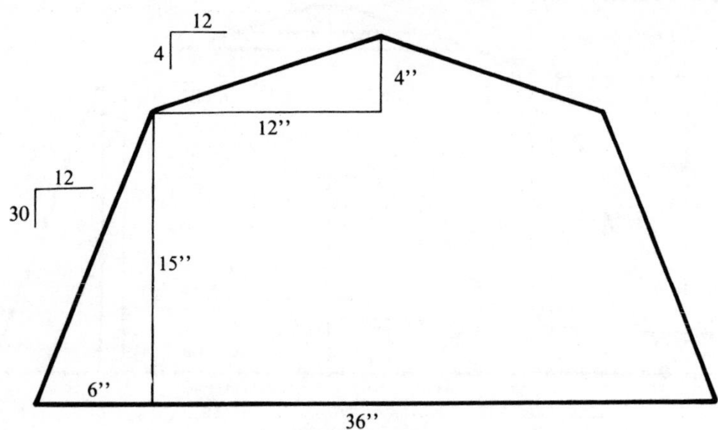

Main model dimensions
Figure 7-64

Let's build the roof in Figure 7-64. The lower pitch will be 30 in 12 and the upper portion will be 4 in 12. The lower framing points will be 6'' in from each side.

Finding the Rafter Lengths

There are two ways to figure the rafter lengths: the framing square method and the secant method.

The Framing Square Method— Figure 7-65 shows the unit triangle and calculation. Since your framing square doesn't have a unit length for 30 in 12 rafters, we have to compute one. The total run is only half a foot. Divide 32.31 inches by 2 to get 16.155 inches, or 16⅛''. This is the mathematical length of the rafter.

Your framing square shows that a 4 and 12 rafter has a unit length of 12.65''. Since the total run is one foot, the mathematical length of the rafter will be 12.65'', or 12⅝''.

The Secant Method— A 30 in 12 rafter has a tangent of 30 divided by 12, or 2.5. The inverse tangent is 68.20 degrees. To find the secant, use [cos] [1/x] and the display will show 2.6926. Multiply the secant by the run of 0.5' and you have 1.3463', or 16⅛''.

The secant for the 4 in 12 rafter is found the same way: 4 divided by 12 gives a tangent of 0.333 and an inverse tangent of 18.4349. Use [cos] [1/x] to find the secant, 1.0541. Since the run is one foot, the rafter length will be the secant of 1.0541 times 1, or 12⅝''.

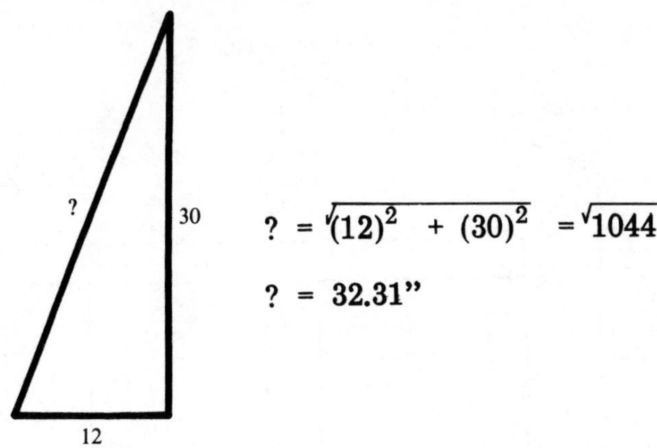

Finding unit length
Figure 7-65

The Model End Wall

Look at Figure 7-66. The model end wall will be framed with a beam pocket at the left side, and a 5¼" post to support the purlin beam which runs the length of the model. The right side will be framed for a bearing wall. A 2 x 6 center ridge beam will bear on top of the plate. On an actual roof, the gable end wall would be framed with a beam pocket provided for the center ridge.

The Rafter Cuts

Now we're ready for the lower rafter plumb cut. Since 30 and 12 won't fit on your framing square, set the gauges at 15 and 6. Even then, the tongue and body have to swap places to set the gauges. Draw the plumb cut and measure its length as 9⁷⁄₁₆".

The plumb cut can also be found mathematically. See Figures 7-52 B and 7-67.

For the upper rafter seat cut, set the stair gauges to 4 and 12 and draw a seat cut line, as in Figure 7-68. Since the model purlin will be 3" thick on top, we can measure back along the seat cut 3" as indicated and make a mark. Through this point draw plumb line (a).

Measuring line (a) gives the proper amount of HAP. A 2¾" HAP will give a bearing of 3". Any smaller HAP will cause part of the seat cut to appear inside the room. That would spoil the line of the ceiling.

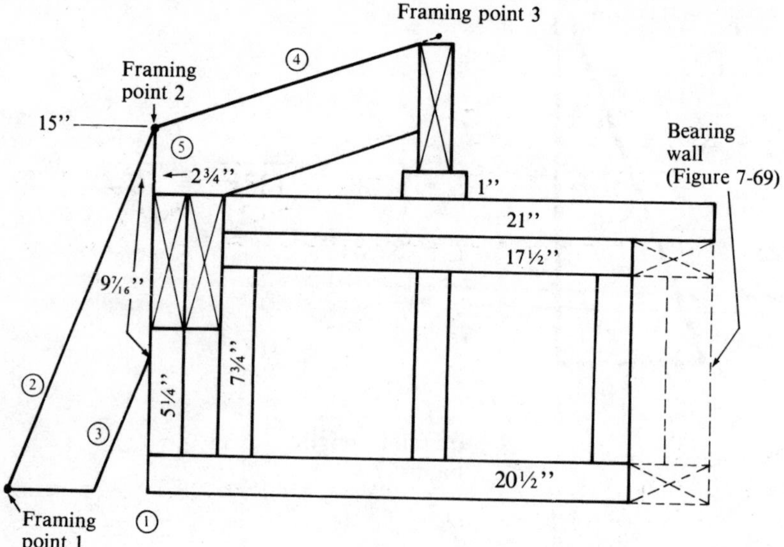

Detailing the end wall
Figure 7-66

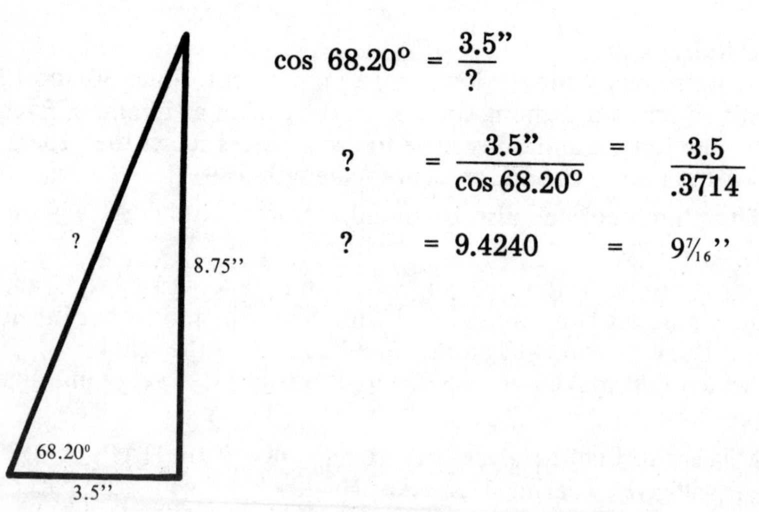

$$\cos 68.20^\circ = \frac{3.5''}{?}$$

$$? = \frac{3.5''}{\cos 68.20^\circ} = \frac{3.5}{.3714}$$

$$? = 9.4240 = 9\tfrac{7}{16}''$$

Finding the plumb cut
Figure 7-67

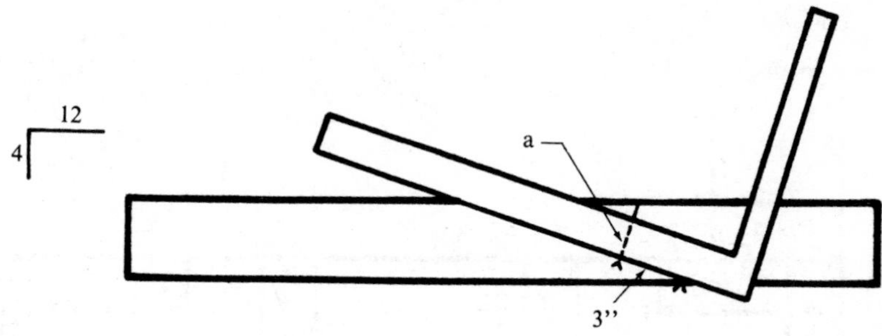

Finding the HAP
Figure 7-68

Figuring the Stud Lengths
Look at Figure 7-66 again. It's drawn in the scale of 3'' to the foot. The plan shows ''framing point 2'' at 15'' above the plate. The run is 6'' to the point where lines (1) and (2) meet. Line (3) indicates the thickness of the 2 x 4 model rafter. Line (4) establishes the slope of the upper roof.

To figure the stud lengths, we have to start at the 15'' height and work down because the HAP of the upper rafter determines the amount of the level cut. Add line (5), the HAP, then add the end-grain picture of the 2 x 6 purlins. Measure your stock to see if your 2 x 6 is 5½'' or 5⅝'' wide, since the milling is often different.

The end wall will have a bottom plate 1½'' thick. Therefore, from the 15'' height subtract the HAP (2¾''), the thickness of the purlin (5½''), and the bottom plate (1½''): 15'' minus 2¾'' minus 5½'' minus 1½'' equals 5¼''.

For the model, cut two 2 x 4's to 5¼'' long. These form the post under the beam pocket. On a full-size roof you would usually cut the post a little shorter. It's much easier to wedge the purlin up to the right height than it is to remove material after the beam is resting in the pocket.

Cut the other members to the length indicated and frame a wall for each end.

Build a purlin by nailing two 5'-long pieces of 2 x 6 together. Be sure to crown both pieces first and put the crowns up. Cut another 2 x 6 for the ridge. Add 1'' of material under each end of the beam as indicated. Frame the purlin and ridge to the model.

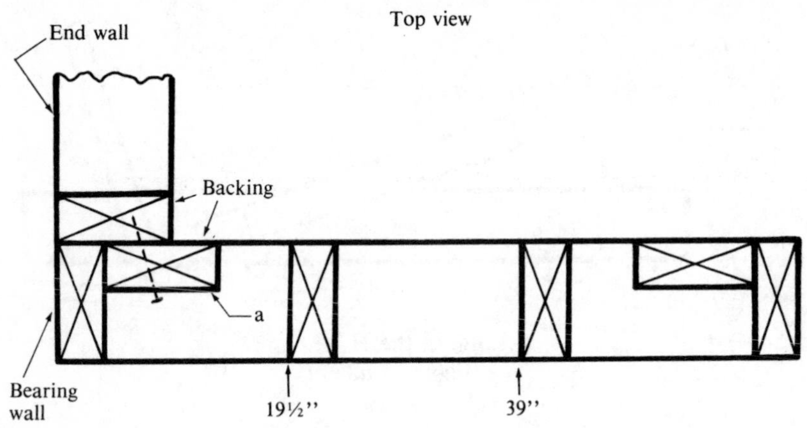

The bearing wall framing
Figure 7-69

The Bearing Wall

Build the bearing wall on the third side of your model. The studs will be 7¾'' long.

Cut six studs and frame them at the intervals shown in Figure 7-69. The stud in position (a) provides two things: backing for the interior wall, and a nailing surface to join the end wall and bearing wall together.

The bottom plate and top plate will be 5' long. We'll hold back the rafter plate for this wall by 3½'' on each end. That makes the rafter plate of each end lap over, tying the three walls together. The rafter plate will be 60'' minus 3½'' minus 3½'', or 53''. Add support for the side wall in the center, over the model plate brace. Refer back to Figure 2-8 A in Chapter 2.

Laying Out the Rafters

Start with the lower roof. Crown a piece of 2 x 4 and lay the crown away from you. Set the stair gauges at 6'' and 15'', using the 15'' side as the plumb cut side. Draw the ridge line, as in Figure 7-70. There will be no shortening, so the ridge line will become the cutting line. Measure down 16⅛''. This is the beginning point of the seat cut. The HAP for this rafter is zero, since there is no overhang in a gambrel roof.

The two lines and two cuts in Figure 7-70 are all you need on this lower rafter. Cut eight of these rafters and lay them out according to the instructions under Figures 2-34 and 2-35 back in Chapter 2.

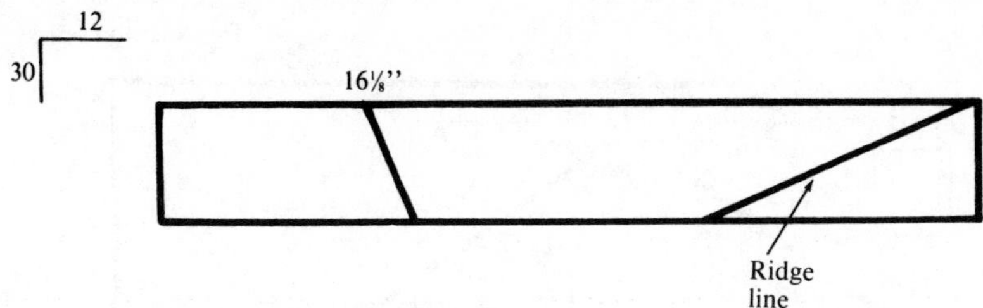

The lines of the lower rafter
Figure 7-70

For the upper rafter, crown a piece of 2 x 4 and lay the crown away from you. Set the stair gauges to 4'' and 12'', returning the square to its accustomed position. Lay out the ridge line and add the 3/4'' shortening (one-half the thickness of the ridge). See Figure 7-71. Mark off the building line as calculated and add the 2¾'' HAP. Cut eight of these rafters and frame them to the model.

For more complex gambrel roofs, you'll need to know that gambrel hip and valley rafters are figured the same way as regular hip and valley rafters. The only difference is that they have no overhang.

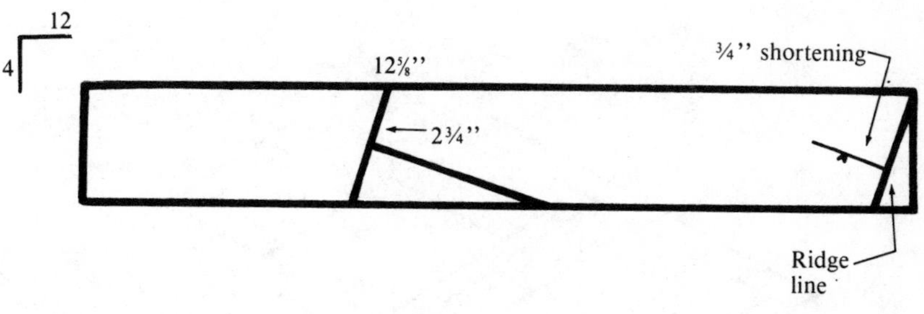

The upper rafter
Figure 7-71

The California

When the span of an addition is greater than the span of the main roof and the pitch is the same, the addition ridge will be higher than the ridge on the main roof.

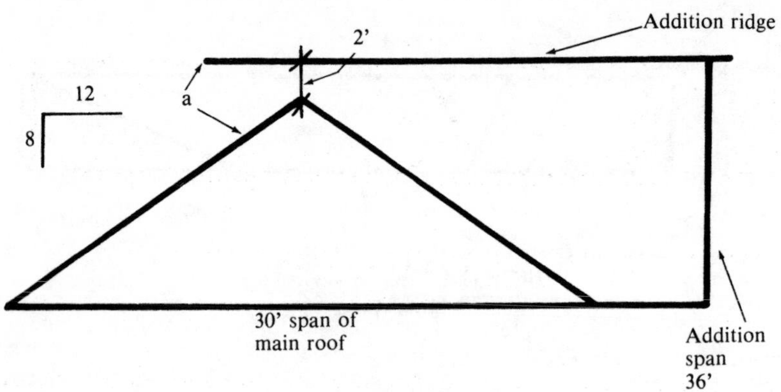

An overshooting ridge
Figure 7-72

In Figure 7-72, the ridge of the addition is 2' above the main roof ridge. Something must be done at point (a) to close the roof surface. A gable end on the side opposite from the addition will close the roof. This construction is called a *California.*

Building a Model California

Attach the 2' addition plate to the main model frame, as in Figure 7-73. Add the common rafters and the 72'' gable ridge to the

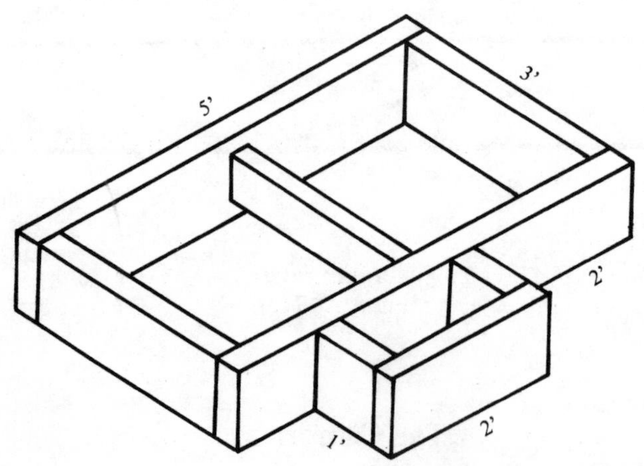

The rafter plate
Figure 7-73

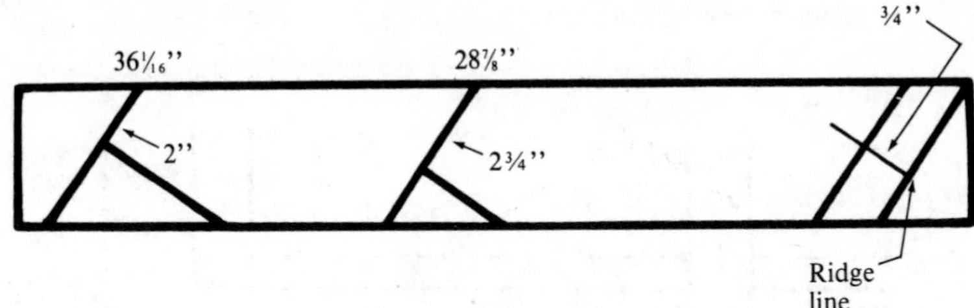

The 2' run common rafter
Figure 7-74

model rafter plate. Then add the sheathing to both slopes of the main roof. The 2' addition side will be used to frame half of the addition, making the total span of the addition 4'.

The Addition Common Rafters

The total run of the addition to the building line is 2'. The run to the fascia line is 2½'.

Rafter Building Line	= Secant x Run	
	= 1.2018 x 2'	= 2.4037
	= 2'4⅞''	= 28⅞''
Rafter Fascia Line	= 1.2018 x 2.5'	= 3.0046
	= 3'0$\frac{1}{16}$''	= 36$\frac{1}{16}$''

Mark the layout as indicated in Figure 7-74 and frame the rafter to the gable end edge of the 2' plate.

The Addition Ridge

Cut a 2 x 4 ridge 48'' long. Mark off 6'' for an overhang at the addition gable end and frame the ridge to the 2' run rafter. See Figure 7-75.

Mark off the main ridge at 53¼'' and the addition ridge at 3'. Brace up the long common rafter so that the ridge is at the correct height. The height is the run of 2' times 8'', or 16''. Add to that the 2¾'' HAP to get 18¾''. Then subtract 1/2'' for loss of ridge height. That gives us 18¼'' for the actual ridge height. Refer back to Figure 7-23 for more information on these calculations.

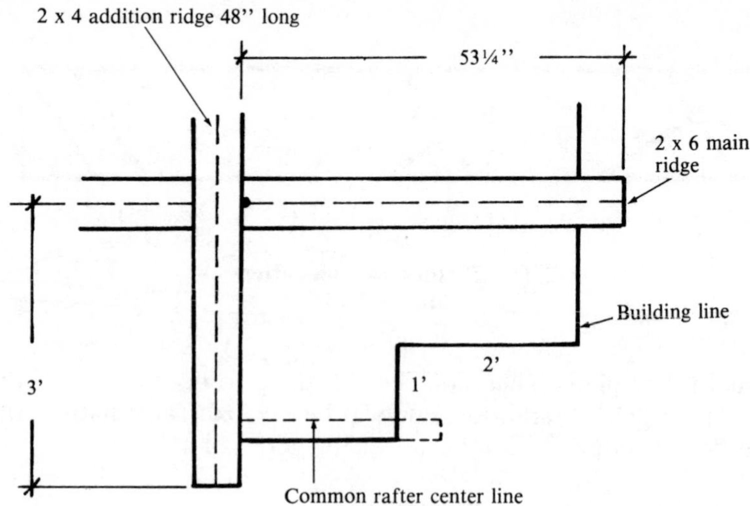

The ridge cross point
Figure 7-75

Now check the ridge to be sure it's level. If the peak of the main roof sheathing is too high, let the sheathing slip down a bit.

The Peak Rafters

To find the length of these rafters, first find the common difference between them. The common difference calculations for these rafters are similar to the common difference in jacks from Chapter 6. Here, however, the ridge and valley do not come together on the main roof. Because of this, you have to add the hypotenuse (total length) of the peak rafter to each common difference. This gives us the correct mathematical length of those rafters.

There are two easy ways to find this mathematical length, which is indicated by the question mark in Figure 7-76.

The Layout Method— Draw an 8 in 12 plumb line as in Figure 7-77. Measure 3½'' and make a mark. Through this mark draw a seat cut line, then measure the mathematical length of the rafter.

The Mathematical Method— Figure 7-78 shows the layout triangle. You have to find the hypotenuse labeled with the question mark.

This mathematical length of 6⁵⁄₁₆'' is added to each of the other rafters to find the correct length. Cut out this rafter and frame it at the peak.

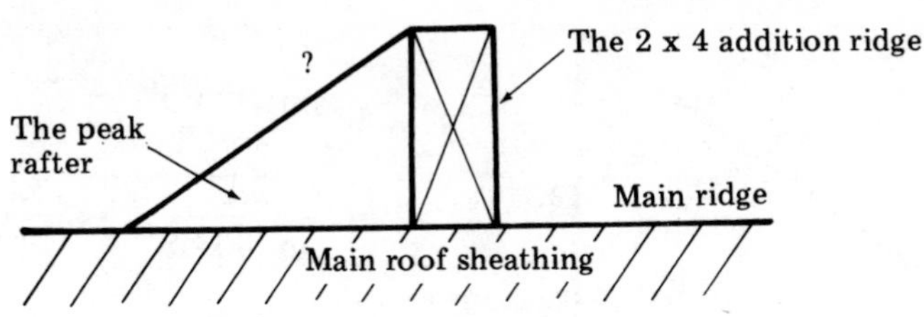

The peak rafter
Figure 7-76

Length of the Other Rafters
The common difference for a regular roof is a function of the total run, as explained in Chapter 6.

Rafter = Secant x Common Difference

The secant of an 8 in 12 roof is 1.2018. The first common difference will be marked at 8¼''. This will be the short point dimension for the seat cut. Multiply 1.2018 by 8.25'' to get 9.9153'', or 9¹⁵⁄₁₆''.

Add the 6⁵⁄₁₆'' of the peak rafter to this 9¹⁵⁄₁₆'' to find the final length of 16¼'' to the seat cut short point.

Look at Figure 7-79 for the layout lines of this rafter. Put in the ridge line and add the shortening. Measure the mathematical length of 16¼''. Transfer the mark to the other side and draw the

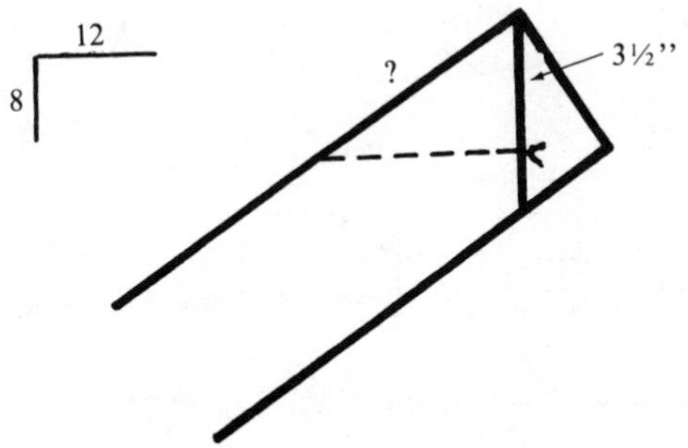

Finding the length by layout
Figure 7-77

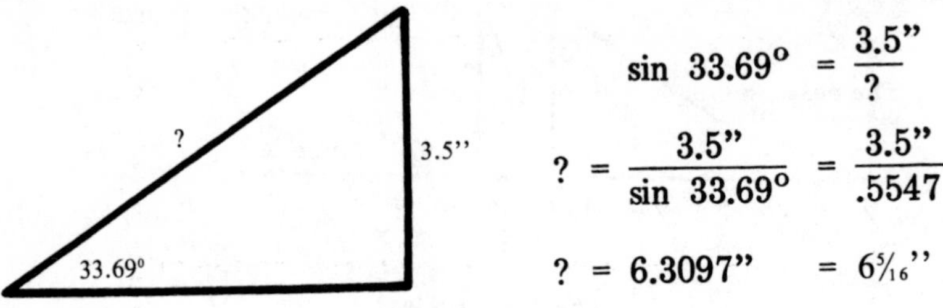

The hypotenuse
Figure 7-78

seat cut. This line is the short point of a 34-degree angle cut. Set your saw to 34 degrees and make the cut. Reset the saw and cut the shortening. This is the way all these rafters are calculated, laid out, and cut.

In Figure 7-80, the rafter at the 16½'' mark will be twice the length of the 8¼'' rafter. Then to each rafter length add the 6 5/16'' of the peak rafter. The second rafter is 19 13/16'' to the short point of the seat cut.

The California End

First calculate the mathematical length for a common difference of 10½''. Multiply 1.2018 by 10.5 to get 12.6194. Then add 6.3097'' and you have 18.9291''. That's 18 15/16'', the length to the short point for the pair of rafters that make up the California. Make two opposite rafters and frame them to the model.

On a full-size roof, the overhang would probably be built as shown in Figure 7-55 and the drawings that follow. On our model, we'll build the California without an overhang. On a full-size roof, you would use valley strips as described in the section on blind valleys.

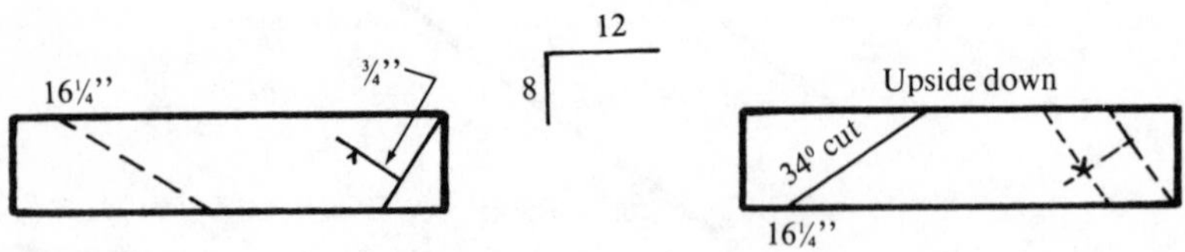

The lines of the valley rafter
Figure 7-79

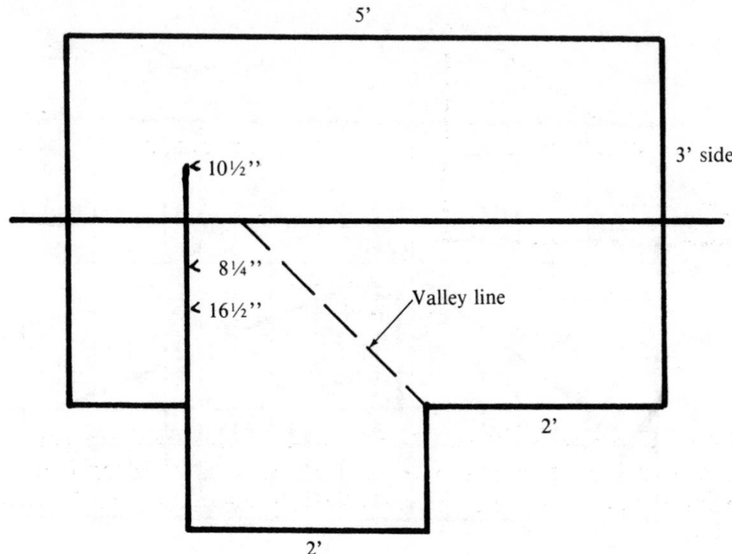

The rafter layout
Figure 7-80

Be sure to mark the shortening for the valley strips before cutting the compound seat cut.

In Chapter 12 we'll cover the irregular California roof. It's like this roof except that the pitch on the addition is different from the pitch of the main roof. This makes the jack rafters between the main roof and the California ridge irregular.

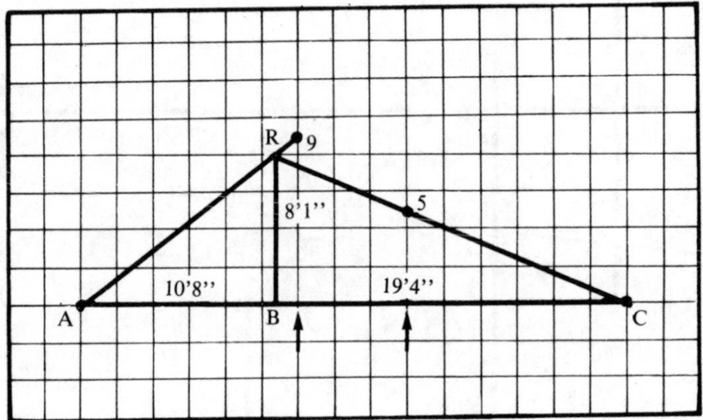

8

Math for Non-Centered Ridges

Occasionally you'll come across a blueprint that does not contain both of the necessary elements for the roof cutter. These elements are total run and unit rise.

In this chapter, we will first solve for these elements on roofs with centered ridges, then go on to roofs with non-centered ridges.

Roofs with Centered Ridges

The examples that follow can be solved with the information from Chapter 1, and serve here as a review of the material in that chapter.

Span and Fractional Pitch Given: Suppose you have the span and the pitch expressed as a fraction, as in Figure 8-1. What's the unit rise? What's the total run?

The total run is one-half of the span. Divide 25' by 2 to get 12.5'. To find the unit rise, multiply 1/4 by 24'' (twice the unit run of 12''). Refer back to Chapter 1, Figure 1-13. One-quarter times 24'' equals 6''. This is the unit rise. The roof in Figure 8-1 has a unit rise of 6 in 12.

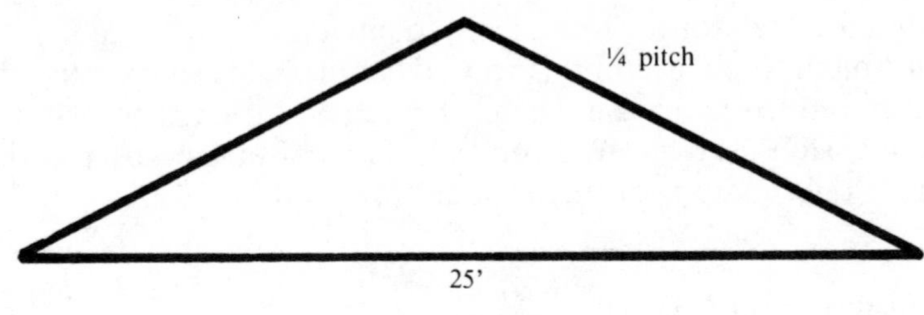

Fractional pitch
Figure 8-1

Total Rise and Span Given: In Figure 8-2, only the total rise and span are given. There isn't a clue to the intended pitch except the notation "pitch to plate". But calculating the unit rise isn't too hard. The total run is one-half the span: 37' divided by 2 equals 18'6". The unit rise is found by the formula:

$$\text{Unit Rise} = \frac{\text{Total Rise (in inches)}}{\text{Total Run (in feet)}}$$

Change the 8' of total rise into inches and the total run into feet. Eight feet becomes 96". One-half of 37' is 18'6", or 18.5'. Dividing 96" by 18.5', we get 5.1892" of rise for each foot of run. This is a 5³⁄₁₆ in 12 roof. Most roof cutters would probably ignore this architectural nonsense and cut a 5 in 12 roof. That would make the total rise 92½" rather than 96". It's possible that the designer really wanted a 5³⁄₁₆ in 12 roof. But much more likely, the designer just didn't understand roof framing practices.

Of course, we could frame it as a 5³⁄₁₆ in 12 roof. No rafter cutting table has rafter lengths for that pitch. But the roof cutter

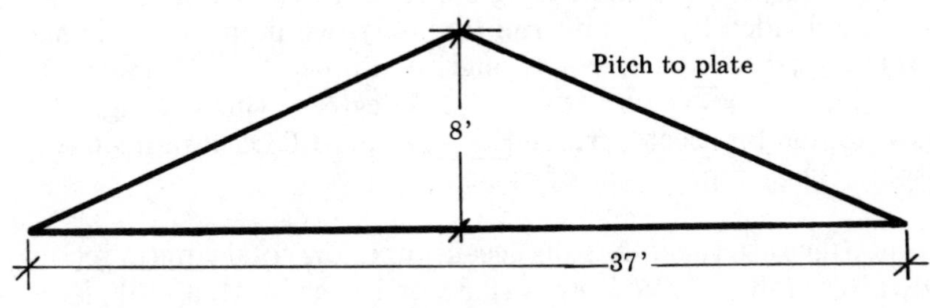

Total rise
Figure 8-2

who has a calculator isn't restricted to pitches in a rafter table. We could "pitch to plate" on this roof. That probably means that the designer wanted us to use whatever pitch was needed to produce the dimensions shown. We can calculate rafter lengths for this pitch by the square root or the secant method.

The Square Root Method

Figure 8-3 A shows how to calculate rafter lengths for the 5³⁄₁₆ in 12 roof. Use the square root method (Pythagorean Theorem) to find the hypotenuse or diagonal in Figure 8-3 B. Use your calculator to find the unit length, 13.0739". The mathematical rafter length is then 18.5' times 13.0739", or 20'1⅞", from the ridge line to the building line. Set your stair gauges on 5³⁄₁₆" and 12". Make sure you are setting the rise on the 16ths scale on the front of the square, not on the back where the 12th scale is located.

Square Root Method:

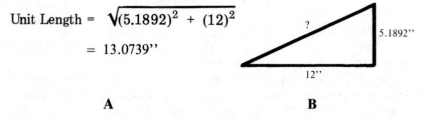

$$\text{Unit Length} = \sqrt{(5.1892)^2 + (12)^2}$$

$$= 13.0739''$$

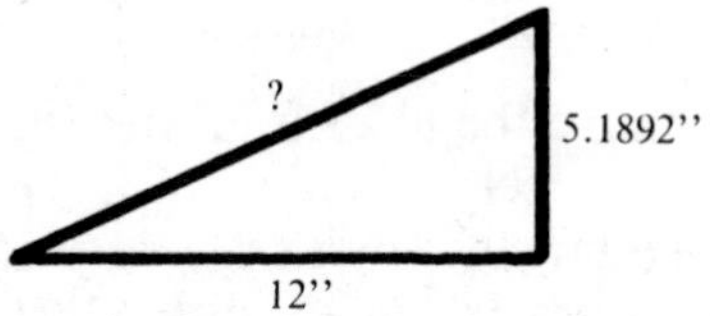

?

5.1892"

12"

A B

Using square root to solve the unit triangle
Figure 8-3

The Secant Method

Let's find the rafter length using the secant, as in Figure 8-4. The unit rise divided by the unit run (the only two numbers available here) is equal to the tangent of angle A. Punch [5] [·] [1] [8] [9] [2] [÷] [1] [2] [=] [INV] [tan] to get 23.38 degrees, which is angle A. Now, to find the secant, punch [cos] [1/x] and 1.0895 appears in the display. This is the secant of angle A.

Since the rafter length is the secant times the total run (in feet), our rafters will be 1.0895 times 18.5', or 20.1557'. That's 20'1⅞", the mathematical rafter length. You won't find any rafter table that has an answer for this problem.

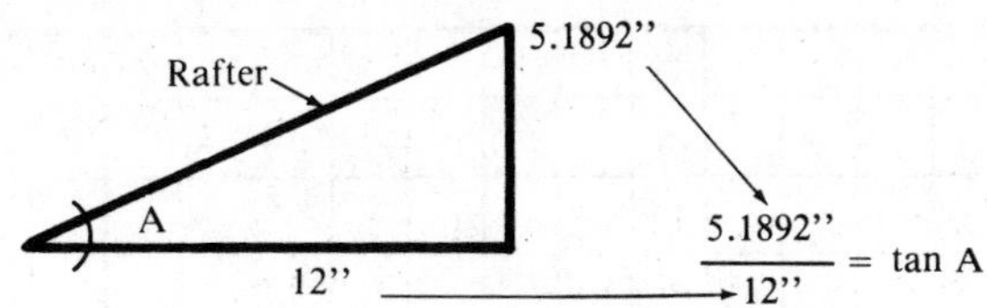

Finding the secant and rafter lengths
Figure 8-4

Roofs with Non-centered Ridges

Look at Figure 8-5. The information on the plan describes this gable roof well enough. But from the roof cutter's point of view, two essential points are missing. What's the total run of each section? What's the total rise? Think about it for a minute and you'll see that this problem is more difficult than it seems. But it can be solved by using a layout on graph paper, or by a math method.

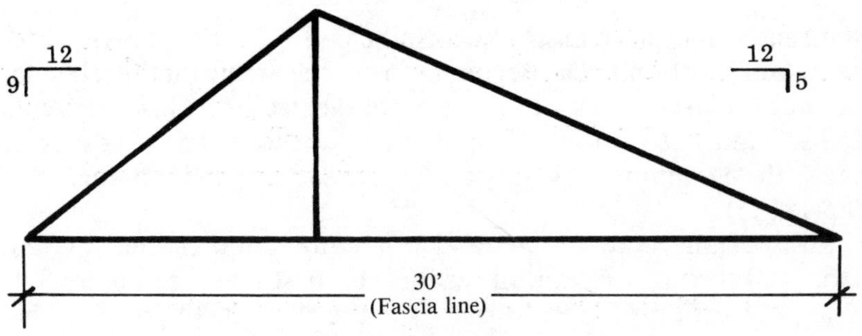

A roof with a non-centered ridge
Figure 8-5

The Layout Method

One way to solve this problem is by drawing it to scale and then scaling off the dimensions. Let's try it. Use Figure 8-6 as a guide, but remember that, because of space limitations, Figure 8-6 was drawn using a scale of 1/4'' equals 2'. It will be easier for you to make each 1/4'' square equal *1'*, so you will use two squares for each square shown in Figure 8-6.

Take out your 1/4'' graph paper and lay out a line 30 squares long, to represent the 30' span. Label the end points of the span (A) and (C). From each of these points, count 12 squares toward

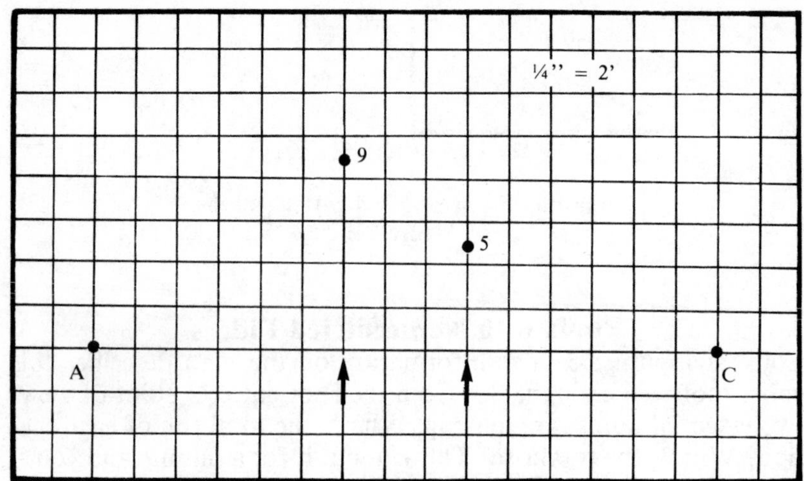

Beginning the layout
Figure 8-6

the center and place an arrow as indicated. The 12 squares represent the 12'' of unit run. Beginning from the arrow on the left side, count up nine squares and place a dot labeled (9). This represents the 9'' unit rise on this side of the roof. Above the other arrow, mark the 5'' unit rise by moving up five squares and making a dot labeled (5).

Draw a line between (A) and (9). Draw another line running from (C) through (5) until it reaches the first line. See Figure 8-7. The intersection of these two lines is the ridge of the roof. Label this point (R). Draw in a perpendicular line (RB) to represent the total rise. This line now divides the 30' span properly. Count the squares or measure carefully with a 1/4'' scale ruler to find each total run and the total rise. The measurements should be close to 8'1'' for total rise, and 10'8'' and 19'4'' for the two total runs.

Let's see if we could use these figures. What would be the rise on each side if we used these total run figures? Converting 10'8'' to a decimal, we get 10.6667'. Multiply this total run by the unit rise (9'') and we have 96''.

On the long span side, the total rise calculation would be 19'4'' (19.3333') times 5'', or 96.6667''. That's 0.6667'' more than we calculated for the short span.

Using the graph paper method, there would be a 11/16'' (0.6667'') difference in height at each side of the ridge. That's too much. We can do better.

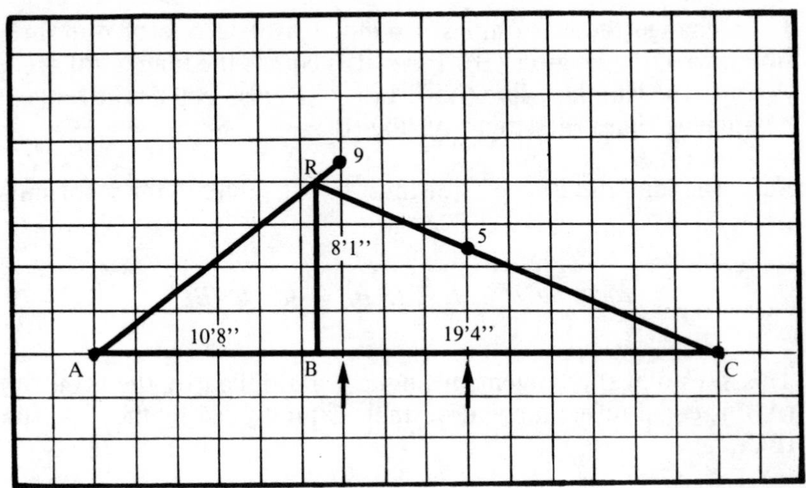

The completed drawing
Figure 8-7

The Math Method

The key to this problem is finding any one of three lengths: the total rise (RB), or either total run (AB or BC) in Figure 8-8. And while we're at it, let's develop a simple formula so we can solve problems like this in a few seconds. Try to follow along as we work out a formula that locates the distance to point (B).

We know that the tangent of an angle, such as angle A or C in Figure 8-8, is the total rise divided by the total run. For the plan in Figure 8-8:

$$\text{Tan } A = \frac{RB}{AB} \qquad\qquad \text{Tan } C = \frac{RB}{BC}$$

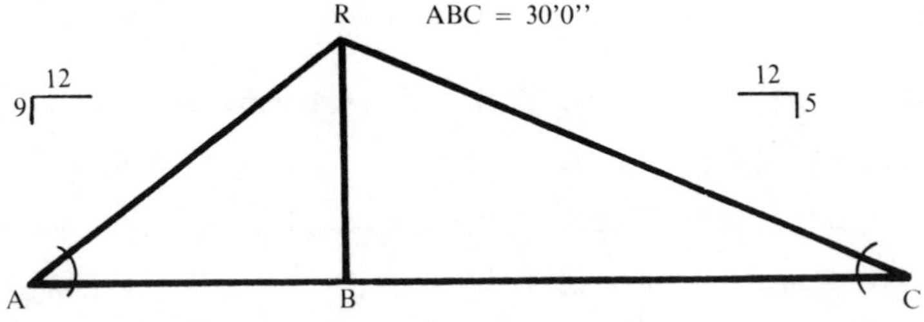

The non-centered ridge
Figure 8-8

Let's change these formulas around to isolate one of our three unknowns, the length (RB). Line (RB) equals the tangent of angle A times the length of line (AB). Line (RB) also equals the tangent of angle C times the length of line (BC).

Since the total rise of each side must be the same, the first formula is:

formula 1: Tan A (AB) = tan C (BC)

This says that the tangent of angle A multiplied by the total run (AB) is equal to the tangent of angle C multiplied by the total run (BC).

We also know that (AB) plus (BC) is equal to the span, therefore:

formula 2: (BC) + (AB) = the span

Using the basic rules of algebra (or common sense), rearrange Formula 2:

formula 3: (BC) = span — (AB)

Formulas 1 and 3 both include (BC). Let's take what's equal to (BC) in Formula 3, and exchange it for (BC) in Formula 1. That gives us:

formula 4: Tan A (AB) = tan C (span — AB)

We still have a little way to go, but it's just multiplying, transferring and dividing:

	$\text{Tan A (A B)} = \text{Tan C (span - A B)}$
Multiplying:	$\text{Tan A (A B)} = \text{Tan C (span)} - \text{Tan C (A B)}$
Transferring:	$\text{Tan A (A B)} + \text{Tan C (A B)} = \text{Tan C (span)}$
Extracting A B:	$\text{A B (Tan A} + \text{Tan C)} = \text{Tan C (span)}$
Dividing:	$\text{A B} = \dfrac{\text{Tan C (span)}}{\text{Tan A} + \text{Tan C}}$

This last formula requires only the information we are given in Figure 8-8. We know the tangent of angle A is 9 divided by 12, or 0.75. The tangent of angle C is 5 divided by 12, or 0.4167. The span is 30'. Substituting numbers into our formula:

$$AB = \frac{(.4167)\ (30')}{.75 + .4167}$$

$$AB = 10'8\tfrac{9}{16}''$$

Multiplying 0.4167 by 30', we get 12.5'. Adding 0.75 and 0.4167, we get 1.1667. Divide 12.5' by 1.1667 and we have 10.7143'. That's about 10'8$\frac{9}{16}$''.

If (AB) is 10.7143', (BC) must be 30 minus 10.7143 or 19.2857'. Converting to a fraction, we get 19'3$\frac{7}{16}$'' for (BC).

We could have calculated (BC) directly. Change Formula 3 to read:

formula 3a: (AB) = span — (BC)

Then substitute this into Formula 1, as before, and we have a formula for finding (BC) directly:

$$BC = \frac{\text{Tan A (span)}}{\text{Tan A + Tan C}}$$

This would yield the same answers as we developed previously.

Now we have the total run and unit rise for each half, so we can calculate the total rise. Use the decimal form in your calculations to be as accurate as possible.

10.7143 times 9'' equals 96.4285'', or 8'0$\frac{7}{16}$'' for total rise.

19.2857 times 5'' equals 96.4285'', or 8'0$\frac{7}{16}$'' for total rise.

That's a much more accurate answer than we got with our graph paper.

Framing the Rafters at the Ridge

A roof with a non-centered ridge can be more difficult to figure, and harder to erect. Figure 8-9 shows the framing problem. The lower pitched rafters have to be held up by the distance (a). You need to calculate how much offset is needed.

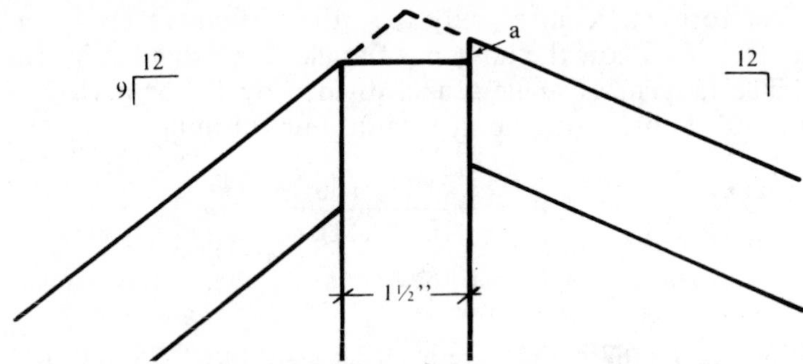

Holding up the lower pitched rafters
Figure 8-9

Figure 7-23, in the previous chapter, explained how to calculate the loss of ridge height. The loss of height is equal to the tangent of the common rafter angle times one-half the thickness of the ridge. Assume that the ridge is 1½" thick. Begin with the higher pitch roof:

$$\text{Loss of Height} = \left(\frac{9}{12}\right)(.75) = .5625 = 9/16"$$

Frame the top of the ridge 9/16" below the total rise.

Here's how to calculate the loss of the lower pitch rafter:

$$\text{Loss of Height} = \left(\frac{5}{12}\right)(.75) = .3125 = 5/16"$$

Subtracting the lower pitched rafter from the higher gives (a), the distance to hold the lower pitched rafter above the ridge.

Distance (a) equals 9/16" minus 5/16", or 1/4".

Problems

We've covered some pretty difficult ground in this chapter. These problems will help you discover any points that need review.

1) A roof has a span of 32'. One side of the roof is 14:12. The other side is 6:12. What's the total run of each side? What's the total rise? How high above the ridge must you hold the 6:12 rafters? The ridge is a rough-sawn 4 x 8 timber.

2) Another roof has a span of 32'. One side of the roof is 16:12 while the other side is 8:12. What's the total run of each side? What's the total rise? How high above the ridge must you hold the 8:12 rafters? This ridge is 1½" thick.

9

Framing a Gazebo

Most experienced roof cutters have framed at least one gazebo. You'll see gazebos used in gardens and backyards to provide shade in summer. They lend a whimsical touch and are common to several architectural styles. Figure 9-1 shows a simple gazebo. The gazebo walls are usually open, and the roof may be at least partially open to the sky.

Gazebos are built in the shape of a regular polygon, usually either an octagon (8 sides) or a hexagon (6 sides). Framing a six or eight-sided structure presents some unique problems that are beyond what most carpenters are able to handle. Until you have a gazebo job, you could skip this chapter. But when the time comes to put a roof on a five, six or eight-sided building, you'll find the information that follows very useful.

Math for a Regular Polygon

A *polygon* is simply an enclosed figure. It can have any number of sides of various lengths. At the left in Figure 9-2 is an irregular polygon. The lengths of the sides vary. You won't frame roofs for many irregular polygons. The regular polygon is much more common. At the right side of that figure is a regular polygon. All sides are the same length.

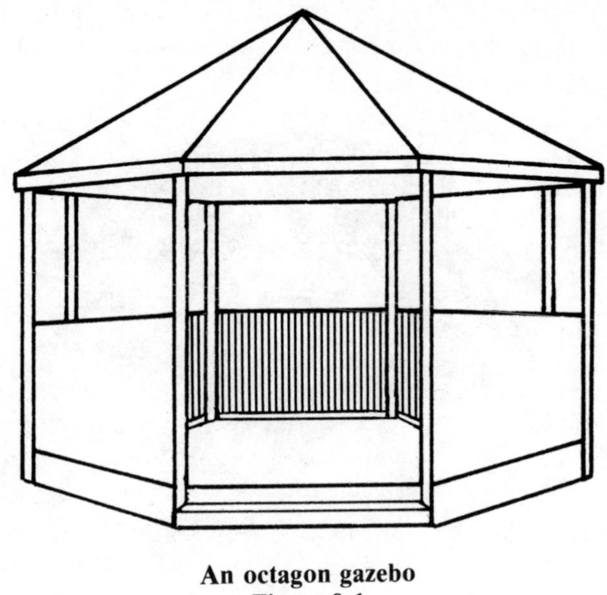

An octagon gazebo
Figure 9-1

Notice that a circle has been drawn around the edge of the hexagon. The circle passes through the meeting points of all sides. These points are called the *vertices* of the polygon. A circle in this position is called a *circumscribed circle*. The vertices of every

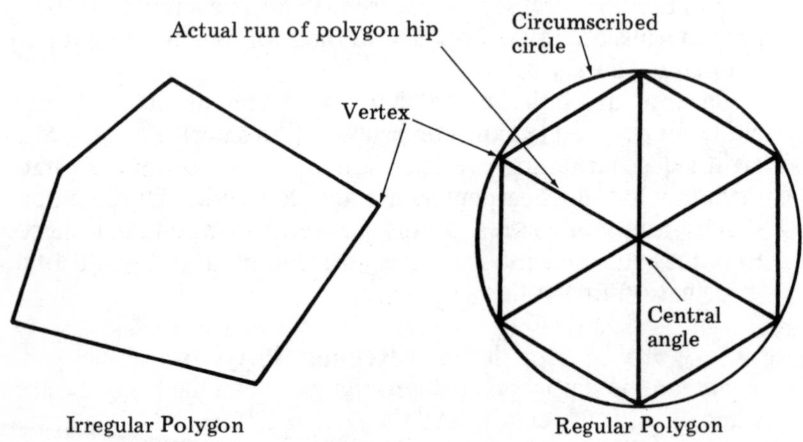

Two types of polygons
Figure 9-2

regular polygon will fall on a circle circumscribed from the central angle. This is the starting point as we set out to cut a roof for our gazebo.

As in the previous chapters, we'll use an example to help make our explanations clear. Our construction plan calls for an octagon gazebo with an 8 in 12 roof and a radius of 19.4831''. (In the second half of this chapter, this same gazebo will be calculated and constructed using the mathematics based on a square.)

Regular Polygon Roofs

Certain things are true about every polygon. That's a big help when you have to frame a gazebo or any building with other than four regular sides. We'll use these common properties to develop a series of 35 steps to follow when cutting the roof for any polygon.

These steps are listed in Figure 9-10, at the end of the first section of this chapter. But before you refer to that, let's explore each of the steps in detail and explain how each is used on our gazebo job.

General Steps

Step 1: Actual Run of the Polygon Hip— Three important points establish the size and shape of the gazebo. These are the radius of the circumscribing circle, the number of sides, and the pitch of the roof. Here are the important facts for our model gazebo:

Radius — 19.4831''
Shape — Octagon
Pitch — 8 in 12

As a roof cutter, you'll recognize the radius of the circumscribing circle as the actual run of the polygon hip. See Figure 9-4.

Step 2: The Central Angle— Figure 9-3 identifies the *central angle* of the polygon. The central angle is formed by extending lines from the center of the figure to each vertex. Notice that there are as many central angles in any regular polygon as there are sides in that polygon.

You know that every circle has 360 degrees of arc. If the circumscribing circle has 360 degrees, the central angle of the regular polygon in the circle has to be 360 divided by the number of sides of the polygon. This will give you the arc of each central angle. For

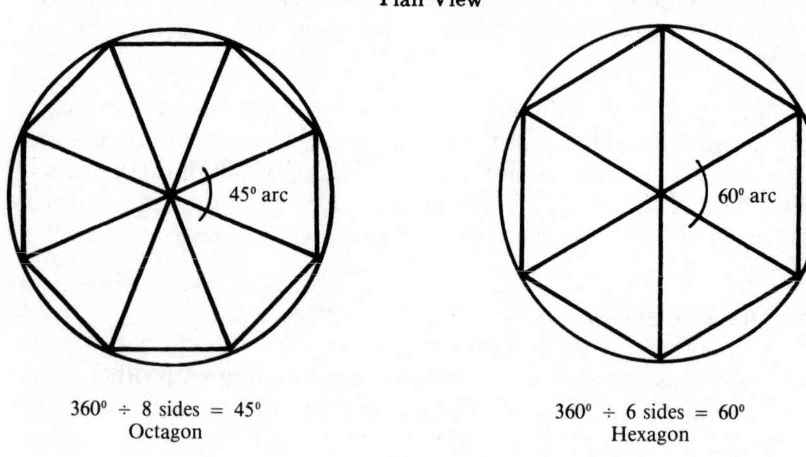

Finding the central angle
Figure 9-3

an octagon, the central angle is 360 divided by eight sides, or 45 degrees. For a hexagon, 360 divided by six sides, or 60 degrees. Remember that:

$$\text{Central Angle} = \frac{360^\circ}{\text{No. of Sides}}$$

You use the central angle when laying out your polygon. Here's how it's done. Draw a circle on a piece of paper. Then draw one line from the center to the edge of the circle. This is the first radius. With a protractor, measure off the degrees of the first central angle. At that point, draw the second radius. Proceed around the circle completing all the radii. Then draw in the sides of the building by connecting the points where the radii meet the circle.

Doing the same thing on a building site is only a little more difficult. The main difference is that you use a builder's transit rather than a protractor.

First, set up up your tripod at the center of the gazebo. Measure out, from the center point, the distance indicated on the plans. Then turn the transit the number of degrees to the next radius and measure out, again from the center point, the same distance. Continue doing that for as many sides as the plans show. When you're finished, measure the length between the points marked at each vertex. They should be the same.

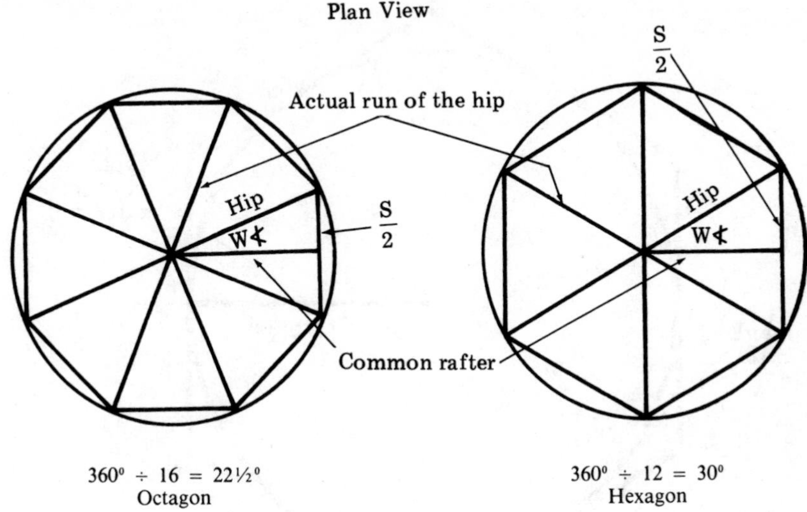

The working angle
Figure 9-4

And be sure that the first wall is aligned with the terrain and is as shown on the plans.

Step 3: The Working Angle— One-half of the central angle is called the *working angle*. The working angle can also be found by dividing 360 degrees by twice the number of sides in any desired regular polygon. See Figure 9-4.

$$\text{Working Angle} = \frac{360^\circ}{\text{Twice the No. of Sides}}$$

You'll use the working angle many times when framing a gazebo roof. On plans and in our examples we'll use the letter "W" and a symbol to indicate the working angle.

Steps for the Common Rafter

Step 4: The Total Run of the Common Rafter to the Building Line— Another fact you need to know about regular polygons is that a circle can be inscribed *within* them. See Figure 9-5. The radius of this inscribed circle is called the *apothem*.

Our inscribed circle just touches the center of each line connecting the vertex points around the octagon. Each of the sides is said

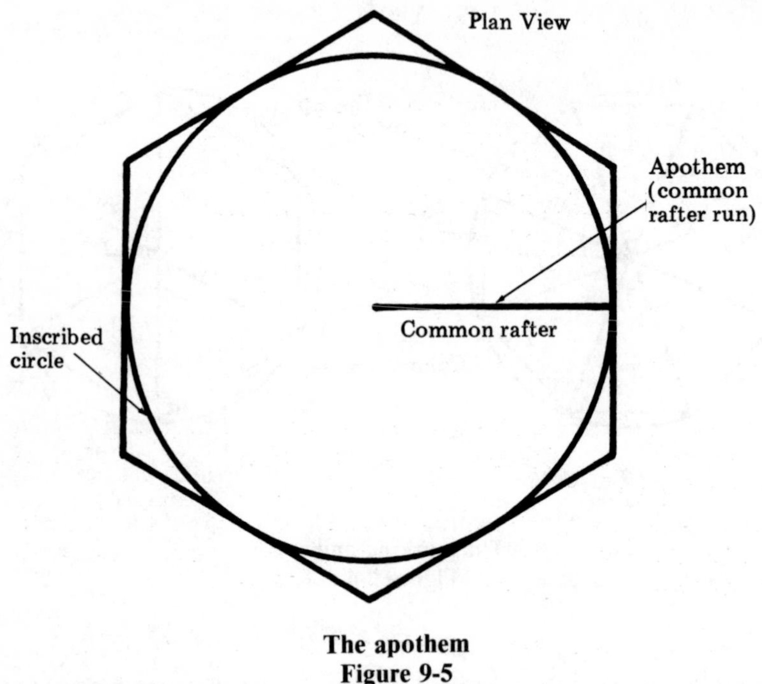

The apothem
Figure 9-5

to be *tangent* to the inscribed circle. The apothem is defined as a perpendicular drawn from the center of one of the sides to the center of the polygon.

As a roof cutter you'll recognize the apothem as the total run of the common rafter for any regular polygon roof.

If we know the radius of the circle that circumscribes the octagon, we can find the radius of the circle that inscribes the octagon. In our case, the circumscribing radius is 19.4831". Let's find the inscribing circle radius, or the run of common rafter.

We know that the actual run of the hip (the radius of the circumscribing circle) is 19.4831 inches. We also know that the central angle is 45 degrees for a regular octagon. Thus our working angle is one-half of 45 degrees, or 22½ degrees. Figure 9-6 shows these relationships. What we want to find is the common rafter run, the radius of the inscribed circle.

Figure B-3 in Appendix B shows the names commonly used to identify sides of a triangle. Compare the drawing at the left side of Figure B-3 with Figure 9-6. What we're looking for is the length of the adjacent side. One of the six functions described in Appendix B is the cosine of angle A, the length of the adjacent side divided

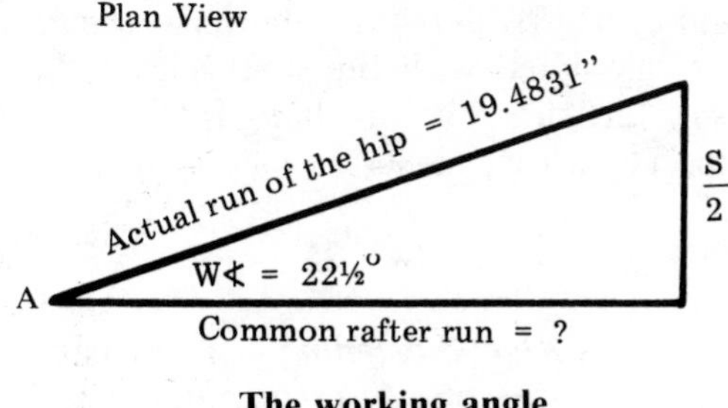

The working angle
Figure 9-6

by the hypotenuse. Since we know both angle A (22½ degrees) and the length of the hypotenuse (19.4831 inches), we can find the common rafter run.

Changing the cosine formula around, we find that the length of the adjacent side (common rafter run) is equal to the cosine of 22½ degrees times the hypotenuse. In our problem:

Common rafter run = Cos W∢ x Actual run of the hip

To solve our problem, first find the cosine of 22½ degrees. On your calculator, punch [2] [2] [.] [5] [cos] and 0.9239 appears. If you multiply this by the radius of the circumscribed circle, you'll come up with the apothem, the run of the common rafter: [×] [1] [9] [.] [4] [8] [3] [1] [=] and you get 18 inches. Now you know why the designer of our gazebo picked 19.4831" for the radius of the circumscribed circle!

Step 5: The Total Run of the Common Rafter to the Fascia Line— Simply add the amount of the overhang to the answer in Step 4. On our model, the overhang is 6 inches.

Step 6: The Common Rafter Mathematical Layout Length for the Building Line— Our octagon building is to have an 8 in 12 roof with an apothem of 18". The building line mathematical length will be:

Building line length = Secant x Actual run

Find the secant from the common rafter secants listed in Figure 1-14. (See Step 26.)

Building line length = 1.2018 x 18" = 21.63" = 21⅝"

Step 7: The Common Rafter Layout Length for the Fascia Line— The fascia line is calculated with the same secant, but using the length of the fascia line run. The overhang is to be 6''. Add 6'' to the 18'' apothem. The answer is 24''.

$$\text{Fascia line length} = 1.2018 \times 24'' = 28.84'' = 28\tfrac{7}{8}''$$

Step 8: The Common Rafter Shortening— There can be only one common rafter on each side of any regular polygon. That would be the rafter at the center of each side. A rafter in any other position on that side would be a jack rafter.

On a polygon roof, the shortening of the common rafter and the shortening of the jack rafter are the same. Both depend on the thickness of the hip. See Figure 9-46 toward the end of the chapter.

$$\text{Shortening} = \frac{\tfrac{1}{2}\text{ the thickness of the hip}}{\text{Sin W}\measuredangle}$$

Step 9: The Common Rafter Setback— The centered common rafter will have a double cheek cut. The amount of setback to be marked off on both sides from the ridge line is found by the formula:

$$\text{Setback} = \frac{\tfrac{1}{2}\text{ the thickness of the common rafter}}{\text{Tan W}\measuredangle}$$

This setback is very similar to Figure 9-47, the setback for the jack rafter.

Step 10: The Setback Cutting Angle— Each cheek cut on the polygon common rafter will be cut with a short point angle setting of:

$$\text{Setback cutting angle} = 90^{\circ} - \text{W}\measuredangle$$

Steps for the Side and Hip Unit Run

Step 11: The Side Of The Building— What's the distance between the two vertices on this octagon? Here's how we find it.

Look at Figure 9-6. Notice that the side of the triangle opposite angle A is identified with *S/2*. This means that the length of this part of the triangle is known to be one-half of the length between two vertices of the circle. It's equal to one-half the length of one side of our gazebo.

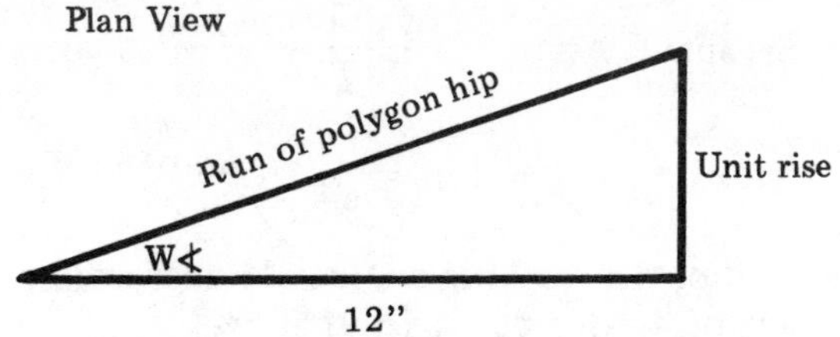

A polygon hip unit run triangle
Figure 9-7

From Appendix B, we know that the sine of the working angle is equal to half the opposite side divided by the hypotenuse.

$$\text{Sin W}\sphericalangle = \frac{\frac{S}{2}}{\text{Actual Run of Hip}}$$

This is half the wall. Therefore, multiplying by 2 and solving for the side gives:

$$\text{Side} = 2 \sin \text{ W}\sphericalangle \times \text{Actual Run of Hip}$$

For our octagon, the side would equal:

$$\text{Side} = 2 \sin 22\tfrac{1}{2}^{\circ} \times 19.4831'' = 14.9117''$$

Step 12: The Polygon Hip Unit Run and Framing Square Setting— We need to find the polygon hip unit run for two reasons. First, we have to calculate a table of secants for our gazebo that lets us convert total run lengths to mathematical rafter lengths. Second, we'll need to know what the framing square setting will be.

A regular common rafter has a unit run of 12''. You'll recall that a 45-degree hip rafter has a unit run of 16.97''. For each regular polygon, the hip will have its own particular run.

To determine the unit run of the polygon hip, let the apothem be equal to 12'' since 12'' is the basic roof framing unit in this country.

Refer to Figure 9-7. If the working angle is known and the run is 12'', then the polygon hip unit run can be found by:

$$\text{Secant } W\measuredangle = \frac{\text{run of polygon hip}}{12''} \quad \text{or:}$$

$$\text{Run of polygon hip} = 12'' \text{ Sec } W\measuredangle$$

For a regular hip, the working angle is 45 degrees. Twelve inches times the secant of 45 degrees is 16.97 inches.

For an octagon hip the working angle is 22½ degrees. Twelve times the secant of 22½ degrees is 12.9887 inches. Let's round that to 13''. This is the unit run for an octagon hip, and you'll use it on the body of the framing square in combination with the unit rise.

Step 13: The Area of Any Regular Polygon—The area of any triangle is one-half the base times the height. For our use, the words apothem and side are more convenient. See Figure 9-8.

The general formula for the area of any regular polygon is:

Area = (½ no. of sides) (Length of side) (Apothem)

Using our steps, this translates to:

Area = (½ no. of sides) (Step 11) (Step 4)

Steps for the Hip Rafter

Step 14: The Secant of the Polygon Hip— In the first thirteen steps, except numbers 6 and 7, we've used plan view mathematics; that is, we've been concerned only with horizontal surfaces. Now we have to start working in three dimensions.

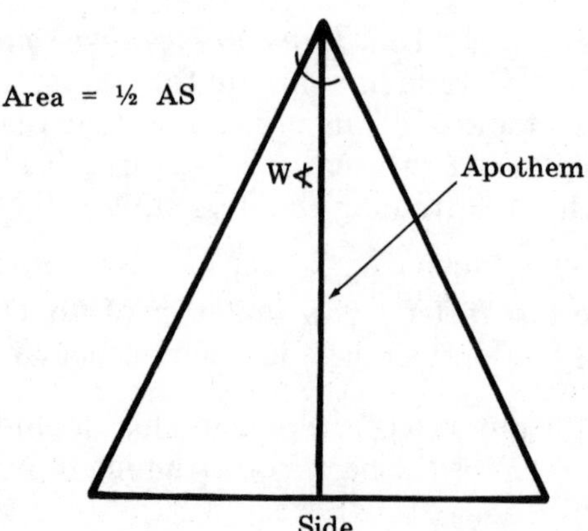

A regular polygon triangle
Figure 9-8

First, let's find the secant of a particular polygon hip so we can multiply that number by the actual run. This will give us the measurement of the building line length and the fascia line length.

We combined 12'' with the rise to find the secant of the common rafter (as in Chapter 2, Figure 2-51 and the text that follows). We combined 16.97'' with the unit rise to find the secant of the regular hip (as in Figure 4-18). Here we will combine the unit run of a particular polygon hip, found under Step 12, with the unit rise to find a particular secant. Let's look at some examples.

Examples - Using the Secant

In Chapter 2, we found that the secant of a common rafter for an 8'' rise roof is 1.2018. Here's how we did it on the calculator:

$$\frac{8''}{12''}\ \boxed{=}\ .6666\ \boxed{\text{INV}}\ \boxed{\tan}\ 33.69^{\circ}\ \boxed{\cos}\ \boxed{1/x}\ = 1.2018$$

You can also look it up. Figure 1-14 lists the secant for an 8 in 12 common rafter as 1.2018.

In Chapter 4, the secant for a regular hip rafter of an 8'' rise roof is 1.1055. Using the calculator:

$$\frac{8''}{16.97''}\ \boxed{=}\ .4714\ \boxed{\text{INV}}\ \boxed{\tan}\ 25.24^{\circ}\ \boxed{\cos}\ \boxed{1/x}\ = 1.1055$$

Figure 4-65 lists the secant for an 8 in 12 regular hip roof as 1.1055.

The general formula for finding secants of any rise of any regular polygon on the calculator is:

$$\frac{\textbf{Unit rise}}{\textbf{The polygon hip unit run}}\ \boxed{=}\ \boxed{\text{INV}}\ \boxed{\tan}\ \boxed{\cos}\ \boxed{1/x}\quad \text{or}$$

$$\frac{\textbf{Unit rise}}{\text{Step 12}}\ \boxed{=}\ \boxed{\text{INV}}\ \boxed{\tan}\ \boxed{\cos}\ \boxed{1/x}$$

If a TI-35 or similar calculator isn't available, use a trigonometry table of natural functions. The unit rise divided by the hip unit run gives the tangent of the angle. Under Step 12, we found the unit run for an octagon hip to be 12.9887''. This divided into 8'' gives 0.6159. This is the tangent of some angle. To find what angle, look for this number in the tangent column. Under 31 degrees 38', you'll find the number 0.61601. Remain at 31 degrees 38' but move over to the column headed "secant" and read 1.1745. Check this by punching it out on your calculator. The number will be the same.

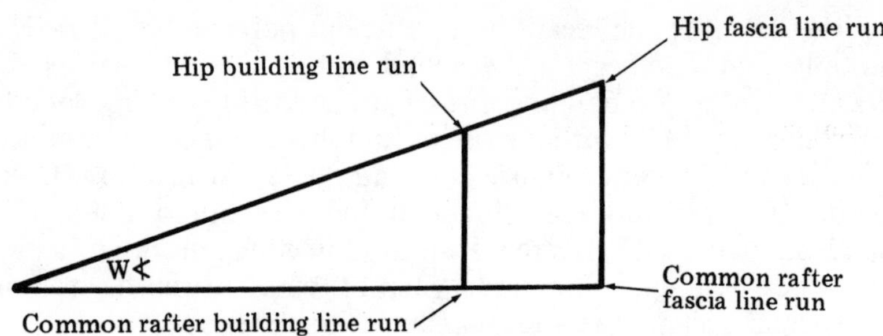

The relationship of the hip run and common run
Figure 9-9

Using either of these methods, a table of secants for different pitches can be generated for any polygon.

Step 15: The Hip Mathematical Layout Length for the Building Line— In Chapter 4, the mathematical layout for the hip was found by multiplying the 45-degree hip secant times the actual run of the hip.

Here, the polygon hip secant will be used along with the radius of the circumscribed circle, which is the same as the actual run of the hip. For our octagon hip rafter then:

Polygon hip math length = Secant x Actual run

Length = (Step 14) (Step 1)

Length = 1.1745 x 19.4831"

Length = 22.8821" = 22⅞"

Step 16: The Polygon Hip Actual Run to the Fascia Line— The length of the overhang is always established by the common rafter side of the building.

The run of the hip and the run of the common rafter have a relationship that's expressed in the cosine of the working angle. See Figure 9-9.

$$\text{Hip run} = \frac{\text{Common run}}{\cos W\measuredangle} \qquad \text{Common run} = (\text{hip run})(\cos W\measuredangle)$$

In our model octagon roof, the total run of the common rafter is 18'' and the overhang is 6''. Therefore, the common rafter total run to the fascia line as found in Step 5 is 24''. Now let's find the actual run of the hip fascia line.

$$\text{Actual run of hip fascia line} = \frac{\text{Step 5}}{\cos W\measuredangle}$$

$$\text{Actual run of hip fascia line} = \frac{24''}{.9239} = 25.9774''$$

Step 17: The Hip Mathematical Layout Length for the Fascia Line— Now it's time to calculate the mathematical layout length for the fascia line.

Mathematical length = Hip secant x Actual run to the fascia line

Mathematical length = (Step 14) x (Step 16)

For our model octagon roof then:

Mathematical length = (1.1745) (25.9774'')

Mathematical length = 30.5''

Step 18: Set Forward for the Hip Birdsmouth and Tail Cut— Figures 9-30 and 9-31, in the second half of this chapter, show the set forward for an octagon figure. This mathematical set-forward formula can be used with any regular polygon.

Set Forward = (Tan W∢) (½ thickness of the hip)

For an octagon, 5/16'' is the proper set forward for a 1½'' thick hip rafter.

For a 45-degree hip of the same thickness, the set forward is:

Set Forward = (1) (¾'') = ¾''

Step 19: Backing or Dropping— Find the backing or dropping for any polygon hip with the method used in Figures 9-37 to 9-39, found later in this chapter. Use the figures appropriate for your polygon.

Using the math outlined here, the backing or dropping would equal:

$$\text{Backing or Dropping} = \frac{\text{(Set forward) (Unit rise)}}{\text{Hip unit run}} = \frac{\text{(Step 18) (Unit rise)}}{\text{Step 12}}$$

Step 20: Angle of the Backing Cut— The angle of the backing cut for the octagon 8 in 13 hip is shown in Figure 9-40.

The general formula for the angle of the backing cut of any regular polygon hip is:

$$\text{Cutting angle} = \frac{\text{Backing}}{\text{½ thickness of hip}} = \boxed{\text{INV}}\ \boxed{\text{tan}} = \frac{\text{Step 19}}{\text{½ thickness of hip}} = \boxed{\text{INV}}\ \boxed{\text{tan}}$$

Step 21: Ridge Line Setback and Framing— If all the hips are to come to the ridge point, then the ridge setback for each hip will be:

$$\text{Ridge setback} = \frac{\text{½ thickness of hip}}{\text{Tan W}\sphericalangle}$$

The saw angle for this cheek cut will be the same as the miter cut for forms and trim. See Step 34.

Step 22: The Radius for the Ridge Block— There are several ways to frame the center point of the gazebo roof. Figure 9-41 shows a different method of construction. Sometimes a polygon-shaped ridge block or hub is made to receive the hip rafters of the gazebo. If so, you have to calculate the radius of the circumscribed circle of that hub. It has to have a radius long enough so that the face area of the hub on each side is equal to the face area of the hip rafter that frames against it.

The correct radius can be found by:

$$\text{Radius} = \frac{\text{Full thickness of hip}}{2\,(\text{Sin W}\sphericalangle)}$$

For an octagon roof with 1½" thick hips, the radius would equal 1.9598".

Step 23: Ridge Line Shortening for Step 22— Now each rafter must be shortened the amount of the apothem for this radius. Multiply the radius of 1.9598" by the cosine of the working angle. That yields the apothem, or shortening. In this case it's 1¹³⁄₁₆". This agrees with the shortening found under Figure 9-41.

There's another way to find the ridge shortening and setback. We make a scale drawing of the plan view and then scale off the shortening and setback. We'll do that later in this chapter (Figure 9-29).

Steps for the Jack Rafter

Step 24: Actual Run of the Jack— Figure 9-51 and the text that follows explain how to calculate the actual run for an 8 in 12 octagon jack rafter.

The general formula for any regular polygon is:

$$\text{Run of polygon jack} = \frac{\text{Distance from corner}}{\text{Tan W}\measuredangle}$$

For a regular 45-degree hip roof, the run of the jack rafter 16'' from the corner is equal to 16'' divided by the tangent of 45 degrees. Since the tangent of 45 degrees is 1, the answer is 16''.

Step 25: The Jack Rafter Actual Run to the Fascia Line— Simply add the amount of the overhang to the answer from Step 24.

Step 26: The Mathematical Length to the Building Line— All jack rafters and common rafters in the regular polygons use a secant based on 12''; therefore, the common rafter roof angle chart from Figure 1-14 can be used.

The secant times the actual run from Step 24 gives the building line measurement.

Step 27: The Mathematical Length to the Fascia Line— Multiply the common rafter secant by the answer from Step 25, the actual run to the fascia line.

Step 28: Shortening the Jack Rafter— Figure 9-46 shows the shortening for an 8 in 12 octagon jack rafter. The general formula for any regular polygon is:

$$\text{Shortening of jack rafter} = \frac{\tfrac{1}{2}\text{ thickness of hip}}{\text{Sin W}\measuredangle}$$

For a regular 45-degree hip, the shortening is 0.75 divided by the sine of 45 degrees, or 1$\frac{1}{16}$''.

Step 29: The Jack Rafter Setback— Figure 9-47 and the text that follows explain how to calculate the setback for an 8 in 12 octagon jack rafter.

The general formula for any regular polygon is:

$$\text{Jack setback} = \frac{\text{½ thickness of jack}}{\text{Tan W}\angle}$$

For a regular 45-degree hip, the setback is 0.75 divided by the tangent of 45 degrees. The tangent of 45 degrees is 1, so the answer is 3/4".

Step 30: The Setback Cutting Angle From the Short Point— The proper cutting angle is the complement of the working angle. To find the complement, subtract the working angle from 90 degrees. For the octagon, this is 67½ degrees as indicated in Figure 9-49.

Steps for the Header Block and Miters

Step 31: Length of Run at the Block— Refer to Figure 9-54.

Step 32: Block Short Point Measurement— Refer to Figure 9-57.

Step 33: Top Bevel Cut Angle— Refer to Figure 9-58.

Step 34: Miter Cut for Forms or Trim— Figure 9-42 shows a mitered form board. The drawing shows the working angle to be the short point angle. However, this angle will also produce a long point angle on the piece that has been cut away from it. You must figure the long point or short point according to the mechanical direction of the beveling angle of your saw. Keep in mind that some saws bevel to the right and some to the left.

Step 35: Growth of the Miter Cut— Figure 9-42 shows a growth of 5/8" at each end for a 2" x 6" octagon form. The general formula for finding this length for any regular polygon is:

$$\text{Length} = (\text{thickness of material})\ (\text{Tan W}\angle)$$

For a square form 1½" thick, the length added at each end would be 1½" times the tangent of 45 degrees, or 1½".

For an octagon form 1½" thick, the length added at each end would be 1½" times the tangent of 22.5 degrees, or 5/8".

A Listing of the Steps

We've come pretty fast in this chapter. But all the information you need is here when you need it. You'll probably want to refer to these pages many times while you're framing that gazebo. To make reference easier, we've listed all 35 steps together in Figure 9-10.

To Find:	Do This:
General	
Step 1 Actual run of the polygon hip (to the building line)	Select a radius (R) for the circumscribed circle[(a)]
Step 2 The central angle	360 ÷ number of sides
Step 3 The working angle	360 ÷ twice the number of sides
The Common Rafter (Figure 9-4)	
Step 4 Total run of the common rafter to the building line (the apothem)	(#1) (cos. W∢)[(b)]
Step 5 Total run of the common rafter to the fascia line	# 4 + the length of the overhang
Step 6 The building line measurement	(#4) (common rafter secant from Figure 1-14)
Step 7 The fascia line measurement	(#5) (common rafter secant from Figure 1-14)
Step 8 Common rafter shortening	(½ thickness of hip) ÷ (sin W∢)
Step 9 Common rafter setback	(½ thickness of common rafter) ÷ (tan W∢)
Step 10 Setback cutting angle	90° − W∢
The Side and The Hip Unit Run:	
Step 11 The side of the building	(#1) (2 sin W∢)
Step 12 The hip unit run and framing square setting	(12") (sec W∢)[(c)]
Step 13 The area of any regular polygon	(#4) (#11) (½ number of sides)[(d)]
The Hip Rafter	
Step 14 Secant of the polygon hip	Unit rise ÷ #12 (not 12") then punch: = [INV] [tan] [cos] [1/x] [(e)]
Step 15 The building line measurement	(#1) (#14)
Step 16 Actual run to the fascia line	# 5 ÷ cos W∢
Step 17 The fascia line measurement	(#14) (#16)

Solving regular polygons
Table 9-10

To Find:	Do This:
Step 18 Set forward at birdsmouth and tail cut	(tan W∢) (½ thickness of hip)
Step 19 Backing or dropping	(#18) (unit rise) ÷ (#12)
Step 20 Angle of the backing cut	#19 ÷ (½ thickness of hip) then punch: [=] [INV] [tan] [e]
Step 21 Ridge line setback if all hips come to center point	(½ thickness of hip) ÷ (tan W∢)
Step 22 The radius for a ridge block	Full thickness of hip ÷ 2 (sin W∢)
Step 23 Ridge line shortening for Step 22	(#22) (cos W∢)
The Jack Rafter:	
Step 24 Jack actual run to the building line	Distance from corner ÷ (tan W∢)
Step 25 Actual run to the fascia line	# 24 + overhang distance
Step 26 The building line measurement	(#24) (common rafter secant from Figure 1-14)
Step 27 The fascia line measurement	(#25) (common rafter secant from Figure 1-14)
Step 28 Shortening	½ thickness of hip ÷ (sin W∢)
Step 29 Setback	½ thickness of jack ÷ (tan W∢)
Step 30 Setback short point cutting angle	90° − W∢
The Header Block and Miters:	
Step 31 Run at the block (see Figure 9-54)	#4 − run of the interrupted common rafter − the thickness of the block
Step 32 Block short point measurement (line (b) of Figure 9-54)	2 (#31) (tan W∢) − (sec W∢) (full thickness of the hip)[c], [f]
Step 33 Top bevel cut angle	(unit rise) ÷ (#12) then punch: [=] [INV] [tan]
Step 34 Miter cut angle	W∢ on the short point side
Step 35 Growth of the miter cut	(thickness of board) (tan W∢)

Solving regular polygons (continued)
Table 9-10

Footnotes: To Table 9-10

(a) If the side of the building has been predetermined, then rearrange step #11 and solve for R:

$$R = \frac{\text{side of building}}{2(\sin W\measuredangle)}$$

(b) For an interrupted common rafter as in (C) of Figure 9-45, decide on the amount of run rather than using #1.

(c) To find sec. on the TI-35 calculator, let the W∢ appear on the display, then punch: [cos] $1/x$.

(d) For the roof area, replace #4 (which yields the area within the building) with #5 (for the flat area under the fascia line). Multiply this by the roof factor found in Appendix E, Table 3, to find the square foot area for your pitch roof.

To find the number of "squares" for roofing material, divide by 100. Also: #4 or #5 and #11 must both be in inches or both be in feet. If inches are used, divide by 144 to convert to square feet.

(e) Or look in a table of trigonometric functions.

(f) Cut from the short point using the W∢.

The Octagon and the Square

The framing square treats the octagon as related to the square, instead of the circle. Since this method is more common to the carpenter, the second part of this chapter will explore this special relationship.

Laying out the Octagon

On a piece of 8½" x 11" paper with a 1/4" grid, lay out an 8" square as shown in Figure 9-11. Draw a vertical and horizontal line through the figure. We're going to draw diagonal lines at 45 degrees like those labeled (a) in Figure 9-11. We don't yet know the exact beginning and end points of these lines. That's what our octagon scale is going to tell us.

The Octagon Scale on the Framing Square

On the front side of the tongue of the framing square is an octagon scale, as shown in Figure 9-12.

The octagon scale has a series of dots separated by numbers. There are four dots in a row, something that looks like a three-digit number, and then another four dots. Actually, the three-digit number is a single-digit number, followed by a vertical line and then another single-digit number. Each vertical line is positioned where a dot should be. The numbers on each side of the vertical line indicate the sequence number for the dot which has been replaced by the vertical line. Some examples will make this clearer.

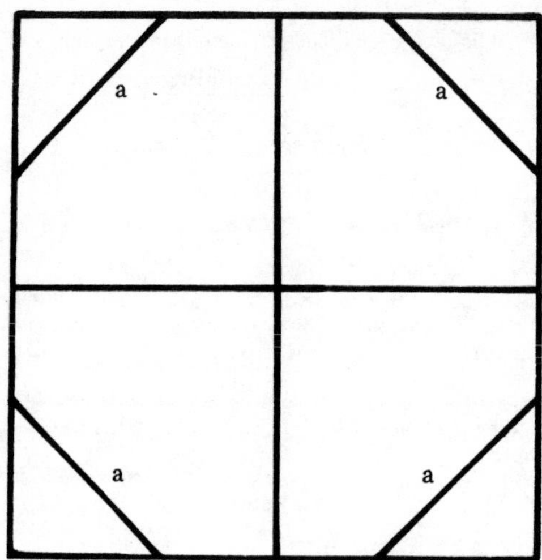

The octagon and the square
Figure 9-11

The number *115* means the vertical center line is the 15th dot on the scale, and *310* means the vertical center line is the 30th dot on the scale.

Here's how the dots are used. If we measure off the space of eight dots on each side of each center line of our drawing, we have the beginning and ending places for the 45-degree lines. The reason you're using eight dots is because this is an 8'' square figure. You use one dot for every inch.

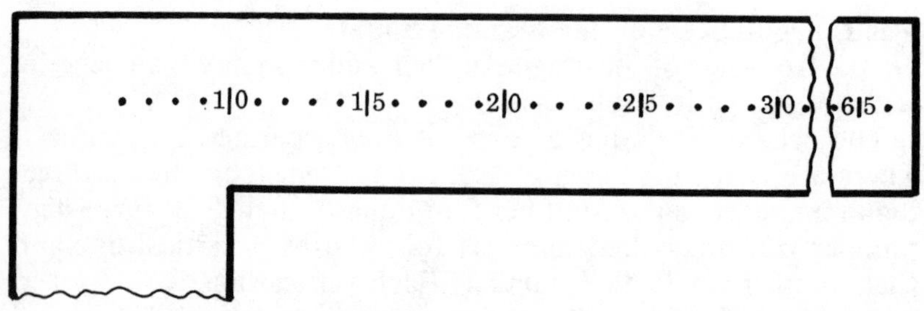

The octagon scale of the framing square
Figure 9-12

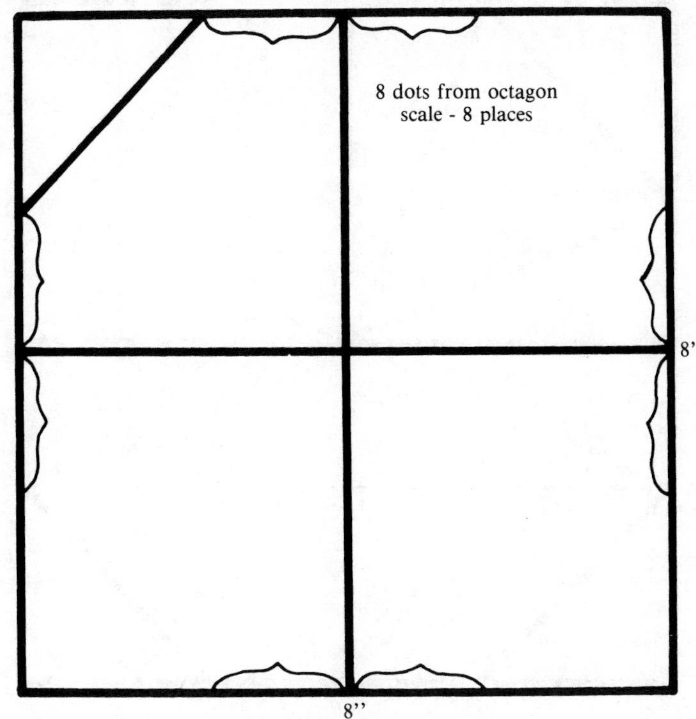

The graph paper drawing
Figure 9-13

On a small piece of paper, mark off the space between the zero point on the octagon scale and dot eight. Then transfer this distance to each side of the points where the vertical and horizontal lines intersect the sides of the box. See Figure 9-13. Now draw in the 45 degree-lines between the points you have just marked. Your drawing should look like Figure 9-11.

The Triangle Inside

Look at Figure 9-14. A dotted line has been extended from point (A) directly across the figure to the 8-dot point on the opposite side. Another dotted line goes from (A) to a second 8-dot point on the opposite side. Use red pencil or pen to draw these same dotted lines in your 8'' square. Now connect the two 8-dot points on the right side of the square so your red lines have created a triangle.

What do we know about this triangle? The two things are obvious immediately. First, it's a right triangle because there's one

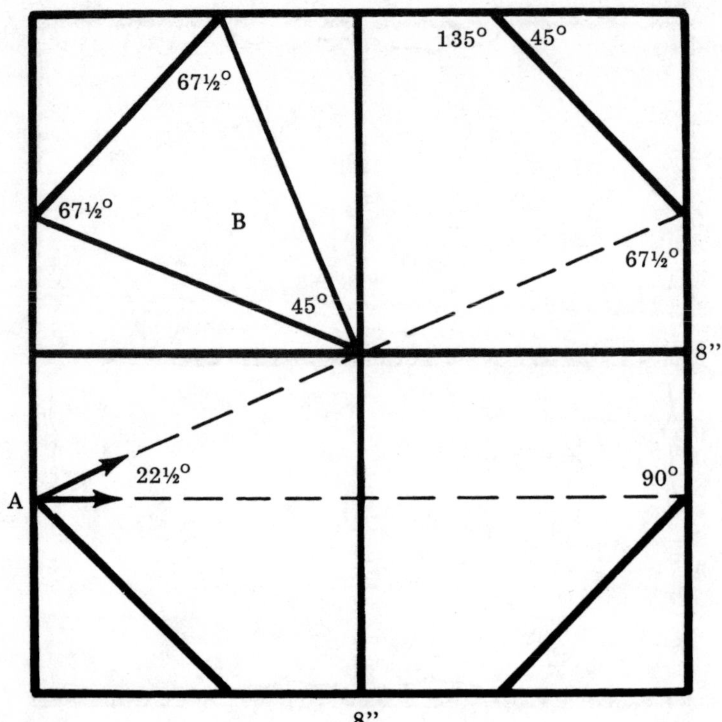

The right triangle inside
Figure 9-14

90-degree angle. That's important because we know many ways to figure lengths and angles on a right triangle. Second, the base line is 8" long — exactly the same width as the side of the square.

Here are some other points that may come to mind about your red triangle. Since we drew the corner lines at 45 degrees, there must be a 135-degree interior angle on the other side of the 45-degree line. Note the 135-degree angle at the top of the square. The diagonal red line bisects one of these corner lines, creating a 67½-degree angle, as noted at the right side of Figure 9-14.

One of the angles in our triangle is 90 degrees. The second is 67½ degrees. Because the angles in a triangle must total 180, the third angle, (A) in Figure 9-14, must be 180 minus 90 minus 67.5, or 22.5 degrees.

In the upper left portion of Figure 9-14, triangle (B), we show another way to get the same answers. If we draw a line from each outer corner to the center, we form eight interior angles. Since there are 360 degrees in a circle, one of the interior angles will be equal to 360 degrees divided by 8, or 45 degrees. Triangle (B) is marked as 45 degrees. This is an isosceles triangle; it has two equal sides. Therefore the other two angles will be equal. Again, 180 degrees minus 45 degrees equals 135 degrees. Divide that by 2 and you have 67½ degrees.

Once again, our upper angle turns out to be 67½ degrees and angle (A) is 22½ degrees.

The Steps in Finding Gazebo Rafter Lengths with a Square
As in the earlier section of this chapter, we'll number the steps in calculating gazebo rafter lengths with a square.

Step 1: The Side to the Square— Figure 9-15 shows our red dotted-line triangle from the last figure. This time, rather than using eight units as its base (or 8'' if you like), we've reduced it down to a unit

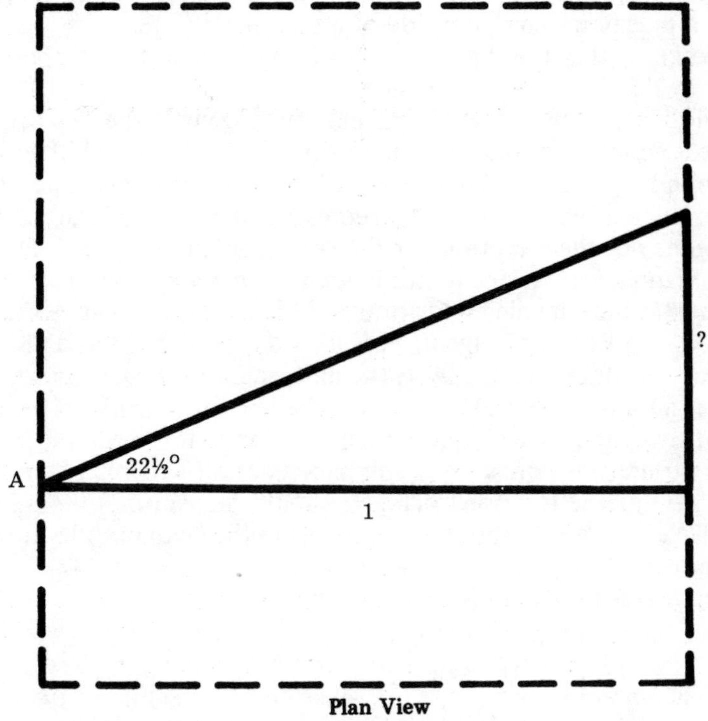

A unit triangle
Figure 9-15

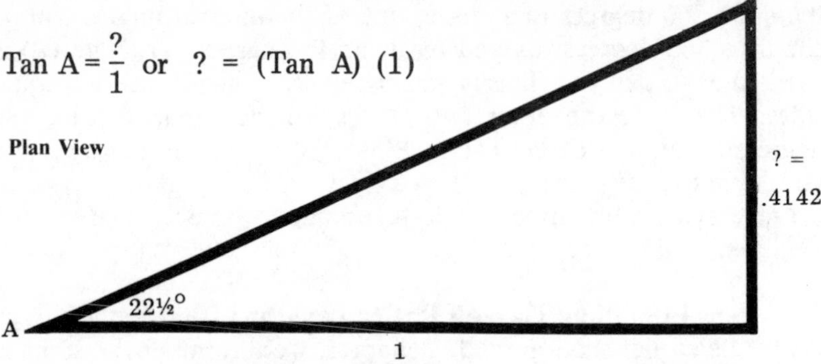

Finding the length of the octagon side
Figure 9-16

triangle whose base is one unit long. This one unit is the same as the width of the surrounding square.

If the base is one unit long, what's the length of the side labeled with a question mark, the side opposite angle (A)?

We know that the tangent of angle (A) is equal to the opposite side divided by the adjacent side.

Another way of saying this is that the tangent of the 22½ degree angle is equal to the question-mark side divided by the width of the surrounding square. We can't call this the run and the rise, however, because our drawing represents the floor of the gazebo octagon, not the elevation for the roof framing members. That's why Figures 9-15 and 9-16 are labeled *Plan View*.

Look at the triangle and formula in Figure 9-16. Angle A is 22½ degrees. To find the tangent, punch [2] [2] [·] [5] [tan] and 0.4142 appears. Multiplying this by 1, the unit length of the triangle base, we get an answer of 0.4142. This is the length of a side of an octagon that fits inside a square whose length is 1.

The triangle we drew on graph paper was 8" square. Therefore each side of the inscribed octagon should be 8" times 0.4142 or 3.3137". That's just about 3⁵⁄₁₆". Measure the octagon side. It will measure 3⁵⁄₁₆".

The two formulas in Step 1 are:

$$\text{The square} = \frac{\text{The side}}{.4142} \qquad \text{The side} = (.4142)\ (\text{the square})$$

These are the plan view formulas only.

Here are some examples. Try answering these questions to test your understanding of what we've covered so far.

1) An octagon is 9' square. What's the length of each wall?

$$\text{Tan } 22\tfrac{1}{2}^{\circ} = \frac{\text{Each wall}}{\text{The square}}$$

$$.4142 = \frac{\text{Each wall}}{9'}$$

$$\text{Each wall} = (.4142)\,(9') = 3.7279'$$

$$\text{Each wall} = 3'8\tfrac{3}{4}''$$

2) We're going to build a gazebo wall with a four-foot opening. There should be at least 6'' of wall on either side of the opening. What is the size of the layout square?

$$\text{Tan } 22\tfrac{1}{2}^{\circ} = \frac{\text{Each wall}}{\text{The square}}$$

$$\text{The square} = \frac{\text{Each wall}}{\text{Tan } 22\tfrac{1}{2}^{\circ}}$$

$$\text{The square} = \frac{5'}{.4142} = 12.07'$$

$$\text{The square} = 12'0\tfrac{7}{8}'' \text{ on each side}$$

3) What's the length of each wall of a gazebo built on a 24' square?

$$\text{Each wall} = (\text{Tan } 22\tfrac{1}{2}^{\circ})\,(\text{side of the square})$$

$$\text{Each wall} = (.4142)\;(24')$$

$$\text{Each wall} = 9.9411''$$

$$\text{Each wall} = 9'11\tfrac{5}{16}'' \text{ long point}$$

Step 2: The Actual Run of the Octagon Hip— So far you've learned how to find the square if only the length of the side is given, or how to find the side dimension if the dimension of the square is given. Now let's find the plan view run of the octagon hip, which is the actual run for that hip.

We've used the secant method to find rafter lengths before. Multiply the secant by the side adjacent to get the hypotenuse. But what's the secant of this 22.5 degree angle? Under 22 degrees 30 minutes in a trigonometry table, you'll find the secant listed as 1.0824. On your calculator, punch [2] [2] [·] [5] [cos] [1/x] and 1.0824 appears.

Multiplying the secant by 8" gives us 8.6591". But this is the actual run length of two rafters, while we need only one: 8.6591" divided by 2 equals 4.3296", or about 4⁵⁄₁₆. If you've made a drawing like Figure 9-17, it will measure just over 4⁵⁄₁₆".

The formula for Step 2 is:

$$\text{Actual run of the octagon hip rafter} = \frac{(1.0824)(\text{the square})}{2} = (.5412)\ (\text{the square})$$

But remember that this formula is for the plan view only.

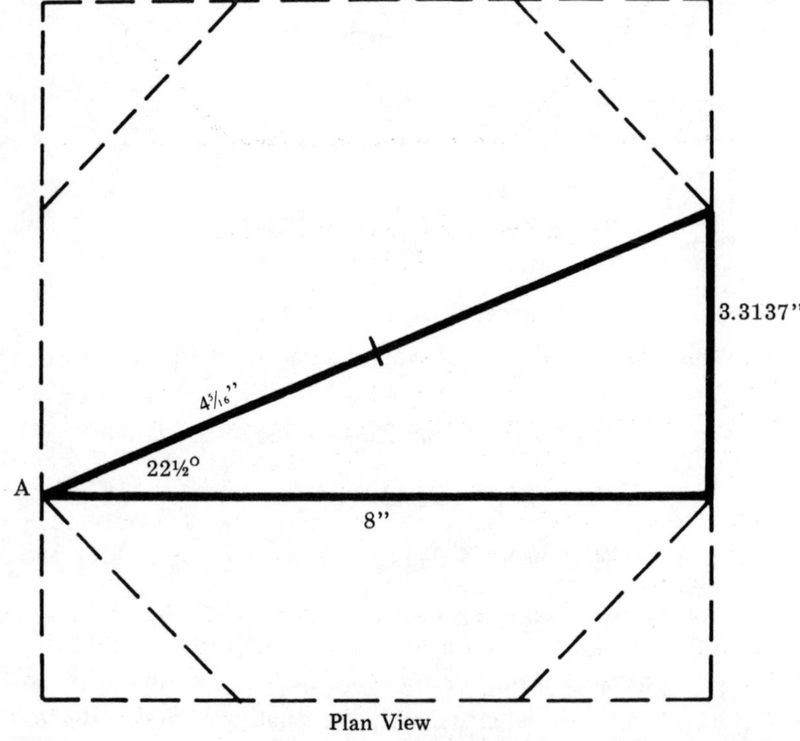

Finding the hip run
Figure 9-17

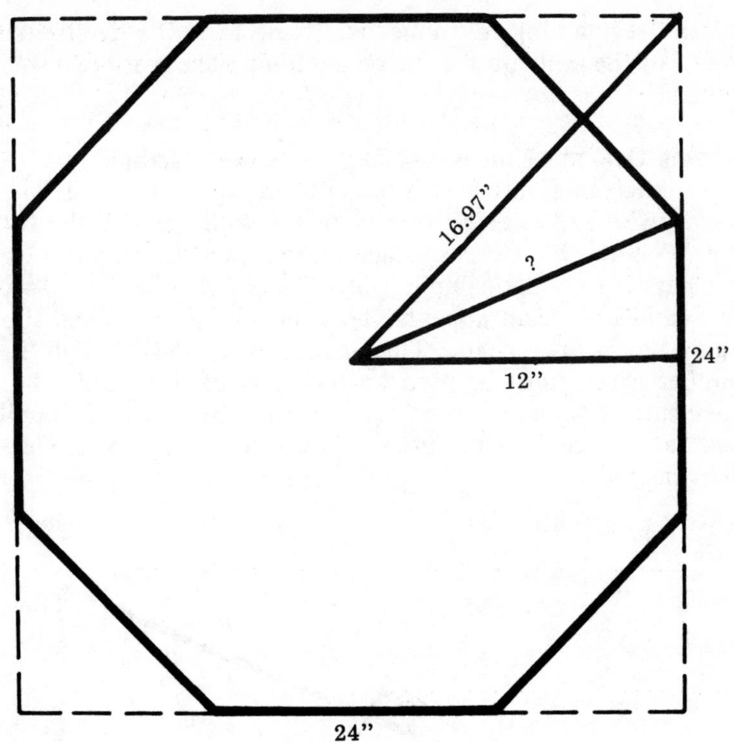

The relationships of unit run
Figure 9-18

The Unit Run of the Octagon Hip

Until now we've been working only with the plan or flat view when we've calculated the actual run of the octagon hip rafter. Now, from this information, we can determine the mathematical length of the rafter necessary for our roof pitch. To do this we'll need to know the value of the octagon secant of that pitch. You'll find this in the octagon rafter roof angle chart, Figure 9-60, at the end of this chapter.

How were the numbers in Figure 9-60 developed? First, we know that the unit run for a common rafter is 12", and this forms the mathematical basis for the regular hip unit run of 16.97". Now, what's the unit run of the octagon hip? Look at Figure 9-18.

If we assume a 24" octagon, we will have a 12" radius. This represents the 12" unit run. Note the line labeled 12" in Figure 9-18. Also, the diagonal to the circumscribed corner will be 16.97" for the regular hip, as indicated. Now we must determine the length of the unit run for the octagon hip. There are two ways to do this.

Finding Octagon Unit Run Using the Secant Method

The triangle in Figure 9-19 is taken from Figure 9-18. The interior angle is still 22½ degrees, or half of the 45 degrees of the 16.97" line. We know by the secant method that the secant times 12" will give us the octagon unit run. Punch [2] [2] [·] [5] [cos] [1/x] and the secant appears. Then punch [×] [1] [2] and the octagon unit run appears. The answer is 12.9887". This is the number which must be used on the body of the framing square when cutting octagon hip rafters. That number should be familiar. You saw it before, calculated in a different way, earlier in this chapter.

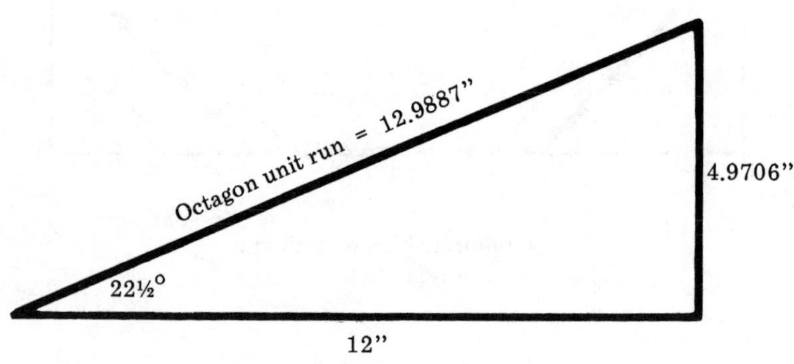

Octagon unit run by secant
Figure 9-19

Finding Octagon Unit Run Using the Square Root Method

We can also find the octagon unit run by the square root method, if we know the length of the other side. In Step 1 of finding gazebo rafter lengths with a square, we learned how to calculate the length of the other side. The length of the side is equal to 0.4142 times the size of the square, which is 24" in this case. Multiplying 0.4142 by 24", we get 9.9411".

Our figure uses only one-half of this length for its side, so dividing by two gives us 4.9706".

Using the square root formula we have:

$$\text{Octagon unit run} = \sqrt{(4.9706)^2 + (12)^2} = 12.9887''$$

This is the same figure we found by the secant method. You'll probably agree that the secant method was quicker and easier.

Step 3: The Octagon Hip Secant by Trig Function Method— Now we have the base number of 12.9887'', which is used to build the table of octagon secants. For an 8'' rise roof, you can find the secant on your calculator by first finding the tangent and going from that to the secant. See Figure 9-20.

$$\text{Tan } \frac{8}{12.9887} = .6159 = \boxed{\text{INV}}\ \boxed{\text{Tan}} = 31.63°$$

$$\boxed{\text{Cos}}\ \boxed{1/x} = \text{Secant} = 1.1745$$

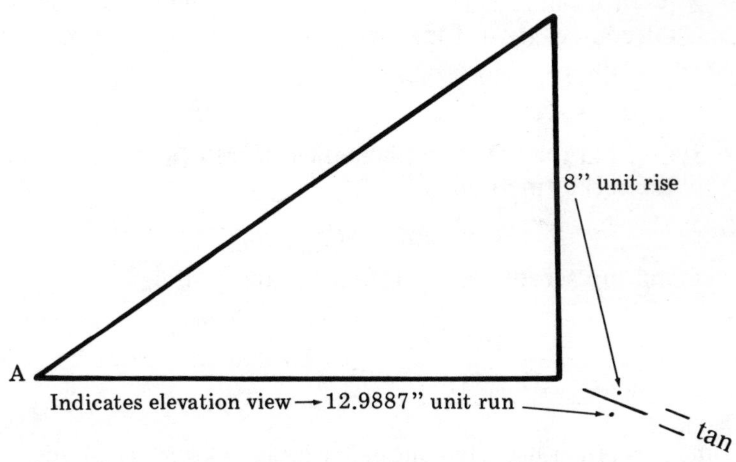

Finding the secant of an 8'' rise octagon hip
Figure 9-20

This is the secant number for an 8'' rise octagon hip. The Step 3 formula is the secant formula we have always used:

Rafter length = Secant x Actual run

So the octagon hip rafter, for an 8 in 12 roof, is 1.1745 times the actual run found under Step 2 above. This is an elevation formula.

Review

We've used three steps to find the mathematical length of octagon hip rafters.

First Step: Find the size of the square. If only the side of the octagon is given, use this formula to find the square:

$$\text{The square} = \frac{\text{The side}}{.4142}$$

Second Step: Find the length of the actual run of the octagon hip. Use the formula:

$$\text{Octagon hip actual run} = \frac{(1.0824)\ (\text{the square})}{2} = (.5412)\ (\text{the square})$$

These two formulas deal with the plan view, the pattern on the ground. Of course, these formulas work no matter whether you use feet or inches.

Third Step: Find the rafter mathematical length (in the elevation view) by the secant method:

$$\text{Rafter} = \text{Secant x Actual run}$$

Any octagon hip secant can be found by first finding:

$$\frac{\text{Inches of rise}}{12.9887} = \text{the tangent}$$

From here, either use a trigonometry table or your calculator. In a trig table, find the page with the degrees you're looking for. Then look across and read the secant. With the calculator, get the tangent in the display. Then punch [INV] [tan] to convert the tangent to the angle. Then push [cos] [1/x] and the secant will appear.

Alternate for Step 3: The Square Root Method— Step 3 may also be done by the unit length method, which is the square root method.

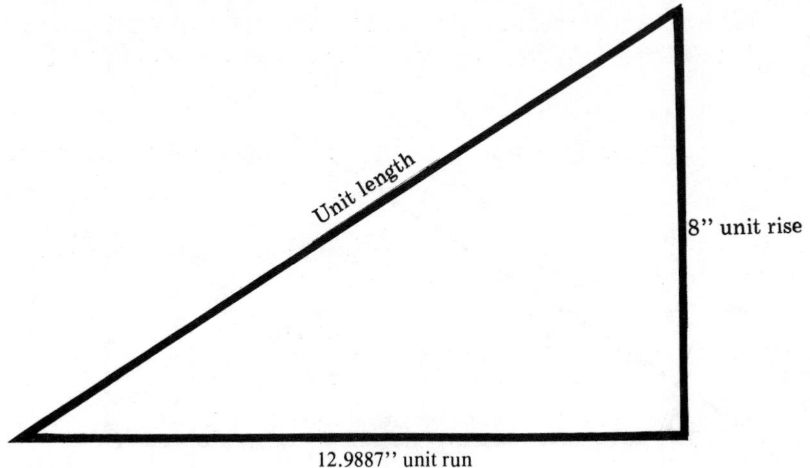

An 8 in 13 unit triangle
Figure 9-21

A unit triangle for an 8" rise octagon roof would look like Figure 9-21. To find the unit length, use the Pythagorean Theorem.

$$\text{Unit length} = \sqrt{(12.9887")^2 + (8")^2}$$

$$\text{Unit length} = 15.2547"$$

This unit length is then multiplied by the number of octagon unit run feet or:

$$\text{Octagon rafter length} = \text{Unit length} \times \frac{\text{Actual run (in feet)}}{12.9887"}$$

This method is much more involved than the secant method given under Step 3 above.

Calculating the Dot Spacing

Back at Figure 9-15, we examined the unit triangle to understand the basic relationship between an octagon side and the circumscribed square. This triangle also helps us understand the basis for the dot spacing of the octagon scale. The octagon scale is printed on the tongue of the framing square.

Look at Figure 9-22. If we assume the unit 1 is one foot, then we would have 0.4142' for the rise side. If we assume the unit 1 to be an inch, we would have 0.4142". Since octagons are laid out to each side of two perpendicular center lines, we would want to space our dots by half the distance of one side. Dividing 0.4142"

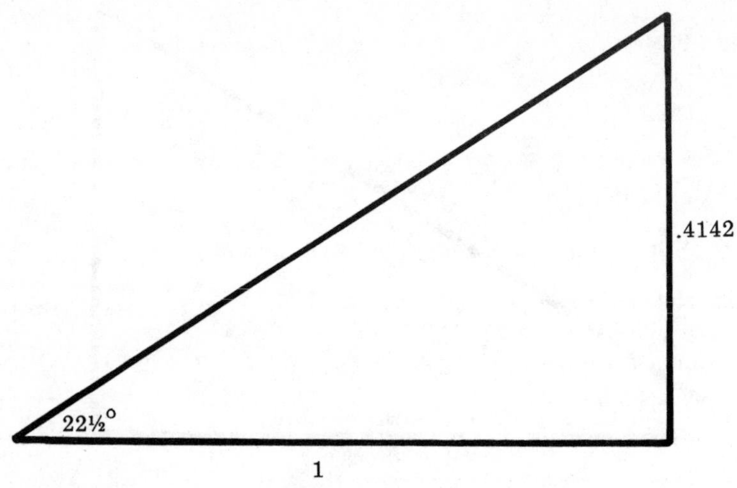

Another look at the unit triangle
Figure 9-22

by 2 gives us 0.2071''. Multiplying that by 16 gives 3.3137 sixteenths. This is the spacing given to each dot on the octagon scale. Measure your framing square to see if that isn't true.

Octagoning a Post

The dots of the octagon scale were put on the square so carpenters could form small planks, a table top, or the end of a square timber into an octagon shape.

There's another way to make an octagon post from a square timber. Lay a ruler across a timber as shown in Figure 9-23 A. The

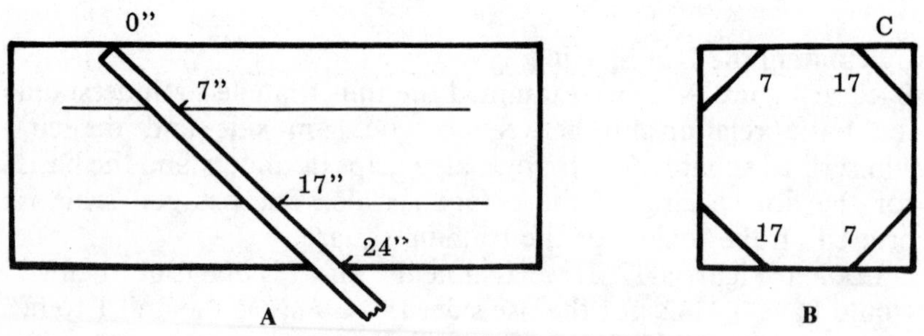

Layout lines for an octagon post
Figure 9-23

zero point on the ruler should be on one edge while 24'' rests on the other edge. Put one mark at 7'' and another at 17''. Draw parallel lines through these points.

These lines become guide lines for the saw. Set the saw at 45 degrees, cutting off the corners of the timber, as shown in Figure 9-23 B. A rip guide and circular saw can be used for three of these cuts. But a table saw gives much better results.

The Relationships of Line Segment (C)

There's another way to determine the amount to cut off when forming an octagon out of a post. Suppose we want to trim off triangle (C) in Figure 9-24.

Knowing the length of line segment (C) is useful from time to time, and octagoning a post is one of these times. What is its relationship to the square and to the side?

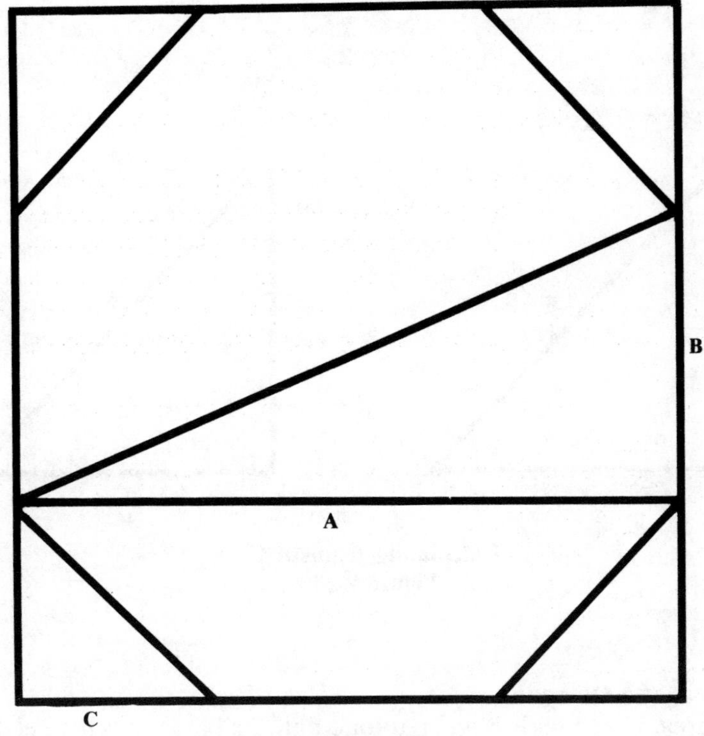

Line segment C
Figure 9-24

Line Segment (C) Related to the Square

If line (A) in Figure 9-24 is equal to 1, then line segment (C) is equal to 1 minus 0.4142 divided by 2. Algebraically, the subtraction must come first. One minus 0.4142 equals 0.5858. Then:

$$C = \frac{1 - .4142}{2}$$

$$C = \frac{.5858}{2} = .2929 \text{ of A}$$

What's the length of (C) in our 8" square? Multiplying 0.2929 times 8" equals 2.3431", or 2 inches and 5.5 sixteenths. Check your 8" square drawing and see if that isn't the length of line segment (C).

Line Segment (C) Related to the Side

In Figure 9-25, the length of side (B) is equal to (C) times the square root of 2. Therefore:

$$C = \frac{B}{\sqrt{2}}$$

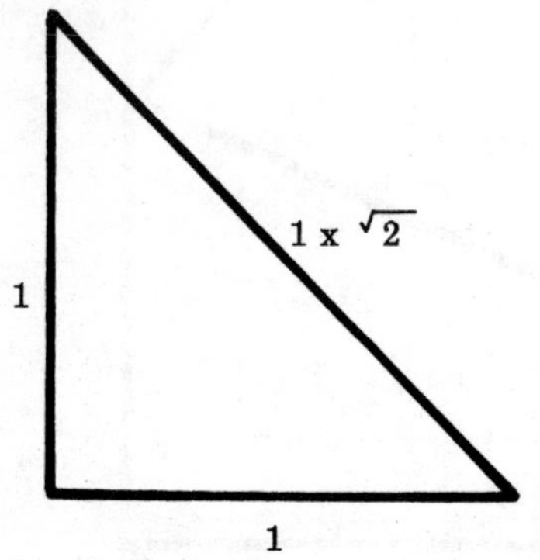

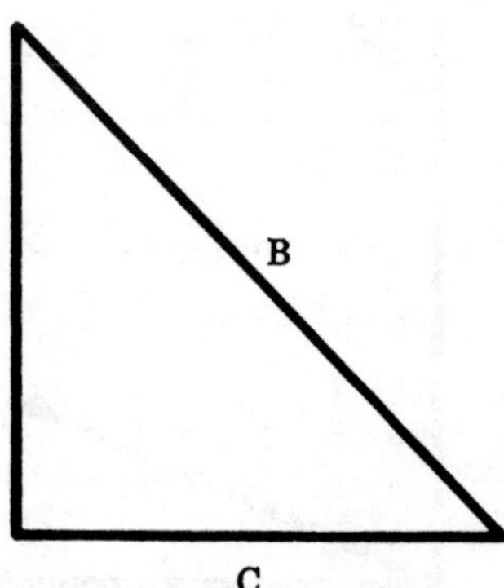

Calculating segment C
Figure 9-25

Area of the Octagon

The area of a triangle is equal to one-half the base times the height. You can think of an octagon as eight triangles. If (B) in Figure 9-26 is the base and (H) is the height of each of these triangles:

$$\text{Area of Octagon} = \frac{8 \times B\,H}{2} = 4\ B\,H$$

$$\text{Area of Octagon} = 4\ B\,H$$

Drafting an Octagon Roof

The mathematics presented so far in this chapter give you all the tools you need to plan and frame a gazebo roof. We'll now apply the concepts you've learned to draft, cut, and build an 8 in 13 model octagon roof for a 3' square base.

Start with paper that measures at least 19 inches square. You'll also need an architect's rule with a 1/2'' scale. Usually you would read each division on this scale as being one foot. But for this job interpret each number on the 1/2'' scale as one inch.

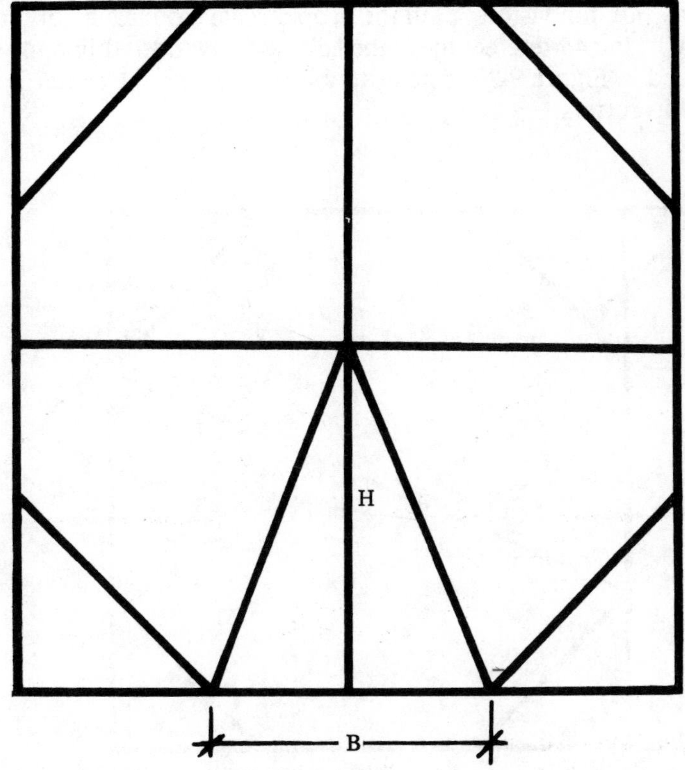

The area of an octagon
Figure 9-26

Use your architect's rule to lay out a square that measures 36 units by 36 units. See Figure 9-27. Of course, the actual size of the square is 18'' by 18'' when measured with a ruler. Place a mark at the center of the square.

Since the dots of the octagon scale are not in 1/2'' scale, they can't be used here.

First we'll calculate the length of side (B) and then side (C):

$$B = \text{Tan } 22\tfrac{1}{2}^\circ \times \text{The square}$$

$$B = .4142 \times 36''$$

$$B = 14.9117 = 14\tfrac{15}{16}''$$

$$C = \frac{B}{\sqrt{2}} = \frac{14.9117''}{1.414} = 10.5442'' = 10\tfrac{9}{16}''$$

Lay out dimension (C) in the proper scale from each corner, and draw in the 45-degree lines labeled (3). Draw dotted line (4) as indicated in Figure 9-27. The dotted lines indicate the center lines of the hip rafters.

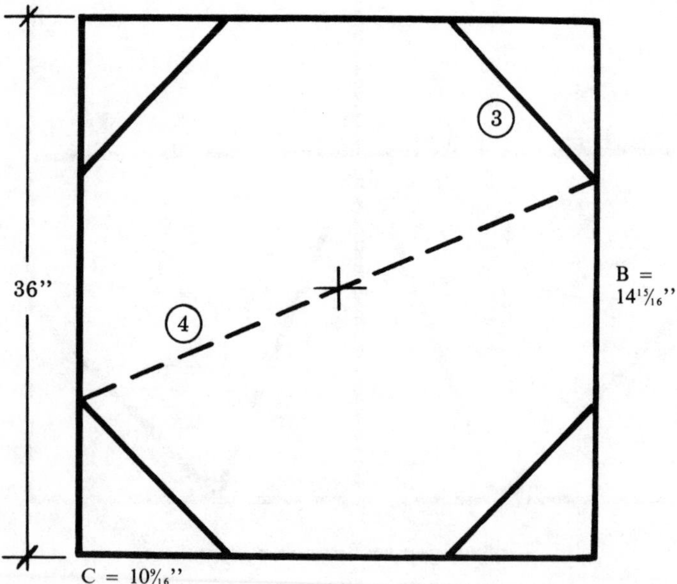

Laying out the square
Figure 9-27

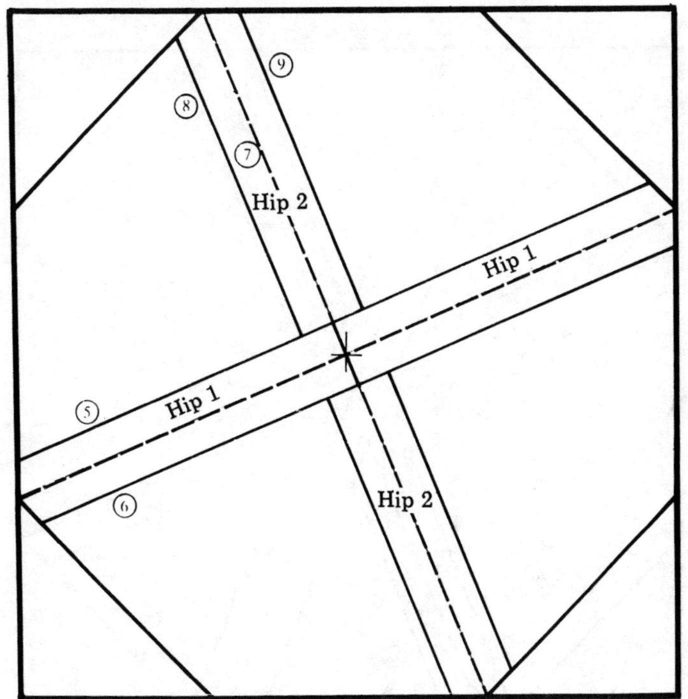

Adding the first 4 hips
Figure 9-28

Now switch to a regular ruler and lay out the hips in full size. See Figure 9-28. Measure 3/4" to each side of the dotted line and draw lines (5) and (6). These lines will be a full 1½" apart. Dotted line (7) should be drawn solid where it goes between (5) and (6). Add lines (8) and (9). Label the first pair of adjacent hips as Hip 1. Label the other hips Hip 2.

Look at Figure 9-29. Part of line (13) is left out and a large segment of line (10) is excluded. Finish lines (11), (12), (14) and (15). The hips formed by these lines are labeled Hip 3. The #1 hips are cut on the ridge line. The #2 hips shorten by half the thickness of the first pair. The #3 hips shorten by half the 45-degree thickness of the first pair, which your drawing indicates as 1$\frac{1}{16}$". Measure the 45-degree thickness of Hip 1 along line (13). It should measure 1$\frac{1}{16}$".

Draw in line (16) across the width of Hip 3. This shows the setback of the #3 hips to be half the thickness of the #3 hip. Measure the setback. It should be 3/4".

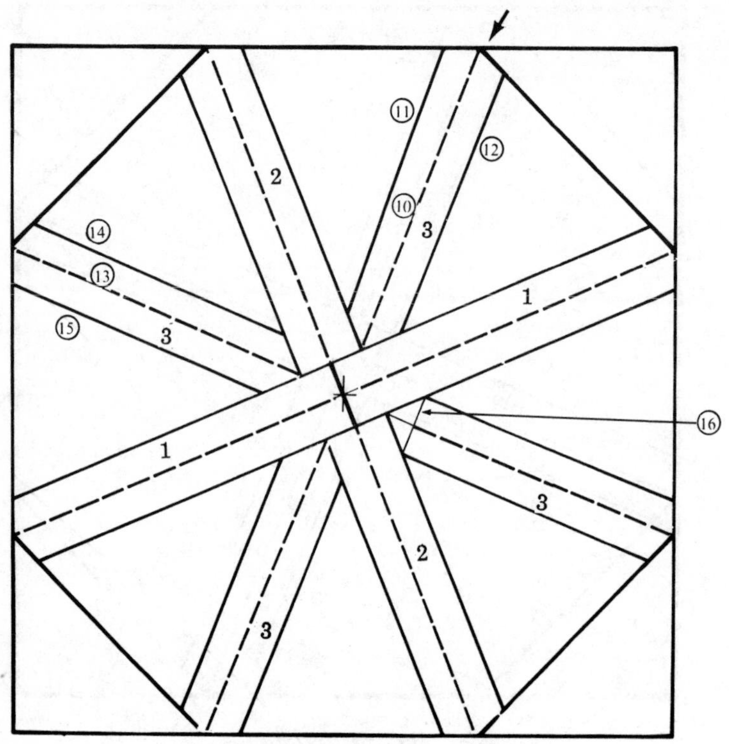

Completing the other hips
Figure 9-29

Set Forward at the Building and Fascia Lines
Look at Figure 9-30. Draw line (17) where the arrow is in Figure 9-29. This shows the amount of set forward necessary for a 1½'' thick octagon rafter. Measure the segment (b). It will be 5/16''.

For a regular 45-degree hip, the cheek cut lines for the birdsmouth and the tail cut were set forward 3/4''. (You might want to check back to Figures 4-42 and 4-49 to verify this.) With the octagon rafter, the set forward will be 5/16'' for 1½'' thick material.

Mathematical Set Forward
Back in Figure 9-14, we determined that the interior angle of an octagon is 22½ degrees. See Figure 9-31. The base of this triangle is half the thickness of the rafter, or 3/4''. We'll use the tangent formula to determine the length of side (b) in Figure 9-31, the set forward at the building fascia line.

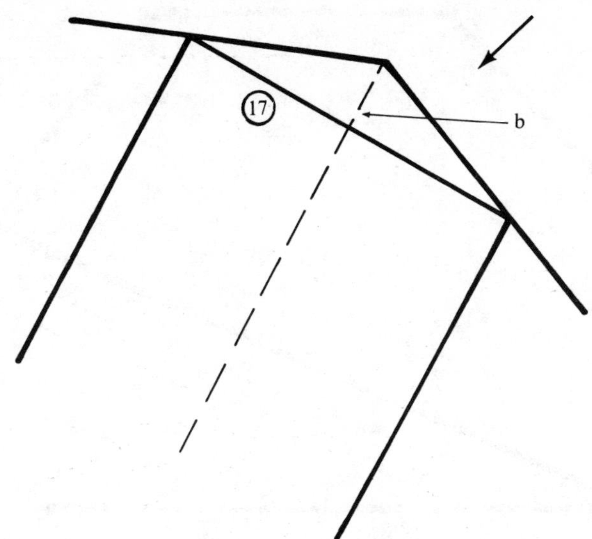

Finding the set forward
Figure 9-30

Using the tangent formula, punch [2] [2] [·] [5] [tan] and 0.4142 appears. Punch [×] [·] [7] [5] [=] and 0.31066 appears. Punch [×] [1] [6] [=] and 4.97 appears. This is the number of sixteenths. So the amount of set forward is 5/16''. This agrees with the distance we measured previously.

$$\text{Tan } 22\tfrac{1}{2}^{\circ} = \frac{b}{\tfrac{3}{4}''}$$

$$b = (\text{Tan } 22\tfrac{1}{2}^{\circ})\,(.75)$$

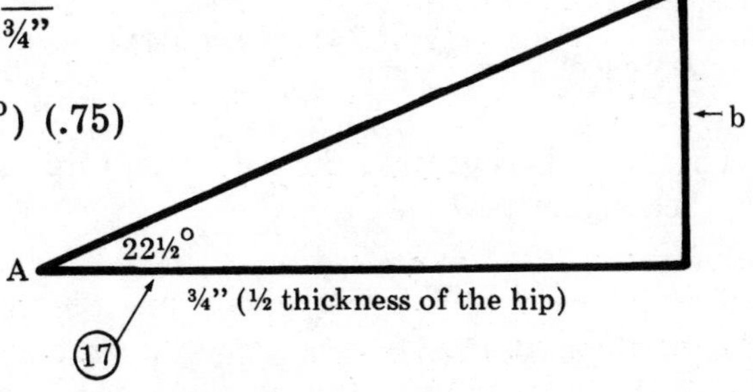

22½°
A
b
¾'' (½ thickness of the hip)
17

Trigonometry set forward
Figure 9-31

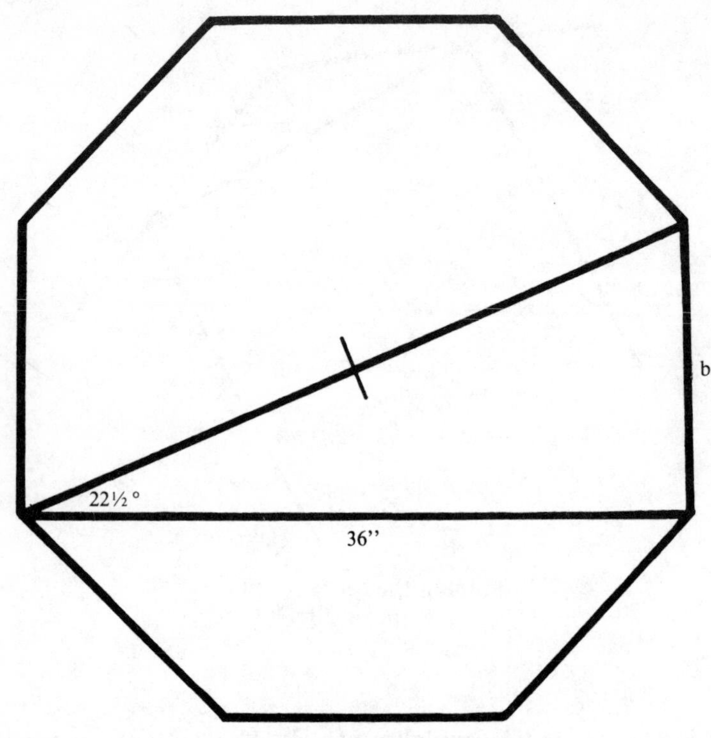

The old secant formula
Figure 9-32

Calculating the Hips

We have been given the square as 36", so we have our Step 1 answer. Now we must find the actual run of the hip. Use the second plan view formula, which is Step 2:

$$\text{Actual run} = \frac{(1.0824)\ (\text{the square})}{2} = (.5412)\ (\text{the square})$$

This formula is just like the old secant formula we have been using. See Figure 9-32.

The old secant formula says:

Rafter length = Secant x Actual run

This formula gives you the rafter length when you know the run. The old formula applies when we are dealing with the rise in a roof triangle. But here we are still dealing with a plan view (flat) dimension. The formula is the same but the two applications are different. Don't get them confused.

The 1.0824 in the Step 2 formula above is the secant of 22½ degrees. Punch [2] [2] [·] [5] [cos] [1/x] and 1.0824 appears. Punch [×] [3] [6] [=] and 38.9661 appears. Punch [÷] [2] [=] and you get 19.4831''. This is the actual run of the hip to the building line.

We must also calculate the run of the hip to the fascia line. A 6'' overhang on each end changes our 36'' square to a 48'' square. This is the answer for Step 1. Running this through the same Step 2 formula gives an actual run of 25.9774''. So:

Building line actual run = 19.4831''

Fascia line actual run = 25.9774''

Now let's get the rafter lengths:

Rafter length = Secant x actual run

Again we use the secant formula. And in this, the third step, we find the actual roof triangle. The secant here is of an angle whose rise is 8'' and whose run is 12.9887''. Punch [8] [÷] [1] [2] [·] [9] [8] [8] [7] [=] [INV] [tan] to find the angle, 31.63 degrees. Then punch [cos] [1/x] and the secant appears: 1.1745. This is the same number you'll find in the octagon roof angle chart at the end of this chapter, Figure 9-60.

Bldg. line math. length = (1.1745) (19.4831'') = 22.88'' = 22⅞''

Fascia line math. length = (1.1745) (25.9774'') = 30.51'' = 30½''

Laying Out the Octagon Hips

Now we can start marking and cutting our rafters. Figure 9-33 shows Hip 1 marked for the fascia line and building line mathematical lengths that we have just calculated.

Crown a piece of 2 x 4 and lay the crown away from you. Set the framing square gauges to just under 13'' on the body and 8'' on the tongue. Draw a ridge line and mark off the mathematical length for the building line and fascia line.

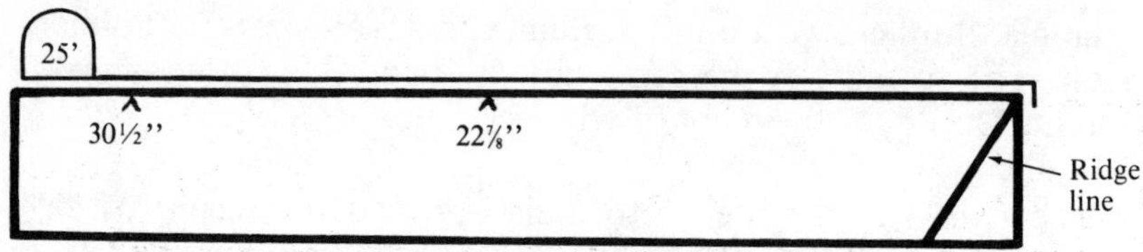

Beginning the layout
Figure 9-33

Using the same HAP and plancher dimensions as before, draw in those lines as in Figure 9-34. Then add the 5/16" set forward found under Figure 9-30. Transfer the set-forward line to the other side. This completes the layout lines for the first pair of hips. These five lines will also apply to the other octagon hips.

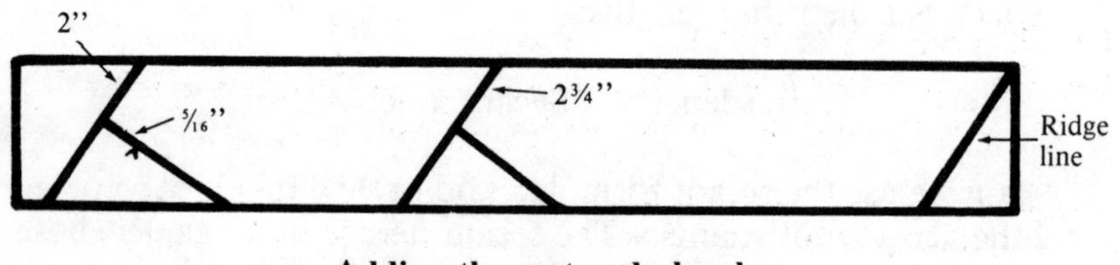

Adding the seat and plancher
Figure 9-34

Cutting the Octagon Hips

On the first pair of rafters, the birdsmouth, plancher and ridge line are cut with the saw at zero degrees, perpendicular to the work. The double cheek at the tail is cut with the saw set on 22½ degrees.

For the second pair, add 3/4" shortening to the ridge line, as shown in Figure 9-35. That's half the thickness of the first pair. Look back to Figure 9-28. Cut the shortening line with the saw on zero degrees.

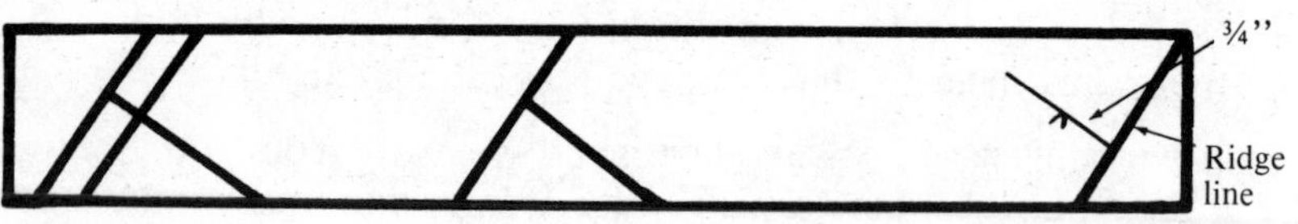

Shortening the second pair
Figure 9-35

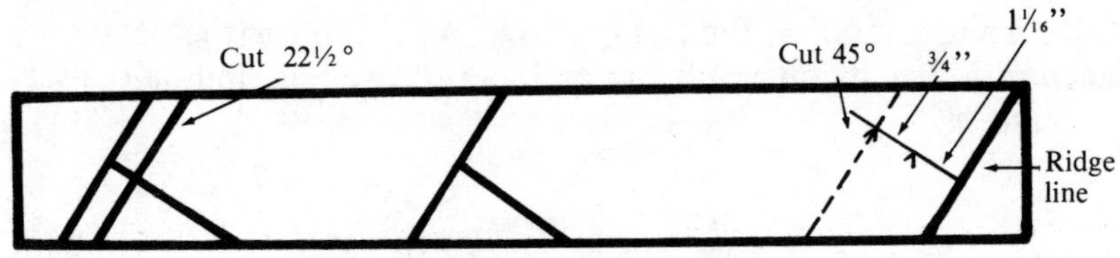

A different shortening and setback
Figure 9-36

The third pair has four similar rafters. Look back to Figure 9-29. The shortening is half the 45-degree thickness of the first pair. This is 1¹⁄₁₆'', and it is marked off on the construction line. See Figure 9-36. The setback for the cheek cut is again 3/4'', and not the 5/16'' at the tail cut. For this ridge cheek cut, set the saw on 45 degrees. You will have to transfer this line to the other side.

The Backing or Dropping
There are three ways to do the backing or dropping: the framing square, the math ratio, or the layout method.

The Framing Square Method— In Figure 9-30, the set forward at the building and fascia line (b) is seen to be 5/16''. Mark a set forward of 5/16'' on the seat cut line as in Figure 9-37.

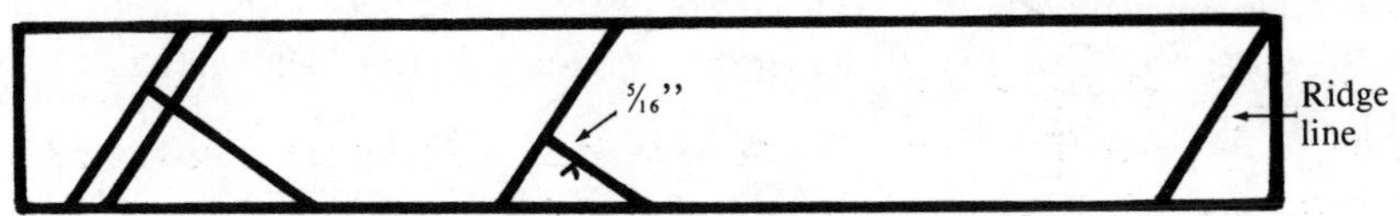

Backing or dropping by construction
Figure 9-37

Follow the procedure for hip backing or dropping in Figures 4-49 through 4-54, Chapter 4. The only difference is in the amount of set forward. The 45-degree hip rafter has a set forward of half the thickness of the hip, while the octagon hip rafter has a set forward of 0.4142 times half the thickness of the hip.

The Math Ratio Method— Here we compare the backing or dropping of a 45-degree hip with the backing or dropping on our octagon hip. The formula at the left in Table 9-38 is from Figure

4-44. The formula at the right is what we'll use for the octagon. Each ratio has its own unit run and its own way of finding the set forward.

$$\frac{16.97}{\tfrac{1}{2}\text{ thickness of hip}} : \frac{\text{Rise}}{?}$$

45° Hip

$$\frac{12.9887}{(.4142)(\tfrac{1}{2}\text{ thickness of hip})} : \frac{\text{Rise}}{?}$$

Octagon Hip

Different backing ratios
Table 9-38

We cross multiply to find the value of the question mark in the octagon formula in Table 9-38. Multiply the tangent of 22½ degrees by one-half the thickness of the hip by the unit rise and divide by the octagon unit run to get the backing or dropping required. Punch [2] [2] [·] [5] [tan] [×] [·] [7] [5] [×] [8] [÷] [1] [2] [·] [9] [8] [8] [7] and you get 0.1913, or 3/16''.

The Drafting or Layout Method— Figure 4-46 in Chapter 4 also shows a drafting method for finding the backing or dropping. The set forward is projected on a unit triangle and the backing is measured off.

Look at Figure 9-39. Here the set forward has been placed at the other side of the triangle, but the result is the same. This position will be used in Chapter 10 to find the backing cuts of an irregular hip rafter, as in Figure 10-22.

The Angle of the Backing Cut

The amount of backing is found from the side view, but the angle of the cut is determined by the end view of the rafter. See Figure 9-40. For this octagon hip, the line (9e) from Chapter 4, Figures 4-53 and 4-54, has been found to be 3/16'', or 0.1913 inches below the top of the hip. Half the thickness of this hip is 0.75''. Use the tangent to find the backing cut:

$$\text{Tan} = \frac{.1913}{.75} = .2551 = 14.31^\circ$$

To cut this angle, place the sole plate of your saw on the side of the rafter as in Figure 4-58. Using a table saw will be much easier.

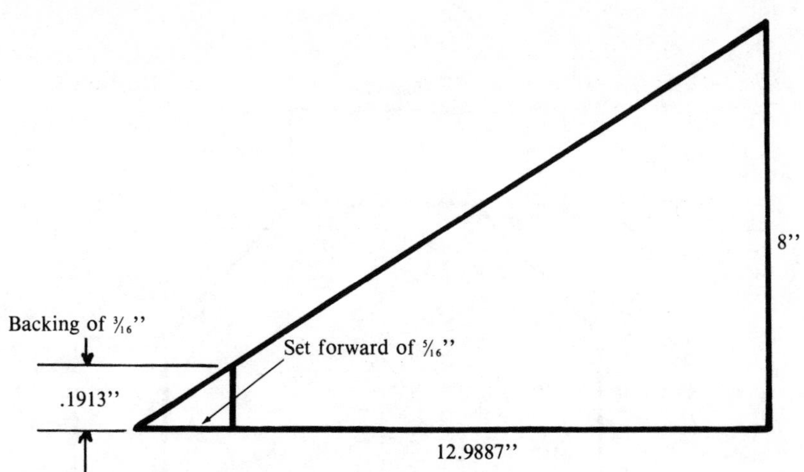

Layout method of backing or dropping
Figure 9-39

Finishing the Octagon Hip Rafters

If you're going to cut a double cheek cut birdsmouth, do that first. To make a double cheek cut birdsmouth, use a set forward of 5/16''. Set the saw to 22½ degrees, and then set the amount of blade showing below the sole plate to half the thickness of the hip. Cut one cheek only with the power saw, then chisel out the other side. Follow the directions under Figures 4-61 and 4-62.

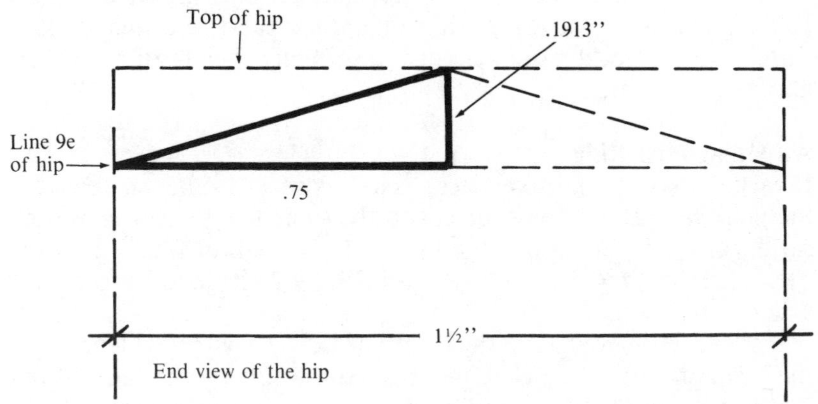

Finding the saw cut angle
Figure 9-40

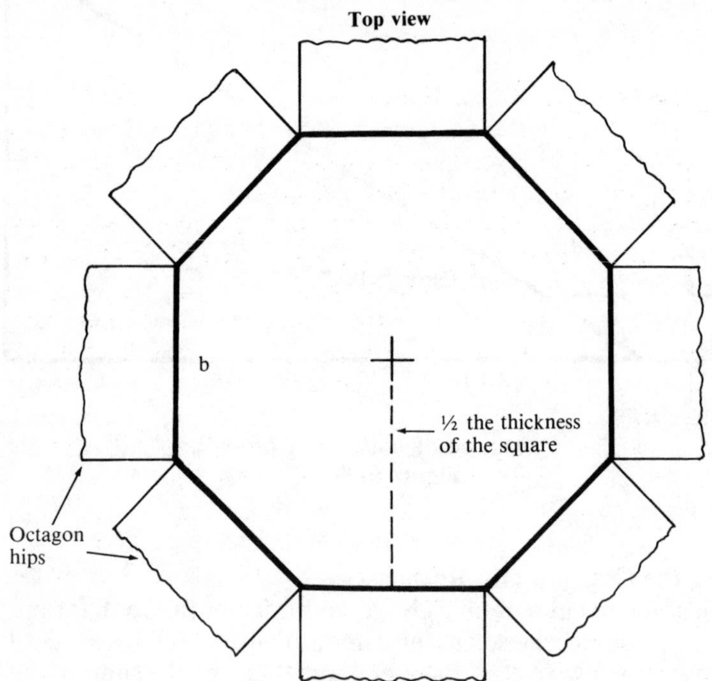

Octagon ridge piece construction
Figure 9-41

Set up 14½ degrees on a table saw and cut backing on all eight rafters. Do not remove the thin line along the top center of the rafter. If you did, it would change the effective height of the HAP and plancher cut.

An Octagoned Ridge

If we cut an octagon ridge piece, as in Figure 9-41, and use that for the peak, all rafters could be cut to the same length. Since all our rafters are 1½'' thick, we would want the side (B) of Figure 9-24 to be 1½''. How long does side (A) in Figure 9-24 have to be to make side (B) 1½'' long?

Under Step 1, we found the length of side (A) when the length of the octagon side was given. We divided the side by the tangent of 22½ degrees or 0.4142. Dividing 1.5 by 0.4142, we get 3.6213'', or 3⅝''.

Square a timber to 3⅝'' and proceed as we did in Figure 9-23.

Since the shortening for these rafters is half the thickness of the square, as indicated in Figure 9-41, shorten each rafter by 3.6213" divided by 2. That's just about 1 13/16".

Building the Octagon Rafter Plate

The side of a 3' octagon was found to be 14.9116". That's just short of 14 15/16". This will be the long-point to long-point measurement of the sole plate. Our angle is 22½ degrees.

It's best to cut the plate on a table saw. Set the blade angle carefully. Cut two pieces with the angle set, then place those two pieces on top of the drawing you have done to see if the angle is correct. Make any necessary adjustments to the saw blade angle. When the two pieces fit correctly on the drawing, you'll have two 22½-degree gauge blocks. Mark one accordingly and keep it handy in your shop.

Cut eight pieces of 2 x 6 to look like Figure 9-42. Measure the long point of each to see if it is just under 14 15/16".

On a piece of plywood, lay out a 36" square. Using the space of 36 dots from the framing square, make a layout as in Figure 9-13. Check these dimensions by measuring both (B) and (C) in Figure 9-24. Line (B) should be 14 15/16". Line (C) should be 10 9/16". Carefully cut out the octagon.

Nail the eight pieces of the rafter plate together, using 8d nails. Place the plywood octagon on top. Measure the four opposite faces. Each reading should be 36". See Figure 9-43.

Tack-nail side (1) in Figure 9-43 and set the distance to the other side at 36". Tack-nail side (2). Check and measure the other sides and adjust the sides as necessary, then nail down the plywood. Turn this over. Now you're ready to frame the roof.

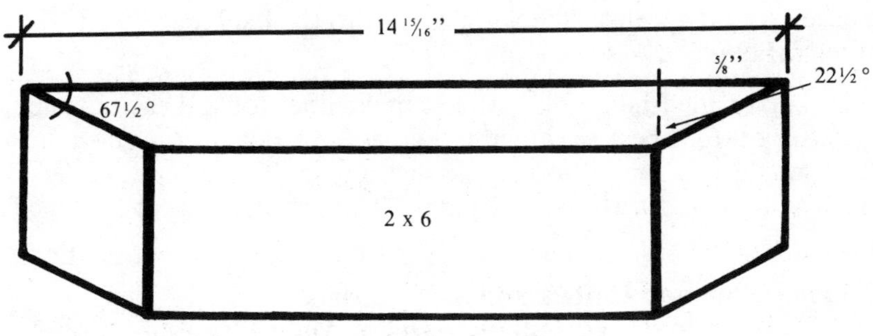

The side of a 3' octagon
Figure 9-42

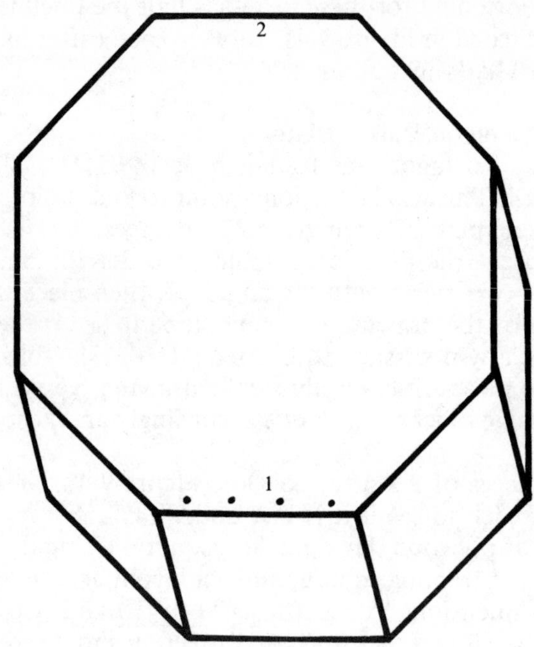

The octagon rafter plate
Figure 9-43

Framing the Octagon

Erect the first pair of rafters and nail them at the birdsmouth and the ridge line. Hold the shoulders of the second pair at the backing level of the first pair. The top part of the first pair will be sticking up above the roof plane. Frame the remaining four rafters on the model. See how well the backing cuts fit to the backing cuts of the other rafters.

On top of the first pair of rafters, make lines toward each of the #3 rafter center lines, as shown in Figure 9-44. This is the shortening triangle you made on your first drawing. These four corners must be removed to finish the plane of the roof.

Octagon Common Rafters and Jack Rafters

We've come a long way with our gazebo. And we're almost done! The last step is filling in the roof with jack rafters, common rafters and blocking.

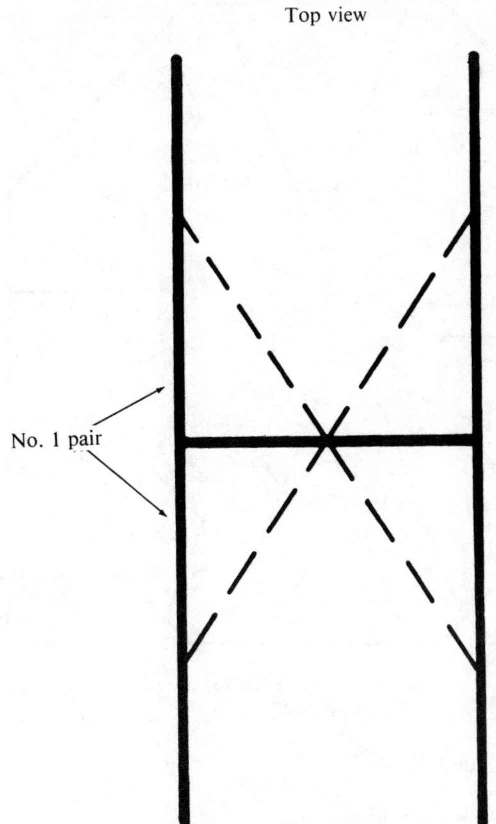

Removing corners
Figure 9-44

Octagon jack rafters fit to the sides of the building; therefore, each jack has a unit run of 12''. Each is laid out with the framing square set on the unit rise and 12''. They differ from regular hip jack rafters in that their run is not equal to their distance from the corner.

Three types of jack and common rafters are possible on an octagon roof. See Figure 9-45. Common rafter (A) is in the center of a side. Jack (B) is at an intermediate position. (In our model the center line of this jack is 3½'' from the corner.) Interrupted common rafter (C) frames against a header block.

Place these jack and common rafters on your drawing at half the size of full scale and color in the shortening and setback.

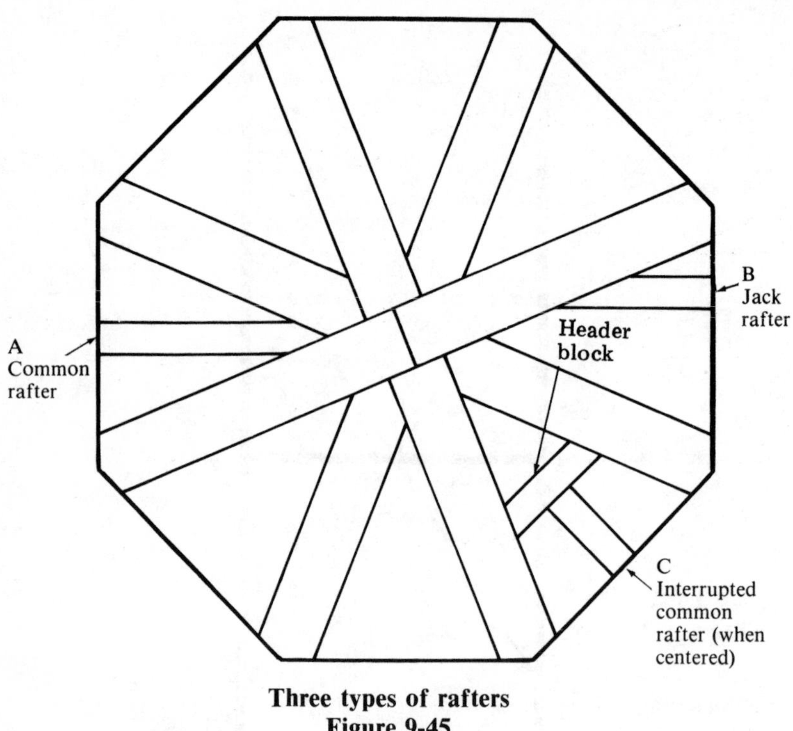

Three types of rafters
Figure 9-45

Octagon Jack Rafter Shortening

Look at the shortening triangle in Figure 9-46. Find this shortening triangle in your drawing. The amount to shorten is the hypotenuse of this triangle. There are three ways to calculate the amount to shorten. Here's the first way:

$$\text{Sin} = \frac{\tfrac{1}{2}\text{ thickness of hip}}{\text{Shortening}} \qquad \text{Shortening} = \frac{\tfrac{1}{2}\text{ thickness of hip}}{\text{Sin } 22\tfrac{1}{2}^{\circ}}$$

On your calculator, punch [.] [7] [5] [÷] [2] [2] [.] [5] [sin] and 0.3827 appears. Then punch [=] and 1.960 appears. That's equal to $1\tfrac{15}{16}$''. This is the shortening for jack (B).

The rule is that the shortening is equal to one-half the thickness of the rafter it fits against, divided by 0.3827.

Here's the second way to calculate the amount to shorten:

$$\text{Secant} = \frac{\text{Shortening}}{.75} \quad \text{or} \quad \text{Shortening} = (.75)\ (\text{Sec } 67.5^{\circ})$$

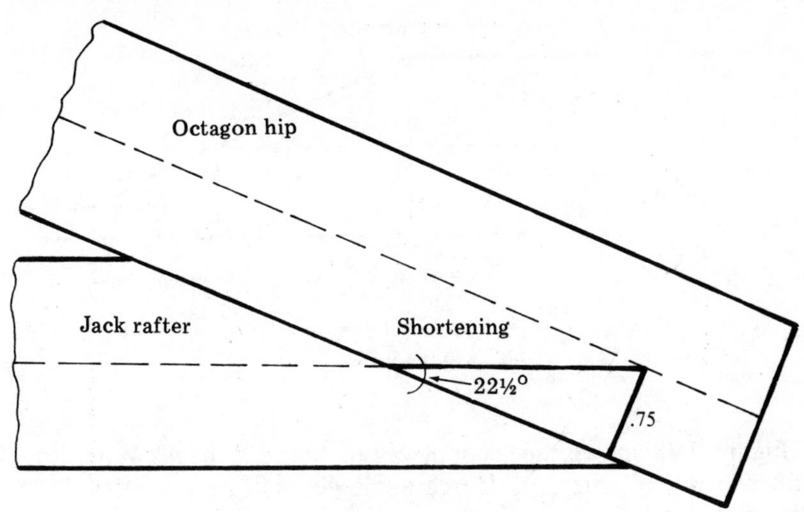

The shortening triangle
Figure 9-46

Punch [.] [7] [5] [×] [6] [7] [.] [5] [cos] [1/x] and 2.6131 appears. Then punch [=] and 1.960 appears again.

The rule could be restated to say that shortening is equal to 2.6131 times half the thickness of the rafter it fits against.

Here's the third way to figure the shortening: Measure it on your drawing. It should be 1 15/16".

Octagon Jack Rafter Setback

You can use either of two ways to figure the rafter setback. Figure 9-47 shows the first.

The shortening was of the hip using the sine function. The setback is of the jack itself and uses the tangent function.

$$\text{Tan} = \frac{\text{½ thickness of jack}}{\text{Setback}}$$

or

$$\text{Setback} = \frac{\text{½ thickness of jack}}{\text{Tan } 22\frac{1}{2}^\circ}$$

In this case, half the thickness of the jack is 0.75", divided by 0.4142 is 1.8107", or just about 1 13/16". Measure your drawing for this amount. Use the 1/2" scale for your half-scale drawing.

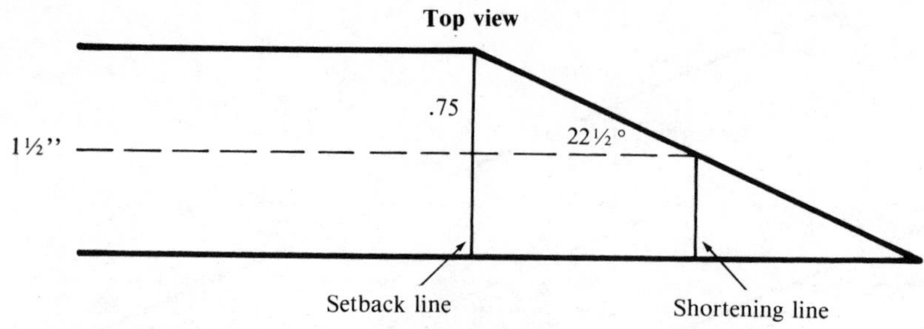

The setback triangle
Figure 9-47

Figure 9-48 shows the second way to figure setback. When the unit run is 12", the 22½-degree angle generates a rise of 4.97". This can be used as a basis for a ratio problem.

$$\frac{4.97''}{.75} : \frac{12''}{\text{run of setback}} \quad \text{or:} \quad \text{run of setback} = \frac{(.75)\ (12'')}{4.97''}$$

The ratio says: as 4.97" is to 12", so half the thickness of the rafter is to the run of the setback. Completing the math gives 1.8109, or $1\frac{13}{16}$" again.

Common rafter (A) and jack (B) in Figure 9-45 both have the same setback. But one is a single cheek cut at 22½ degrees, while the other is a double cheek cut at 67½ degrees.

The Run of Common Rafter (A) and its Layout

Since common rafter (A) in Figure 9-45 would actually extend to the center point, the run for our model common rafter will be half the span of 3', or 1½'.

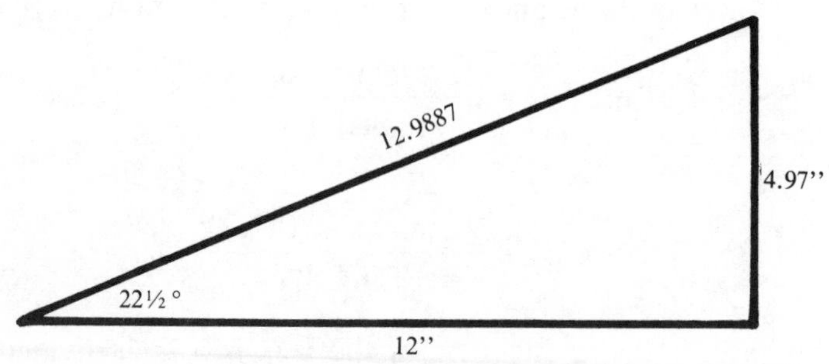

Run relationship triangle
Figure 9-48

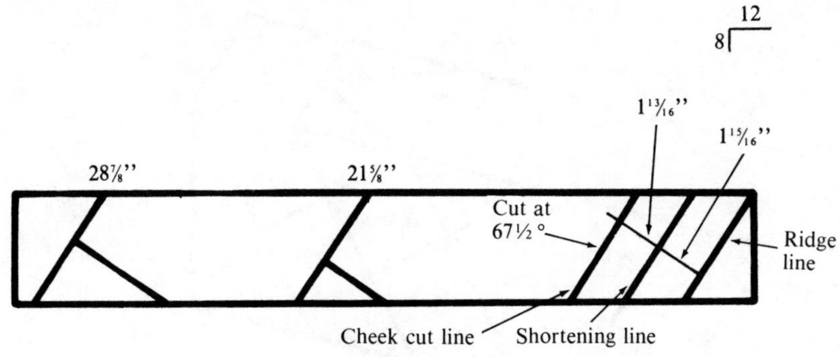

Laying out common rafter A
Figure 9-49

Layout for this rafter will be like laying out an 8 in 12 roof with a unit length of 14.42".

Crown a piece of 2 x 4 and lay the crown away from you. Set the stair gauges on 8" and 12".

Under Figure 2-17, we laid out the model common rafter for 1½' of run and a 6" overhang. Figure 9-49 shows layout for the same rafter. Add the same HAP and plancher. At the ridge line, add the octagon shortening and octagon setback, then transfer the cheek cut line to the far side of the rafter.

Making the Cheek Cuts

These cheek cuts must be made at 67½ degrees. Your power saw can't be set to this angle, so you'll have to make the cut by hand.

Square the shortening and cheek cut lines across the top of the rafter and place a dot in the center of the shortening line as in Figure 9-50. From each corner of the shortening, draw the two diagonal lines on top. With a hand saw, follow the angled line on top and the cheek cut line.

This procedure is often needed for cutting irregular rafters.

The Run of Jack (B)

For a regular hip roof, the jack's distance from the corner is the actual run of the hip jack. Jacks on an octagon roof are different. You have to divide this apparent distance from the corner by 0.4142 to find the actual run of an octagon hip jack rafter.

$$\text{Actual Run of Octagon Jack} = \frac{\text{Distance from corner}}{.4142}$$

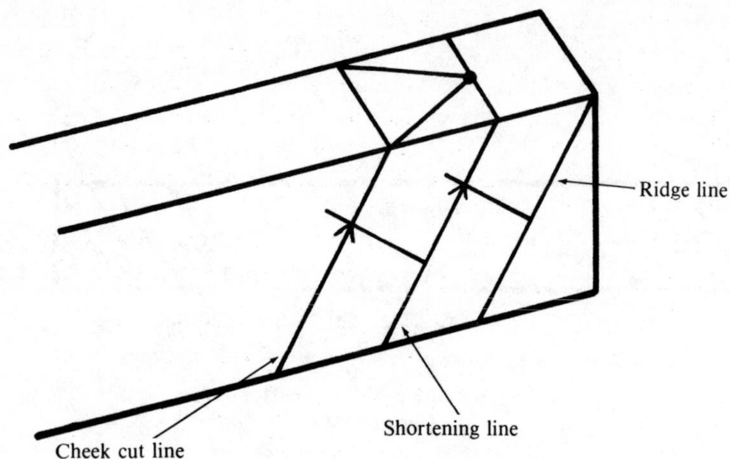

Marking for a large angle cut
Figure 9-50

On our model, the hip is to be framed 3½'' from the corner. See Figure 9-51. What is the run? Divide 3.5 by 0.4142 and you get 8.4497'', or 8$\frac{7}{16}$''. This is only the plan view run, not the length of the rafter.

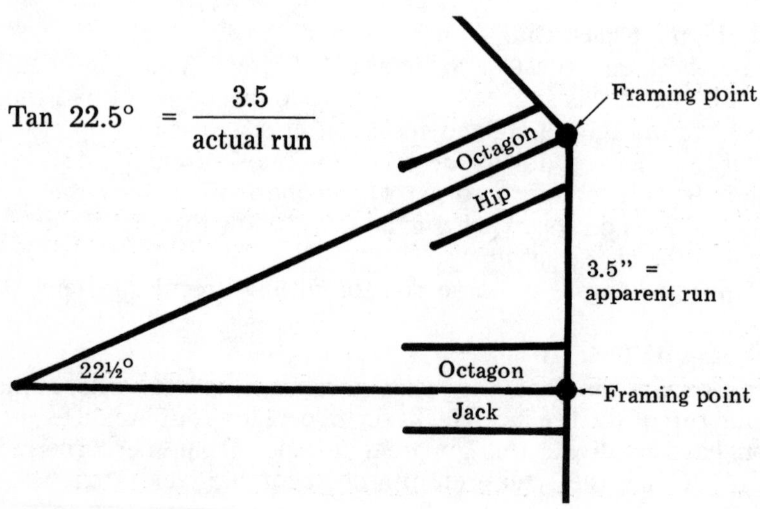

Actual run of the octagon jack
Figure 9-51

Mathematical Length of Jack (B)

Now we must figure the mathematical length of the jack. One way would be to look up the answer in a rafter length table, if you have one. Under "8 in 12 common rafters" for a rafter with 8¼" run, you'll find the number 9¹⁵⁄₁₆". Under 8½" run, the figure 10³⁄₁₆" is given. The mathematical length of our rafter lies somewhere between 9¹⁵⁄₁₆" and 10³⁄₁₆". Probably 10⅛".

Let's figure it ourselves rather than rely on the table. There are two ways to do it. Here's the first way, using the unit length and total run:

$$\text{Rafter} = \text{unit length} \times \text{total run in feet}$$

$$= 14.42 \times \frac{8.4497''}{12''}$$

$$= 10.1537'' = 10\tfrac{1}{8}''$$

Here's the second way, using the secant and actual run:

$$\text{Rafter} = \text{secant} \times \text{actual run}$$

$$= 1.2018 \times 8.4497''$$

$$= 10.1553'' \times 10\tfrac{1}{8}''$$

Laying Out Jack Rafter (B)

Crown a piece of 2 x 4 and lay the crown away from you. Set the stair gauges to 8" and 12". Mark off the length of 10⅛" as in Figure 9-52.

Since the overhang for the octagon is the same as the overhang for the gable model and both roofs are 8 in 12, both mathematical

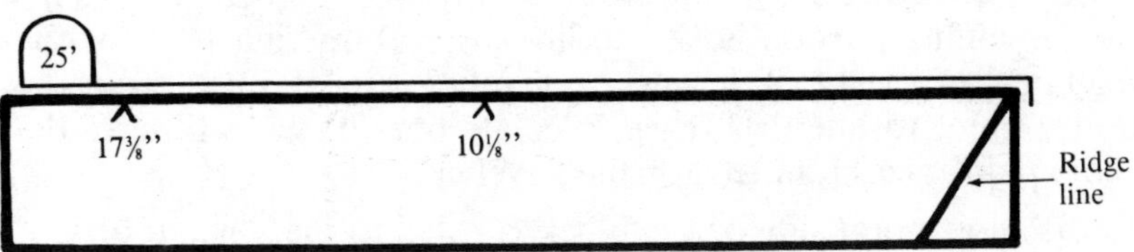

Laying out jack B
Figure 9-52

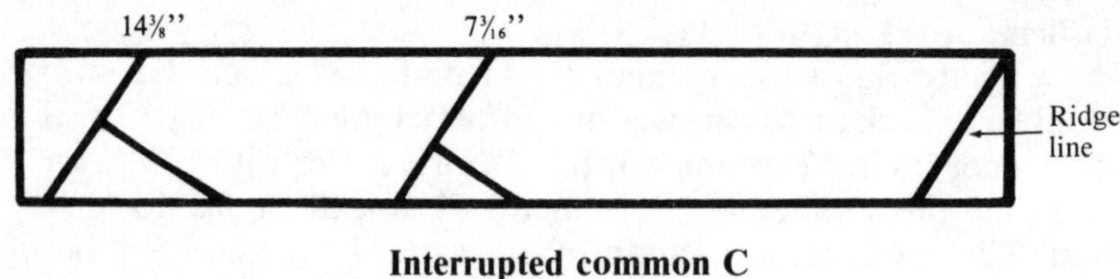

Interrupted common C
Figure 9-53

lengths will be the same. Under Figure 2-16 in Chapter 2, the mathematical length of the overhang was calculated as 7.21''. Adding this amount to the building line run of 10.1553'' gives us 17.3653'', or 17⅜''.

Mark the fascia line at 17⅜''. Add the seat cut and fascia cut. Then add the shortening and setback as in Figure 9-49. Since this is a single cheek cut jack rafter, you will not have to transfer the cheek cut line to the back for the rafter to fit in position.

Interrupted Common Rafter (C)

The run of this rafter hasn't been given, so let's say that it's 6 inches.

Since this is a common rafter at 8 in 12, the mathematical length will be found by multiplying 14.42 by 0.5'. That's 7.21'', or 7³⁄₁₆''. The overhang is also 0.5'. Therefore, the fascia cut line will be twice 7.21, or 14⅜''. Mark it as shown in Figure 9-53.

Crown a piece of 2 x 4 and lay the crown away from you. Set the stair gauges at 8'' and 12''. This rafter looks exactly like a common rafter, but there is no shortening at the ridge line. Lay out the same HAP and plancher and cut this rafter out.

Length of the Header Block

Notice the header block in Figure 9-45. What's the length of this block?

Look at Figure 9-54. Since the total run of the octagon is 1½', the remaining portion is 12'' to line (a). Subtracting 1½'' for the thickness of the block leaves a remaining run of 10½'' from the center point to line (b). If we calculate line (b), we will have the short point measurement for the bevel cuts.

The short point side of the block is equal to the tangent of 22½ degrees times the run times 2. See Figure 9-55. This dimension is line (b) in Figure 9-54, from hip center line to hip center line. Of course, the shortening on each side still has to be subtracted.

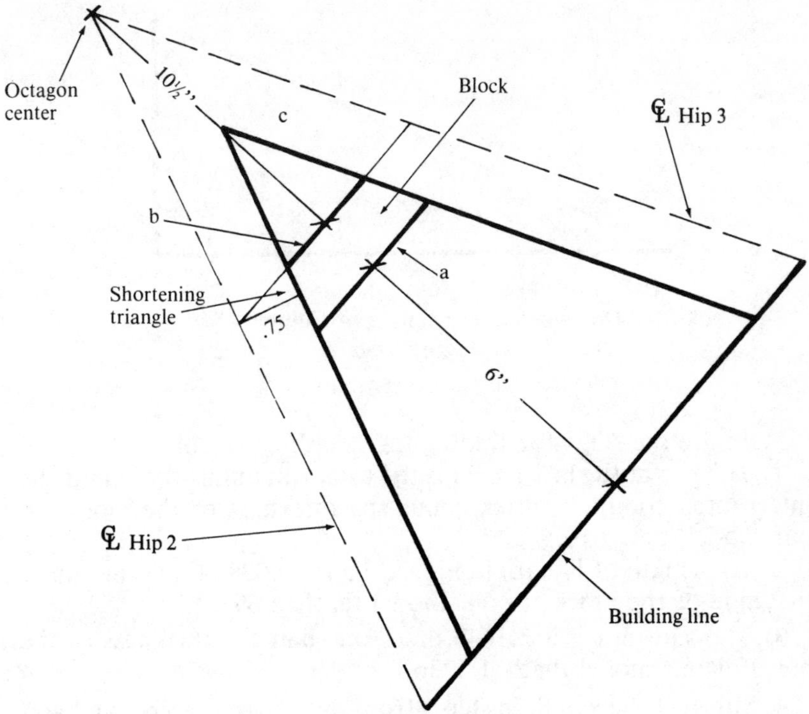

Calculating the header block
Figure 9-54

The Block Shortening Triangle

Figure 9-56 shows the shortening triangle from Figure 9-54. The shortening for both sides is equal to the secant of 22.5 degrees times 0.75 times 2.

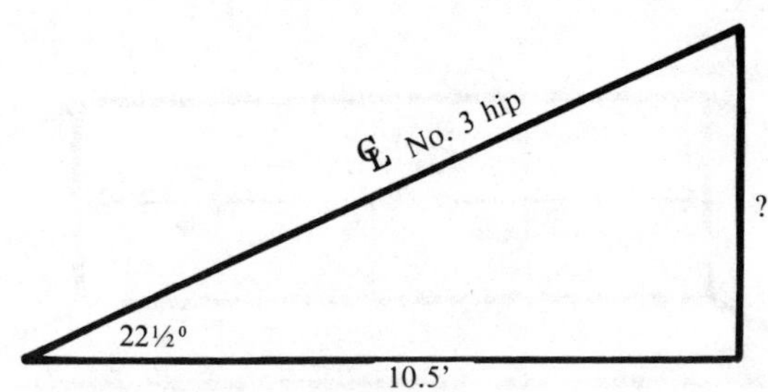

Triangle C of Figure 9-54
Figure 9-55

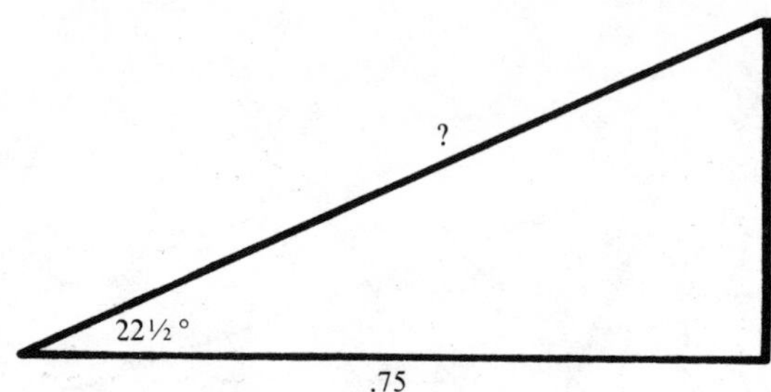

The shortening triangle of Figure 9-54
Figure 9-56

Here are the steps for finding the length of the block:

1) The run at the block equals the total run minus the run of the interrupted common rafter minus the thickness of the block. In our model, that's 10.5".

2) The length of line (b) in Figure 9-54 is 0.8284 times the run at that side of the block. In our model that's 8.6985".

3) The shortening is 2.1648 times one-half the thickness of the hip. For our model that's 1.6236".

4) Subtract the length in step 3 from the length in step 2 and you have 7.0749". That's about 7 1/16", as shown in Figure 9-57. That's your short point measurement. Cut each end at 22½ degrees.

The Final Cut

Cut the top of the block to the same angle as this 8 in 12.9887 roof. See Figure 9-58. The angle is just past 31½ degrees. Frame the block along with the interrupted common rafter (C). Congratulations. You've just become an experienced octagon roof builder!

Study Figure 9-59, and remember these basic relationships any time you're framing an octagon by this method.

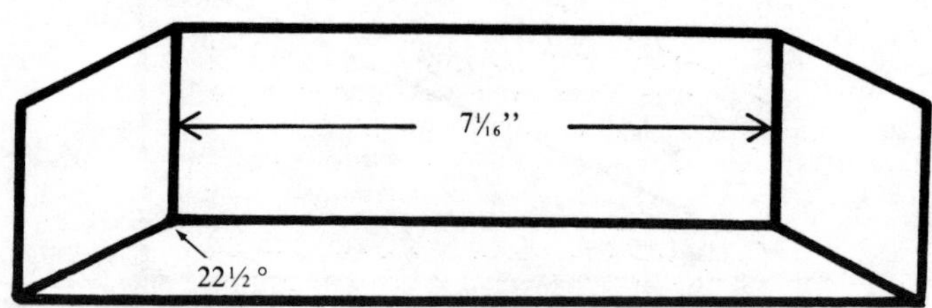

The header block
Figure 9-57

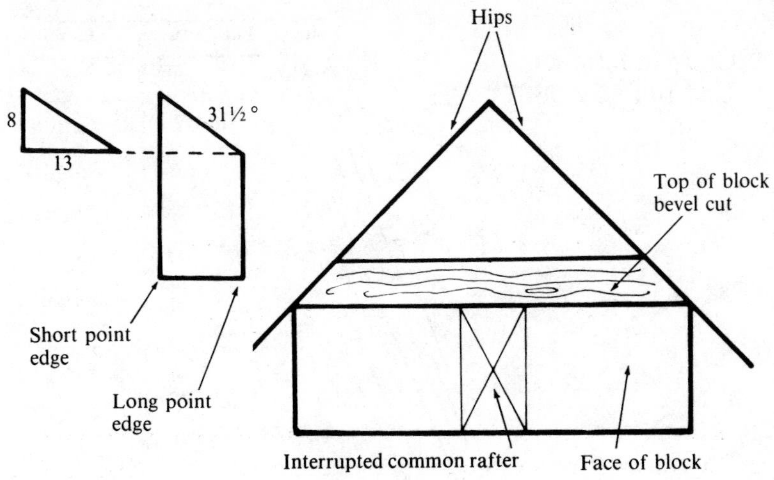

The top cut
Figure 9-58

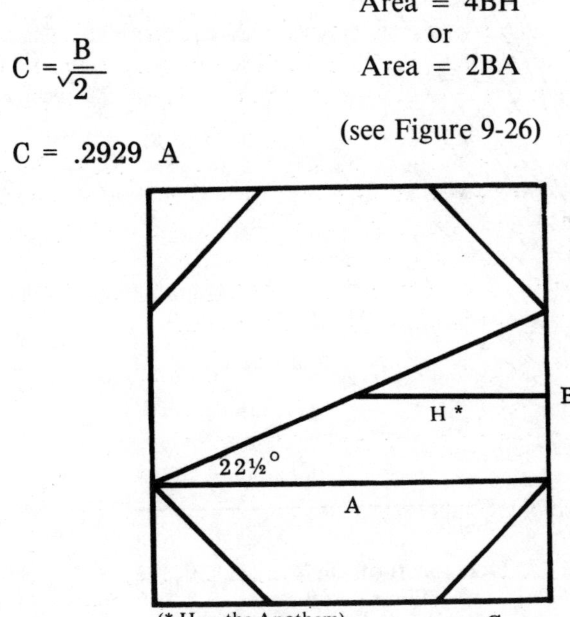

The basic relationships
Figure 9-59

Octagon rafters
Unit run = 12.9887"

Rise	Pitch	Degrees	Secant
24	1	61.58°	2.1010
23	23/24	60.55°	2.0336
22	11/12	59.44°	1.9669
21	7/8	58.26°	1.9010
20	5/6	57.00°	1.8360
19	19/24	55.64°	1.7720
18	3/4	54.19°	1.7089
17	17/24	52.62°	1.6471
16	2/3	50.93°	1.5866
15	5/8	49.11°	1.5276
14	7/12	47.15°	1.4703
13	13/24	45.02°	1.4148
12	1/2	42.73°	1.3615
11	11/24	40.26°	1.3104
10	5/12	37.59°	1.2620
9	3/8	34.72°	1.2166
8	1/3	31.63°	1.1745
7	7/24	28.32°	1.1360
6	1/4	24.79°	1.1015
5	5/24	21.05°	1.0715
4	1/6	17.12°	1.0463
3	1/8	13.01°	1.0263
2	1/12	8.75°	1.0118
1	1/24	4.40°	1.0030

12.9887"

0"

Octagon roof angle chart
Figure 9-60

10

A Simple Irregular Roof

A builder you haven't worked for before called yesterday and asked that you return his call. You call back, and he explains that he has a problem with the roof on one of his jobs. The lead carpenter on the crew that does most of his framing can't come up with a solution. But someone on the carpentry crew mentioned your name while they were discussing the problem. You agree to have a look at the plans in the builder's office the next day.

When you arrive at his office in the morning, he explains the problem: He's building a small cabin on a hillside lot overlooking the lake. Because the lot is narrow, the maximum width of the cabin is 21' by 35'. The entire development has a covenant that guarantees a view to each owner. Because of this covenant, the pitch of the roof can't exceed 8 in 12. The plans show a hip roof with a uniform overhang of 3'6''. Up to this point, it was just a standard hip roof job. And that's the way the builder bid it.

But here's the catch. As they were starting to frame the roof, they discovered that a regular 8 in 12 roof would limit the attic space where solar hot water equipment has to be installed. The heating equipment needs full attic space under the ridge at least three and a half feet beyond where the end of the ridge would normally fall!

Here's the builder's question. Can you design an irregular hip roof that gives him enough space for the solar heating system?

You agree to have a look at the problem and answer the question by tomorrow. The builder makes a copy of the plan sheets you'll need and notes space requirements for the solar heater on the plan copies.

Now you have to figure a way to fit that heating system under an 8 in 12 hip roof!

Drafting the Model and Scaling Measurements

You start out by making a drawing of the roof, Figure 10-1. The 6' by 4' dimensions are 1/7th of the actual roof dimensions. The mathematical length of the ridge is 5' minus 3', or 2'. That's 14' on the actual roof, not quite enough space for the solar tanks. After considering the space lost from the thickness of the rafters, you calculate that the ridge has to be 3½' longer, 17'6" overall. If the ridge could be stretched out another 3½', everything would fit fine.

In Figure 10-2, we've extended the ridge to the right an extra 6". On our 1/7th drawing, 6" equals 3'6". That creates enough room for the solar equipment. But now the hips at the right aren't at 45 degrees to the building lines. Notice also that these hips don't pass over the corners of the building. Three sides of this roof will be 8 in 12, but the fourth side is irregular. We don't know the pitch yet, but it isn't 8 in 12. And we won't find lengths for these rafters in any rafter table.

Let's start by putting a roof like the one shown in Figure 10-2 on our 3' by 5' model. We'll avoid almost all mathematics by simply drawing a plan view (top view) of the roof, and some sectional or side views of the rafters. Then we'll scale measurements directly off the drawings.

Near the end of the chapter you'll find two blank cutting lists, Tables 10-47 and 10-48. Use lists like this when calculating rafter lengths and shortening for an irregular roof. Filled out cutting lists for this chapter are at the end of the book.

The Principle of this Drafting Method

To understand this principle, let's say we have gone to a remote location to cut a roof. Inadvertently we left at home our framing square, our calculator, and our rafter tables. Luckily enough we brought our wits.

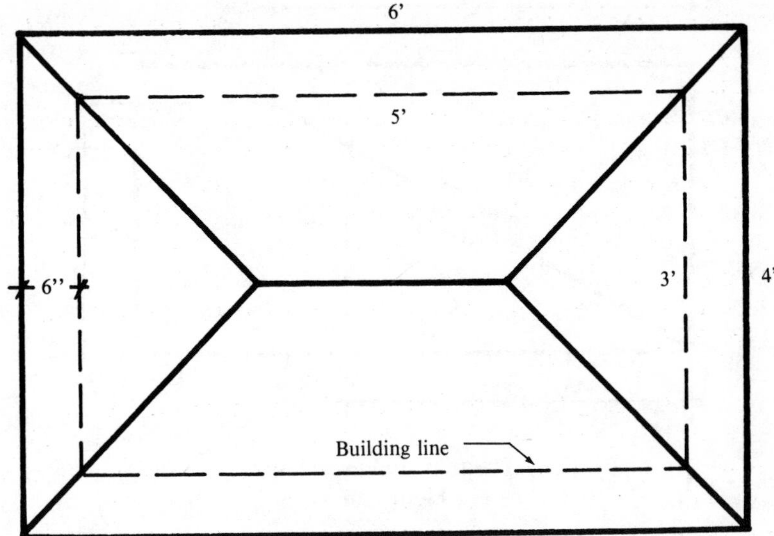

A 1/7 scale model regular hip roof
Figure 10-1

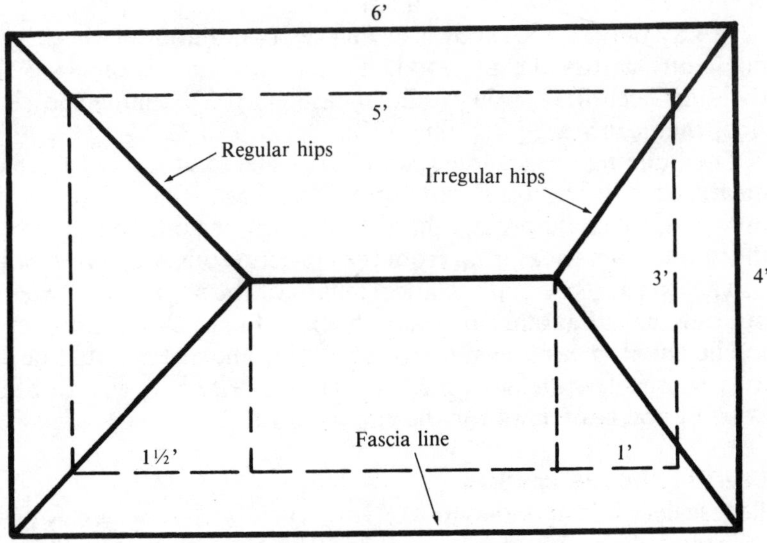

The model with an irregular roof
Figure 10-2

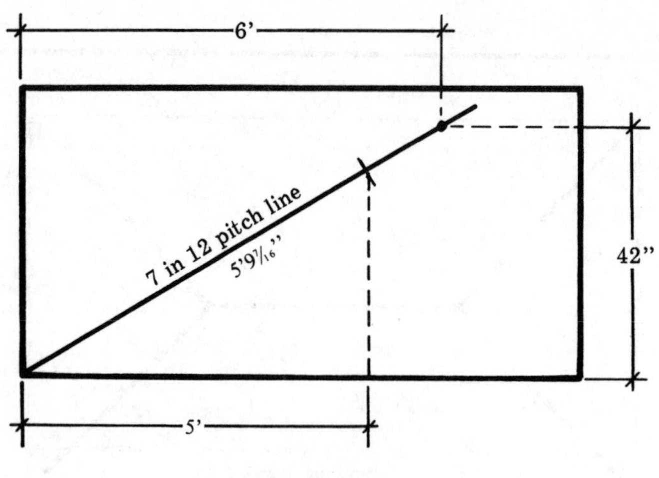

The plywood calculator
Figure 10-3

There are some 4' x 8' sheets of plywood on the job site. One of them will make a fine calculator. For this example, we'll use a simpler roof to demonstrate how the "plywood calculator" works. Follow along on Figure 10-3.

The building in this example is to have a 7 in 12 pitch gable roof. A 6' run at 7:12 would have a rise of 42'', since 6 times 7'' equals 42''. Lay out a dot as shown in Figure 10-3, and draw the 7 in 12 pitch line across the plywood. This is our calculator. Had the building been 4:12, you would measure up 32'' along the right side, then draw the pitch line. (8' at 4'' equals 32''.)

The building has a gable span of 20'. The total run is 10'. If we measured half the total run (or 5') out from the corner of the plywood, we could project this distance upward until it intersected the 7 in 12 line. Measuring from the corner to this projection point will give us *half* the mathematical length of the rafter, since we laid out only half the total run. Refer back to Figure 2-5 in Chapter 2.

The length measured should be 5'9 7/16''. The rafter would be cut to twice this length, or 11'6 7/8''. Using this principle, we can begin drafting the two views for the cabin roof.

Drafting the Rafter Plate

The scale of 3'' to the foot gives very good accuracy for the carpenter. Since our model is only 3' by 5', each view will fit on 24'' x 18'' paper. Some parts of this drawing will be done in the 3'' scale and some parts will be drawn full scale.

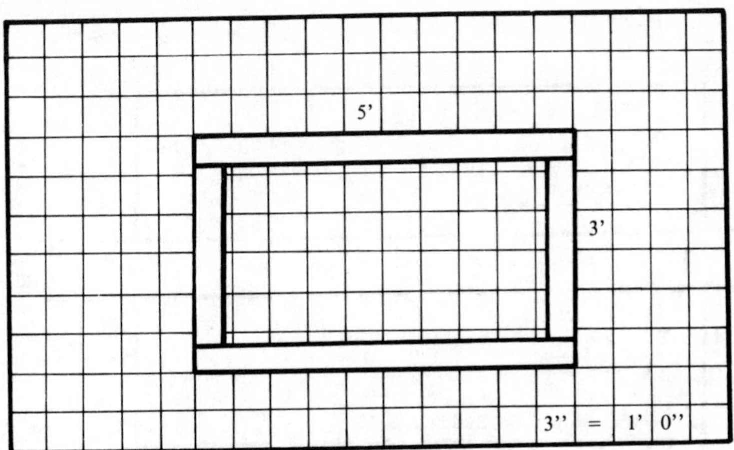

Beginning the layout
Figure 10-4

Begin by laying out the perimeter of the rafter plate in the center of the paper. The square will measure 15'' by 9''. See Figure 10-4.

Although the actual model rafter plate is only 1½'' wide, make the drawing with a normal 3½''-wide rafter plate. On a 3'' scale, each mark on the architect's ruler equals 1/8''. The numbers in the center are for both the 3'' and 1½'' scale. Cover the numbers that don't apply with masking tape to make the ruler easier to read.

The Regular Hips and Framing Point

The regular hip end will be on the left side of the paper, as in Figure 10-2. Measure to locate the framing point on your drawing and then draw the dotted center line of each regular hip.

Drawing the Ridge and End Rafters

Through the framing point, draw the dotted center line of the ridge and end commons. Extend this line 6'' (in the 3'' scale) to either side of the rafter plate to represent the same 6'' overhang used in previous models. Now change scales and draw in the full 1½'' thickness of these members, as shown in Figure 10-5.

Draw another dotted line, perpendicular to the first one, through the regular framing point to represent the center line of the common rafters at that end. See Figure 10-6. Extend this line for the 6'' overhang. Add the full-scale 1½'' thickness to these commons also. Be careful not to draw across the ridge on the inside of the commons. But draw straight through on the hip side to establish the end of the ridge.

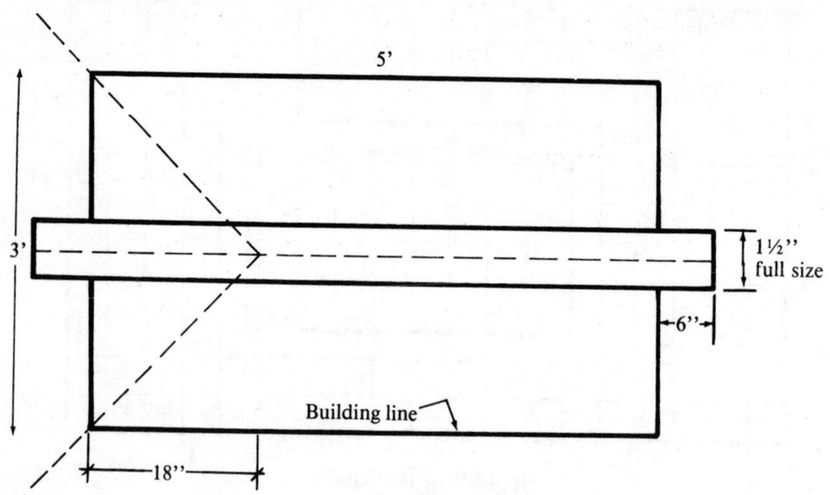

Drawing the ridge and end commons
Figure 10-5

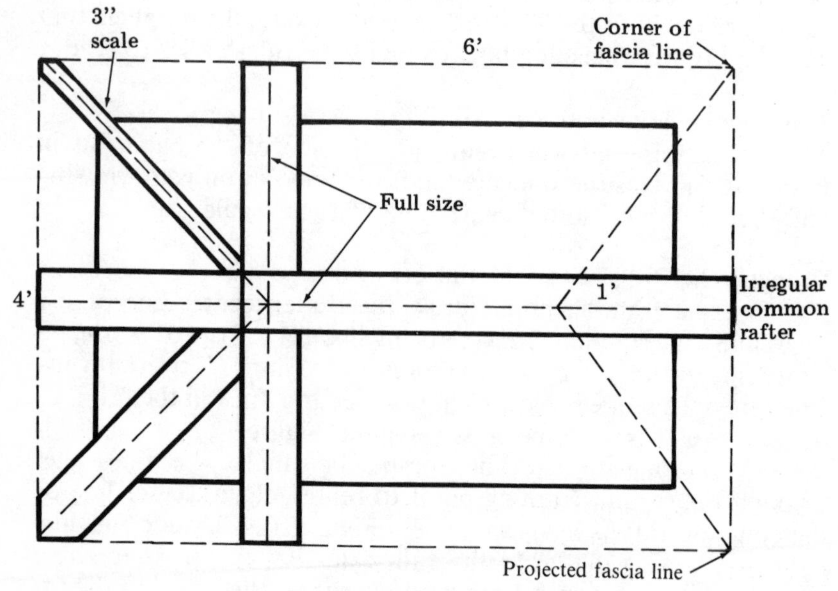

Locating the projected irregular hips
Figure 10-6

Also draw the lower regular hip in full scale. The upper hip can be drawn to the 3" scale. Project the fascia line, as in Figure 10-6, to complete the tail cut of these regular hip rafters.

Finding the Irregular Framing Point

Look back at Figure 10-2. The regular hip is 1½' from the building line on the left side of the building. The irregular hip is 1' from the other end. That makes the spacing between the two framing points 2½'. Measure from the building line, either 1' on the irregular side or 4' on the regular side, and place the framing point on the ridge center line.

Drawing the Irregular Hips

Project the fascia line to both remaining corners. Now draw a dotted line from each corner of the fascia line to the irregular framing point, as in Figure 10-6.

Notice that these lines do not pass over the corner points of the building. In unequal pitch roofs, the unequal hips bear on the corners of the building only if there is no overhang.

When an overhang is involved, the fascia board must be held at a constant level, even on the side of the irregular hip. Here's how to do this. Move the steeper pitched irregular hip off of 45 degrees, so that it meets the fascia board at the right level.

One important function of the sectional views is to establish a level line for the plate, a level line for the peak, and a level line for the fascia. All of the framing members must fit within these lines because all of these points must be at one established level for the roof. That's why an irregular hip roof isn't most roof cutters' cup of tea.

Indicate the thickness of the lower irregular hip in full scale and the thickness of the upper hip in 3" scale. Draw line (a) of Figure 10-7. It shows the separation between the ridge and the irregular common rafter.

Locating the Other Members

Back at Figure 2-34, in Chapter 2, we had to divide space into even parts. We found the material remaining between center lines and then divided by the number of parts needed. The figures used weren't center lines of rafters. They were layout lines, or lines showing the beginning space the rafter was to occupy.

On this model we must calculate the length of jack rafters from the center line measurements, which are the framing point lengths

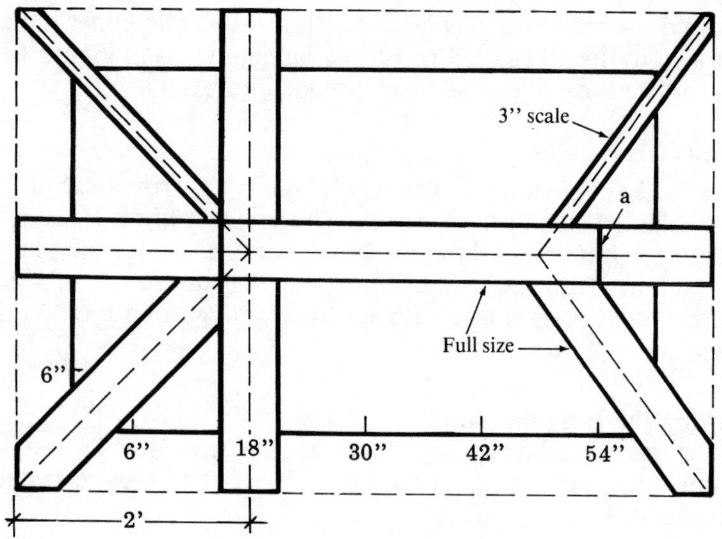

Laying out the plate
Figure 10-7

of the rafter. Here the center line measurements are more useful to us than the layout lines of Figure 2-35.

Look at the common rafter in Figure 10-7. Notice that the center line of the common rafter and the end of the regular hip are 2' apart. Draw a center line for the jack rafter 6'' in from the building line, as shown in Figure 10-7. That equally divides the space between those rafter tails.

The spacing between the center line of the common rafter and the center line of the hip jack will be 12''. Adding 12'' to the center line of the common rafter, which is at 18'', gives a measurement of 30''. Lay out this center line at 30'', and add center lines at 42'' and 54''. These are all in 3'' scale on your plan.

Draw in the other two hip jack rafter center lines at the span of the regular hip end. You can do this either by projecting the intersection at the hip center line, or by measuring over 6'' from each side.

At the irregular end, also place center lines 6'' from each building line. At the regular end, with the hips at 45 degrees, the jack rafters line up opposite each other on the hip. But at the irregular end, the two jacks will not meet on opposite sides of the hip.

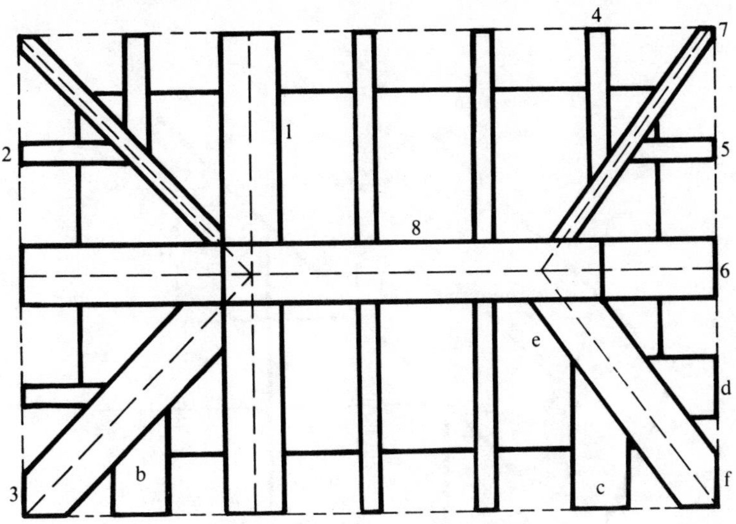

The scale of the various members
Figure 10-8

Look at Figure 10-8. All but three of the rafters on the lower side of the ridge are drawn in full size. The other members are drawn in 3'' scale. Erase the plate lines from inside the rafters to make your drawing more realistic.

Finding the Cheek Cuts and Shortenings

At (b), (c), (d), (e) and (f), construct a setback line as shown in Figure 10-12. We will measure the shortening and setback for each of these rafters.

There are two cutting lists near the end of this chapter (Figures 10-47 and 10-48). The line numbers used in these lists correspond to the rafter numbers in Figure 10-8. As we calculate rafter lengths for this roof, enter your measurements on the cutting lists.

Rafters 1 and 3

The lengths of rafters 1 and 3 are easy to calculate. You mastered this information back in Chapter 2 and Chapter 4.

Rafter 2

Rafter 2 is a regular hip jack rafter. There are four of them on this roof. Using the 3'' scale, measure the building line and the fascia line runs. Enter the lengths on your cutting list.

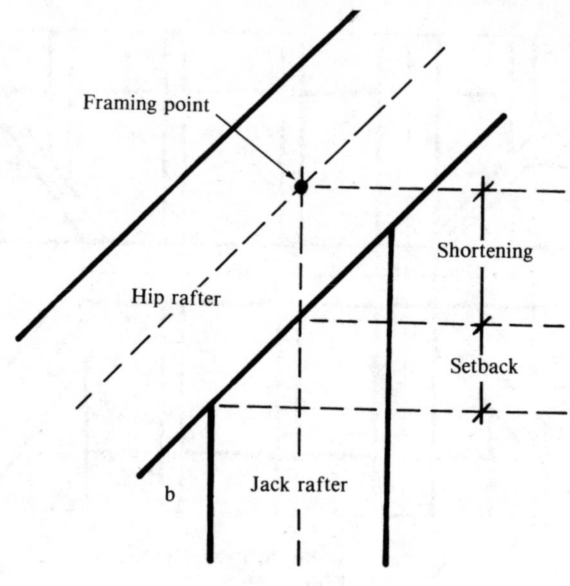

Drawing the setback
Figure 10-9

Measure the shortening and setback at (b) as shown in Figure 10-9. This part of the drawing is in full scale. Use these dimensions directly without adjusting for scale.

Rafter 4

Rafter 4 is a regular 8 in 12 hip jack rafter in some respects. But it differs in actual run, shortening, setback, and the angle of the cheek cut. Using the 3" scale, measure the building line and fascia line runs. Then, using the full scale, measure the shortening and setback.

Using a protractor, measure the angle of the cheek cut. See Figure 10-10. Enter these figures on line 4 of the cutting list.

Rafter 5

Rafter 5 is the irregular side jack rafter. As jack 4 is related to the common rafter, so this jack is related to the irregular common rafter.

Measure the building line run in the 3" scale. Add 6" for the fascia run. Using the full scale, measure the shortening and setback. Using the protractor, find the angle for the cheek cut. Enter these figures on line 5 of the cutting lists.

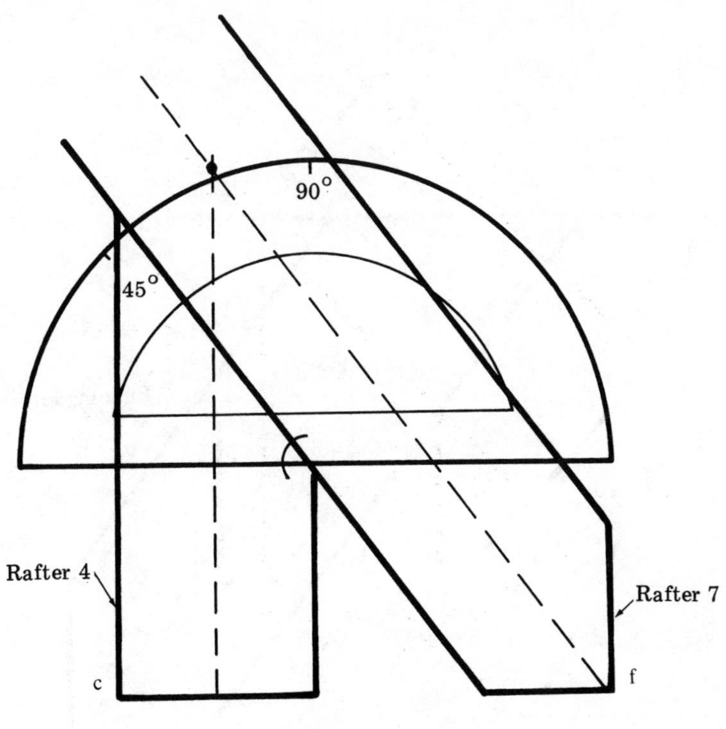

Using the protractor
Figure 10-10

Rafter 6

The run of the irregular common rafter is obvious when looking at Figure 10-2. The shortening is the distance from the irregular framing point to line (a) in Figure 10-7. Measure this distance in the 3'' scale. Enter the amounts on line 6 of the cutting lists.

Rafter 7

Measure this irregular hip rafter length from the framing point to the fascia, using the 3'' scale. Remember to measure to the intersection of the building line and the rafter center line.

Look at Figure 10-11. At (e), measure the shortening and setback for the ridge cut. At (f), measure the setbacks for the fascia cuts. Use full scale for these measurements.

Measure the cutting angle for each of these cuts with your protractor, and enter all figures on your cutting list. List all tail cut measurements on the line with the asterisk at the bottom of Figure 10-47.

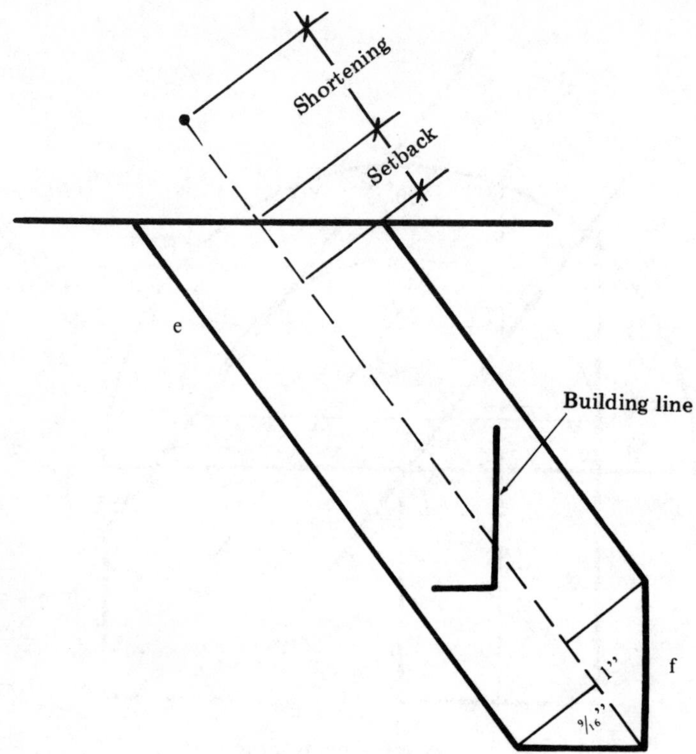

The irregular hip
Figure 10-11

The Sectional Lines
Figure 10-12 shows three section lines. These lines cut the plan at critical points where more information is needed. From the viewpoint of each section line, we'll construct section views. These three section views will help us calculate the dimensions needed to cut this roof.

Label the three lines "Section A-A: Regular Common Rafter", "Section B-B: Irregular Common Rafter" and "Section C-C: Irregular Hip" as in Figure 10-12.

Drafting the Sectional View: A Synopsis
The completed section view is given in Figure 10-13. All the lines are numbered so that it will be easy to construct this drawing. We'll use numbers in parentheses in the text to refer to circled numbers in the figures that follow. Line (1) represents the rafter

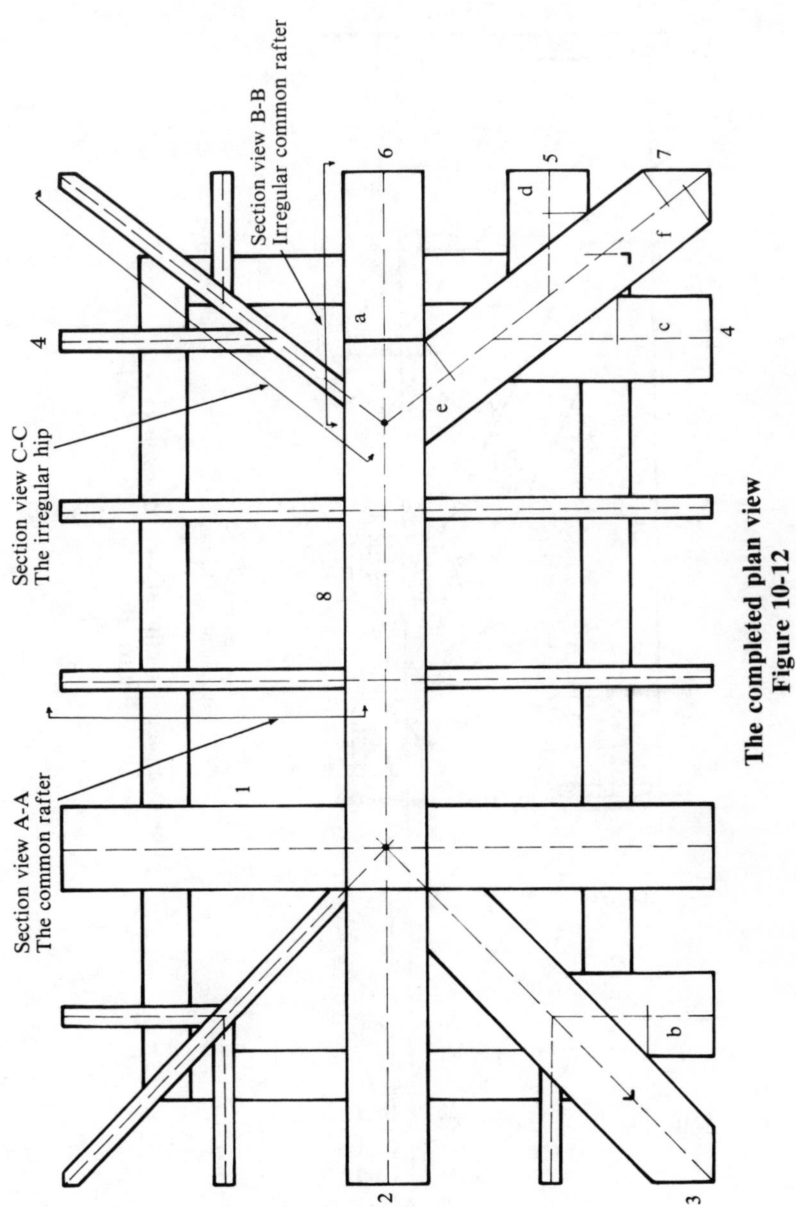

The completed plan view
Figure 10-12

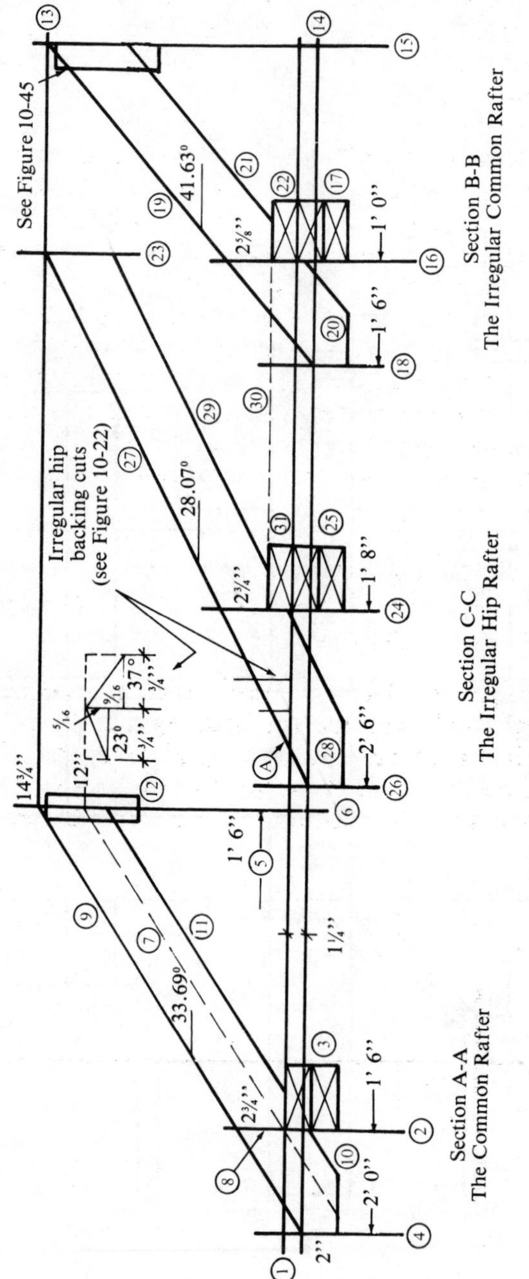

The completed sectional view drawings
Figure 10-13

plate. It's the main reference line. When a common rafter is drawn, as in Section A-A, then the ridge line (13) and the top edge of the fascia line (14) are established. These three lines (1), (13) and (14), are then drawn across the page and control the dimensions of the irregular parts of this roof.

Each irregular member must now be fit between the ridge line (13) and the fascia line (14) in order to close the roof and to keep the fascia board at the same horizontal level on all four sides.

To accomplish this, a ridge center line for each type of irregular member is arbitrarily established. These lines are (15) and (23) in Figure 10-13. Next: from Figure 10-12, the plan view run to the fascia line of each type of irregular member is found and this distance is laid off toward the left from line (15) and (23) respectively. This distance must be laid out along the fascia line (14). The top line of each irregular rafter, line (19) and (27), is then drawn from the fascia line intersect to the ridge intersect. This establishes the irregular members. The next sections will explain all this in detail.

Section A-A: The Common Rafter

Make a drawing of your own, using the same scale of 3'' equals 1'. You'll need 24'' x 10'' paper.

Lines (1) through (4)— Draw the rafter plate line (1) all the way across the page, and then add the building line (2) as in Figure 10-14. Draw the rafter plate and the top plate (3) to scale.

For line (4), measure back the amount of the overhang and draw a perpendicular line. The common rafter tail will end where it

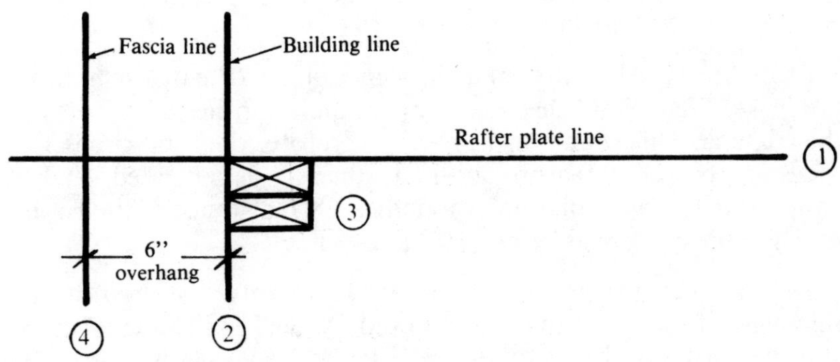

Section A-A: Lines (1) through (4)
Figure 10-14

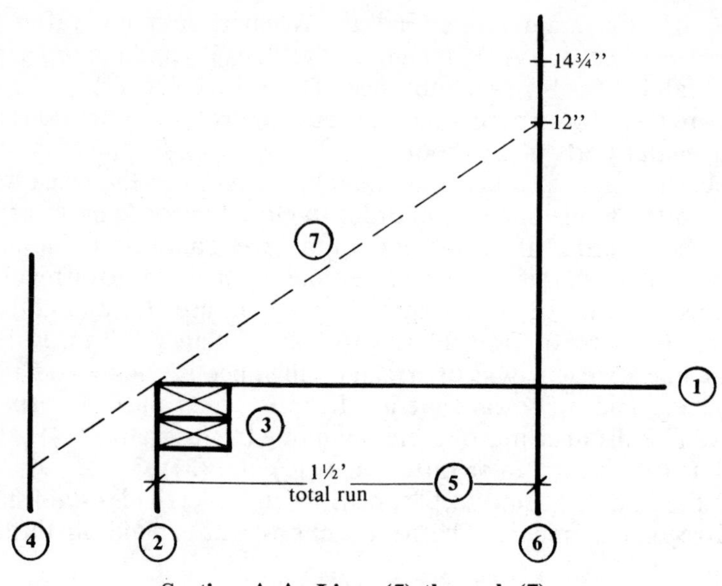

Section A-A: Lines (5) through (7)
Figure 10-15

meets the fascia. That's below the level of the rafter plate, line (1). That intersection will establish the fascia line (14).

Lines (5) through (7)— Measure to the right 1½' from the building line. That's the run of the common rafter. In Figure 10-15, we've measured along line (5) 1½' from the building line to locate ridge line (6). That's the center of the building and the ridge. This is the line we call the ridge line ..... the first line we put on the corner of the 2 x 4 during the layout of the rafter.

From line (1), the rafter plate line, measure up the mathematical total rise. That establishes the slope or angle for the rafter part of the triangle. This is an 8 in 12 roof. Therefore, the total rise is 1½ times 8'', or 12''. Measure and mark the 12'' as indicated. Make another mark above that point to indicate the distance of the HAP (2¾''). This mark will be at 14¾'' above line (1).

Dotted line (7) in Figure 10-15 represents the two-thirds thickness of the rafter that was allowed for the HAP. Lines (7), (1) and (6) form the basic total-length triangle. Therefore, line (7) must connect to the mathematical total rise of 12'' and the corner of the rafter plate. This is true only for this first section view.

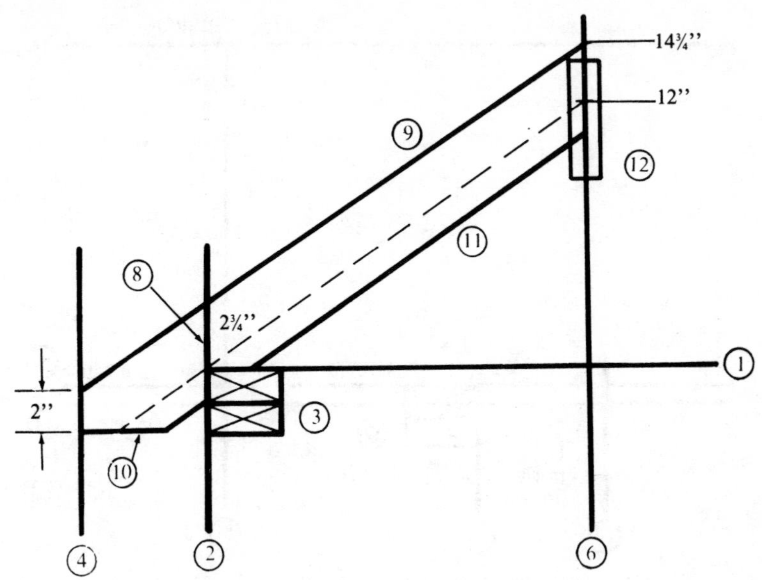

Section A-A: Lines (8) through (12)
Figure 10-16

Lines (8) through (11)— Notice in Figure 10-16 that line (8) is an extension of line (2). This extension of 2¾'' is for the HAP. You've already allowed for the HAP along line (6). Connect these two HAP points to form line (9). Be sure to extend line (9) all the way to line (4), the fascia line. The intersection of lines (9) and (4) is the beginning point for line (14), the fascia level line.

From the point where lines (9) and (4) intersect, measure down 2''. That's the plancher cut. Then draw line (10), the plancher line.

Measure perpendicular from line (9) the thickness of the rafter, in this case 3½''. Use the 3'' scale. Draw in line (11) as indicated.

The process we've used here is the same process you would use to begin any irregular roof sectional drawing. We'll use it again in the next chapter.

Line (12): The Ridge— Draw the thickness of the ridge in 3'' scale. Under Figure 7-23, in Chapter 7, we calculated the loss in ridge height for an 8 in 12 roof. Here we'll leave the top of the ridge a half-inch below the ridge line.

Lines (13) and (14)— Project the ridge line across the page and label it line (13). Note that the ridge line isn't the same as the actual

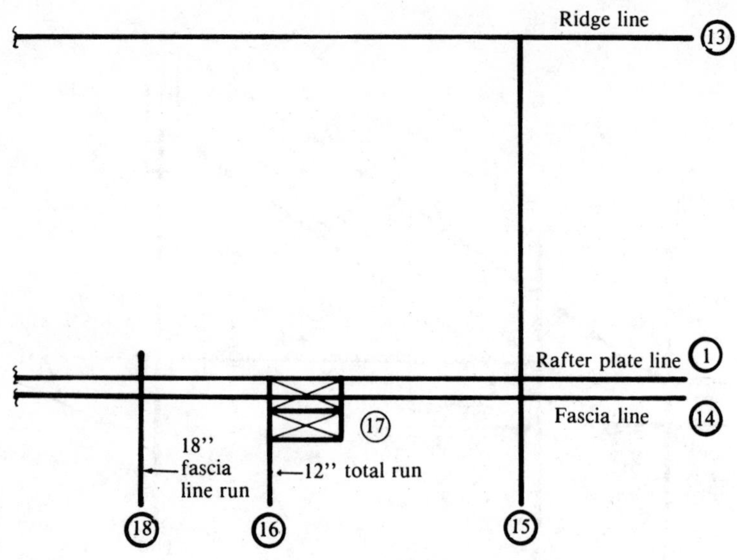

Section B-B: Lines (15) through (18)
Figure 10-17

height of the ridge. Then with a red pencil, project line (14), the fascia line, across the page, as in Figure 10-13.

Section B-B: The Irregular Common

All irregular members must now fit between the projected lines (13) and (14). That makes our layout possible. Each irregular member will be placed between these lines according to the length of its actual run to the fascia line.

Lines (15) through (18)— Look at Figure 10-17. We've drawn reference line (15) to represent the ridge center line. Draw this line, in scale, 4'6" from line (6). Along the rafter plate line (1), measure back from line (15) the amount of the building line run of the irregular common rafter. Referring to Figure 10-2, the building line run of the irregular common rafter is one foot. This locates the building line for this rafter. Add the two plates (17) as before. Line (18) is the fascia line. It's 6" away from the building line (16).

Measure 1½' from line (15) and draw line (18) so that it crosses the fascia level line (14). We'll use the intersection of lines (14) and (18) to establish one end of line (19). See Figure 10-18.

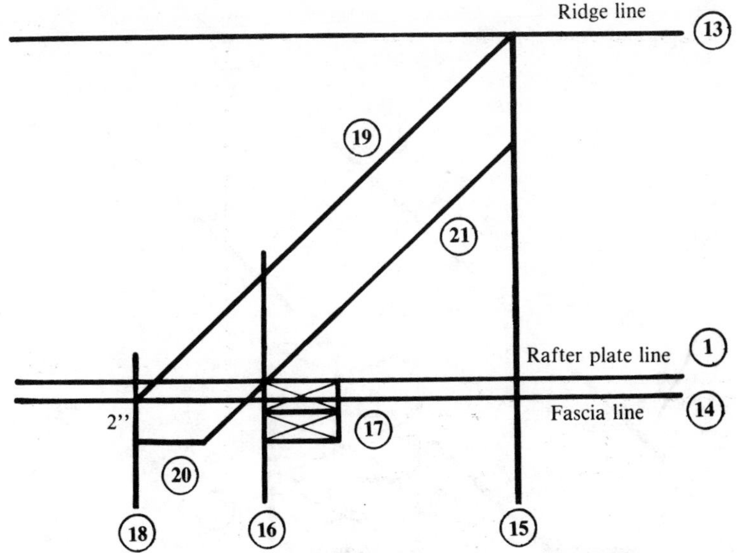

Section B-B: Lines (19) through (21)
Figure 10-18

Lines (19) through (21)— The other end of line (19) is at the intersection of (13) and (15). Draw in line (19). In Figure 10-15, line (7), the dotted line, connected the mathematical total rise and the rafter plate line. In Figure 10-18, line (19) connects the fascia and ridge intersections. All remaining views will also connect the ridge and fascia intersections.

Measure along line (18) the amount of the plancher cut and draw in (20), the plancher line. A plancher line could have been projected across the paper, but that would have cluttered the drawing unnecessarily.

Draw in line (21) using the same procedure as you did for line (11).

Line (22): Raising the Plate— Notice in Figure 10-19 that line (21) hardly touches the rafter plate. You can't make a birdsmouth on a rafter like that. But if we added a 2 x 4 to the top of the rafter plate, we could cut a birdsmouth and still leave about two-thirds of the rafter intact for good strength.

Draw in the third 2 x 4 at (22). This is called the *raised plate,* which provides a birdsmouth bite to these irregular rafters.

On a regular hip roof, there's only one HAP for all framing members on that roof. That's also true of an irregular hip roof.

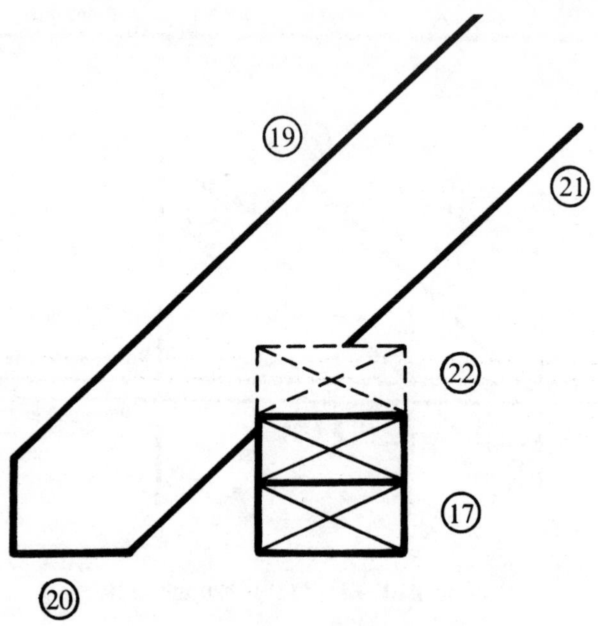

Section B-B: Lines (22) raising the plate
Figure 10-19

But sometimes the HAP of irregular rafters has to be adjusted to fit properly with the rest of the roof. A raised plate changes the HAP on irregular rafters.

Section C-C: The Irregular Hip

Look at Figure 10-20. We've drawn in line (23), a ridge center line, just as we did for line (15) back in Figure 10-18. Draw line (23) 2'10'' in scale from line (6). This measurement lets the view fit well on your drawing.

The building line distance (24) was measured as 20''. The fascia line was measured as 30''. Measure and draw these lines.

Lines (27) through (31)— Line (27) connects the intersection of lines (14) and (26) with the intersection of lines (13) and (23). Then measure down line (26) 2''. That's the amount of the plancher cut. Draw line (28) in Figure 10-21.

Line (29) is drawn as usual. Notice there is no room to cut a birdsmouth here either. Therefore the raised plate will have to extend under both irregular hip rafters. Extend the raised plate line

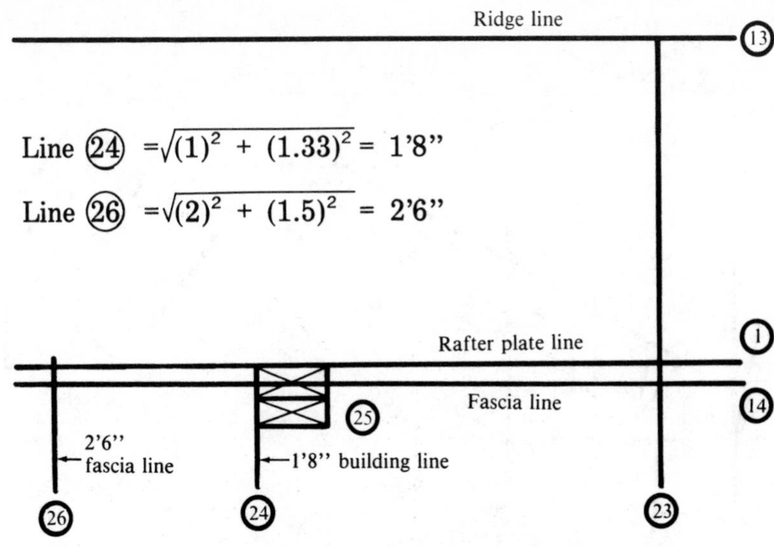

Section C-C: Lines (23) through (26)
Figure 10-20

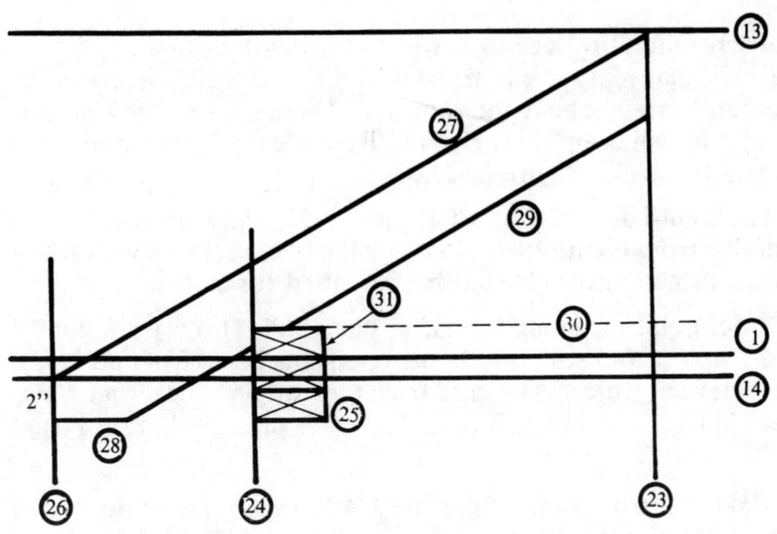

Section C-C: Lines (27) through (31)
Figure 10-21

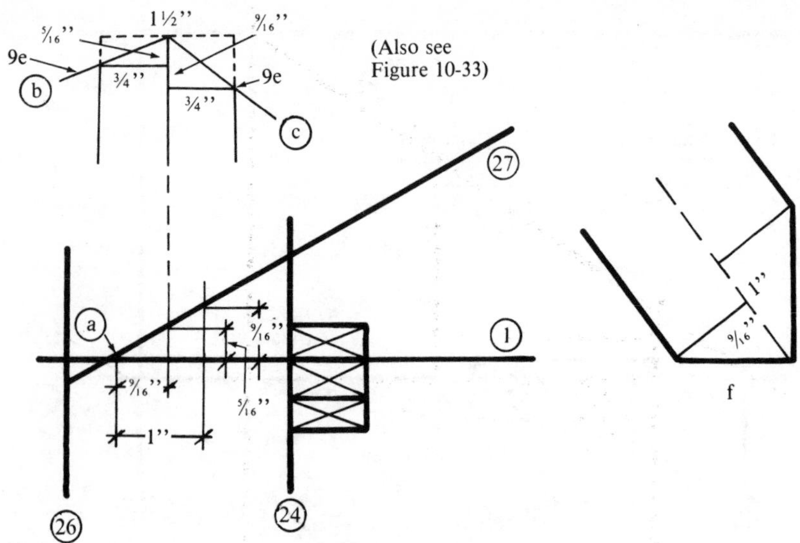

Section C-C: Projecting the backing cuts
Figure 10-22

(30). This shows that the same plate extends under these members. Draw in the raised plate (31).

The Irregular Hip Backing Cuts

The irregular hips have to be backed slightly so the corners of the hip don't extend above the roof line. There are two backing cuts for this irregular hip roof. Here we'll use the method of projection to find them. See Figure 10-22.

The right-hand portion of Figure 10-22, showing point (f), is redrawn from Figure 10-11. It shows the fascia setback for each of the cheek cuts. Line (14) has been omitted for clarity.

From point (a) along the rafter plate line (1), measure out 1'' and 9/16'' in full scale. Both measurements begin from point (a). The lines are projected upward from their origin. The distance between line (1) and line (27) in each place is measured, as in Figure 10-22.

Make the projection a full scale 3/4'' to each side of the dotted center line. See the upper left drawing in Figure 10-22. Lay out the projected amounts as indicated. In this case it will be 5/16'' and 9/16''.

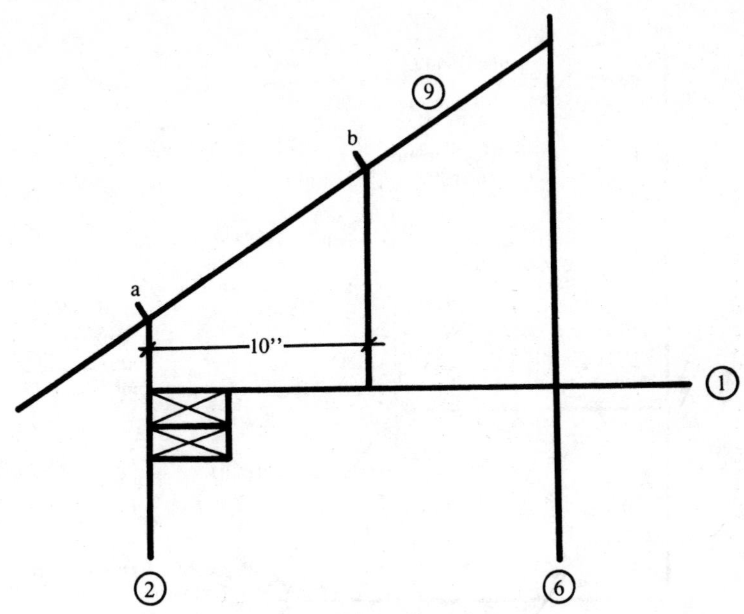

The mathematical length of rafter 4
Figure 10-23

Form triangles by drawing lines (b) and (c), giving an actual cross-sectional view of the backed irregular hip. Add 5/16'' and 9/16'' to line 7 on the cutting list. Measure the angles of each cut with a protractor.

Measurements on the Sectional View Drawings

Now you can measure lines (9), (27) and (19) directly in the 3'' scale to find the mathematical length of the building lines and fascia lines. You can also measure the distance of the HAP for rafters 5, 6 and 7. The HAP will be the same on rafters 5 and 6.

Rafter 4 had a measured building line length of 10''. Lay out 10'' from line (2), going toward line (6). Project this up to point (b) of Figure 10-23. The measured distance (ab) along line (9) is the mathematical length of jack rafter 4. This is the length before shortening and setback.

Lay out jack rafter 5 the same way, except that you lay it out on "Section B-B: The irregular common rafter."

Layout with Mathematical Calculations

We've laid out the whole roof using very little math. Roof cutters have been making layouts like this for many years because there

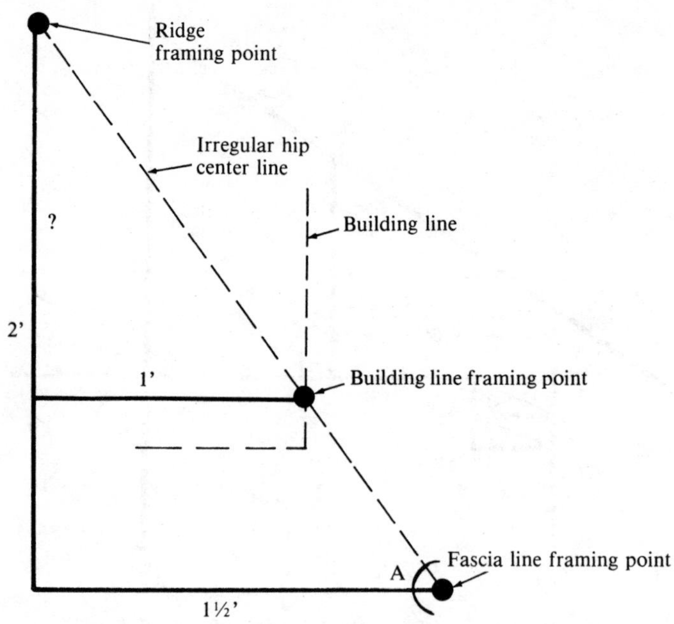

Run of the irregular hip
Figure 10-24

was no simple way to calculate all the lengths required. Before hand-held calculators became commonplace, it was easiest to make a scale drawing and measure the rafter lengths required.

Now, let's see if we can calculate these rafter lengths without a scale drawing. When we've finished, you decide which method you want to use.

Section C-C: The Irregular Hip

On this roof, we know the actual run to the building line for the irregular common rafter. But we have to calculate this distance for the irregular hips. There are two ways to make this calculation.

The First Way— From Figure 10-12, we have been able to draw the triangle of Figure 10-24. The building line framing point lies somewhere on the irregular hip center line. The tangent of angle A is equal to [2] [÷] [1] [·] [5] [=]. 1.3333 shows up in the display. Then punch [INV] [tan] and you get 53.13 degrees. Remember that number.

Now, we need to find the building line segment marked with a question mark in Figure 10-24. Use a ratio to find that length:

$$\frac{1.5'}{1} : \frac{2'}{?} \qquad ? = 1.3333'$$

This ratio says that a run of 1½' is to a rise of 2' as a 1' run would be to the question mark. The answer is 1.3333'.

Now use the Pythagorean Theorem to find the length of the irregular hip run from the ridge to the building line. This is the hypotenuse of the triangle.

$$\text{Irregular hip run from ridge to building line} = \sqrt{(1)^2 + (1.3333)^2}$$

$$= 1.666'$$

$$= 1'8''$$

1'8'' is the same distance as we measured on the plan view when we were using the scale drawing.

The Second Way— The cosine expresses the relationship of the hypotenuse to the run. We know the tangent is 1.3333 for this hip. With the tangent showing in the calculator window, punch [INV] [tan] [cos] and 0.6 appears. Dividing 0.6 into 12'' gives you an answer of 20'', or 1'8'', the same as we found with the ratio. See Figure 10-25.

And again we have the actual run of the irregular hip to the building line. The calculator method seems easier to me. Maybe you'll agree.

The Irregular Hip Fascia Line

The fascia line length of the irregular hip lies in a rectangle that is 1½' by 2'. Look again at Figure 10-24.

Use the Pythagorean Theorem to find the fascia line length.

$$\text{Full run of irregular hip} = \sqrt{(1.5)^2 + (2)^2}$$

$$= 2.5'$$

$$= 2'6''$$

Setback at the Tail

In "The First Way," we found the tangent of the center line angle to be 1.3333. Now look at Figure 10-26. For calculating the short setback, the tangent formula is:

$$1.3333 = \frac{.75"}{?} \qquad ? = .5625 = 9/16"$$

The short side angle is 53.13 degrees. That's the inverse tangent of 1.3333. For the long side, we have to find the complementary angle. That's the angle that, when added to 53.13 degrees, equals 90 degrees. 90 degrees minus 53.13 degrees is 36.87 degrees. Its tangent is 0.75''. Therefore:

$$\text{Tan } 36.87° = \frac{.75"}{?} \qquad ? = \frac{.75"}{\text{Tan } 36.87°} \qquad ? = 1"$$

Both of these lengths have been written on the cutting list.

The Saw Angle for the Tail Cuts

Notice the complementary angles in Figure 10-26. The angle calculated at (a) is used to cut the long setback side, and the angle calculated at (b) is used to cut the short setback side.

Angles and Settings of the Various Members

The angle of the common rafters will be 33.69 degrees for an 8 in 12 roof. Your stair gauges will be set on 8'' and 12''. The angle for the regular hips is also known, 25.24 degrees, with the stair gauges

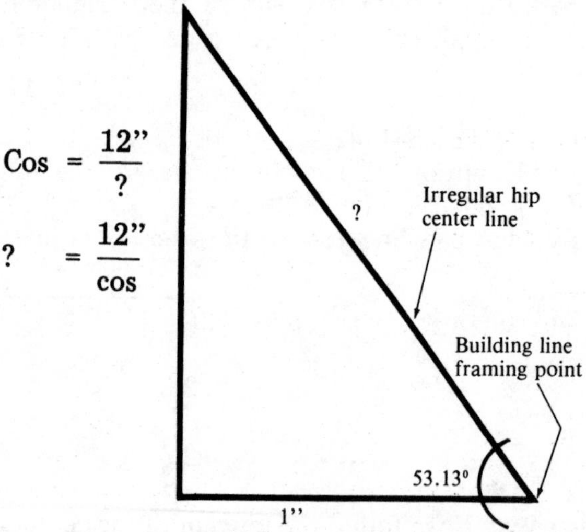

Cosine and the run
Figure 10-25

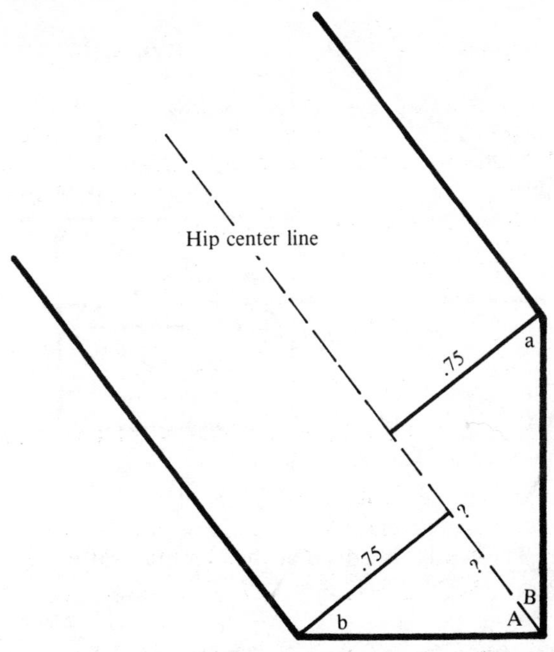

Calculating setback
Figure 10-26

set on 8'' and 16.97''. Now we must find the angles of the irregular hips and the irregular common rafter, and the framing square settings for each.

Irregular Hip Angle and Fascia Line Drop

Look back to Figure 10-13. Line (27) shows the pitch of the irregular hip rafter. Let line (27) represent the hypotenuse of a triangle whose run has been calculated as 2'6''. This is the distance from line (23) to line (26). The rise of this triangle is the full rise of line (23). That's identical to line (6), or 14¾'', *plus* the distance of the fascia line between lines (1) and (14).

The common rafter has an angle of 33.69 degrees and a tangent of 0.6667. Since the run of the overhang is known as 6'', we can find that the rise is 4''. See Figure 10-27. The HAP is 2¾'' of this. Therefore, the distance between lines (1) and (14) is 1¼''. The 14¾'' of line (23) plus this 1¼'' gives a total rise of 16''.

Look at Figure 10-28. If the total rise is 16'' and the run is 30'', it's easy to find the angle for this irregular hip. [1] [6] [÷] [3] [0]

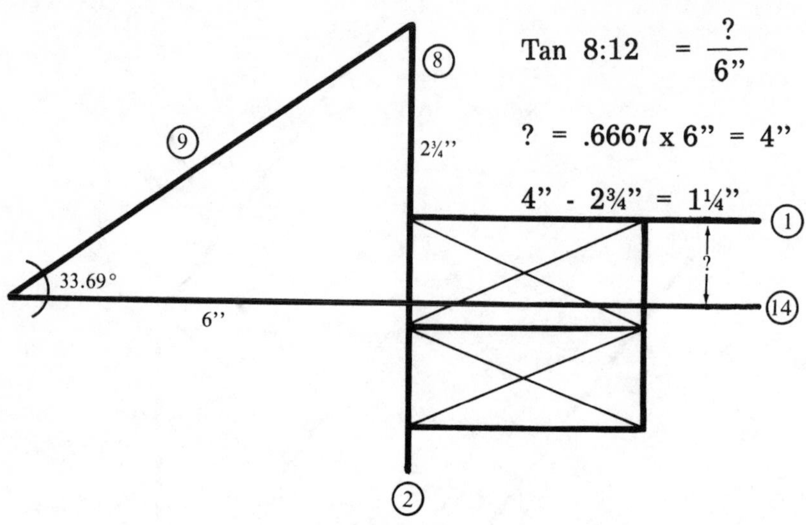

The fascia line drop at the common rafter
Figure 10-27

[=] 0.533333. To find the angle, [INV] [tan] [=] and you get 28.07 degrees. With this figure showing in the calculator, punch [cos] [1/x] and 1.1333 appears. This is the secant that we'll use to calculate the mathematical lengths of the hips.

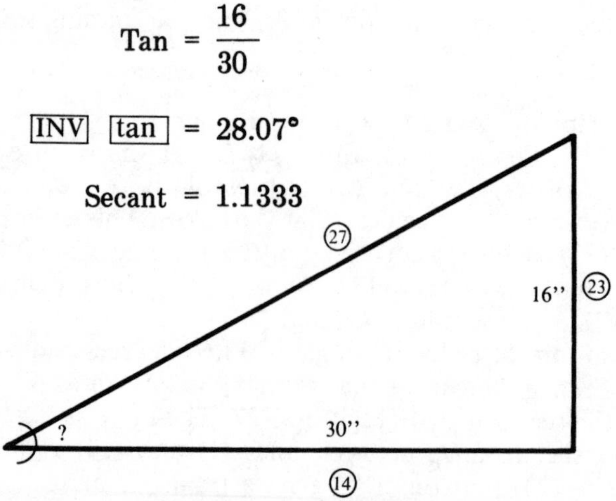

$$\text{Tan} = \frac{16}{30}$$

[INV] [tan] = 28.07°

Secant = 1.1333

27

16" 23

?

30"

14

The irregular hip rafter angle
Figure 10-28

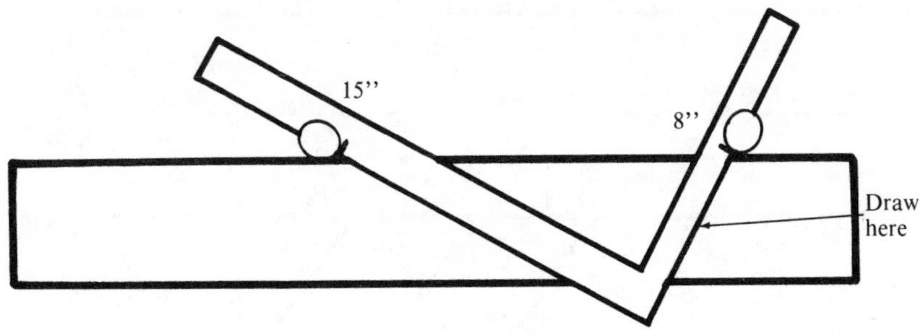

Establishing the line
Figure 10-29

The Stair Gauge Settings

The stair gauge settings for this rafter must be some fractional part of 30'' and 16''. This rafter could be cut using 15'' and 8''.

Since this is a hip rafter, we could set the body to 16.97'' as usual and then find the setting for the rise. We can calculate this setting in either of two ways.

Finding the Settings by Layout— Set the stair gauges to 15'' and 8'' and draw a plumb line, as in Figure 10-29. Now move the stair gauge on the body to 16.97'' and loosen the other gauge. Let the body stair gauge rest against the top of the 2 x 4. Position the square so that its tongue lies along the line you just drew. Move the loosened tongue stair gauge against the 2 x 4 and tighten it. The tongue should read 9$\frac{1}{16}$'' on the front side of the tongue.

Finding the Settings by Ratio—

$$\frac{15}{8} : \frac{16.97}{?} \qquad ? = \frac{(8)(16.97)}{15} = 9\tfrac{1}{16}''$$

The ratio above says 15'' is to 16.97'' as 8'' is to ?. The answer is 9$\frac{1}{16}$''. But don't try to set this on the 12th scale.

The Irregular Hip Shortening and Setback

Since the fascia line and ridge line are parallel, all angles A in Figure 10-30 are equal. Back in Figure 10-24 we found the tangent of this angle to be 1.3333. The angle is 53.13 degrees.

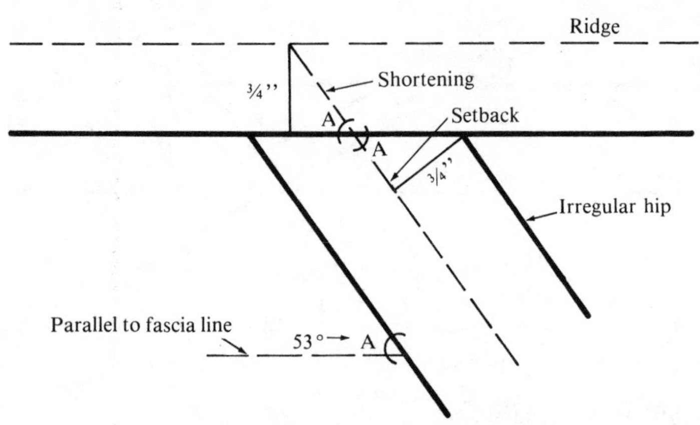

The shortening and setback triangles
Figure 10-30

To calculate the shortening, we have to find the hypotenuse from the rise and the angle.

$$\text{Sin } 53.13° = \frac{.75}{?} \qquad ? = \frac{.75}{\text{Sin } 53.13°} \qquad \frac{.75}{.8} = .9375$$

The sine of 53.13 degrees is 0.8. Dividing 0.75 inches by 0.8 gives us 0.9375'', or 15/16'' for the shortening.

The setback distance is the run. This is found by the tangent relationship. See Figure 10-31. Divide 0.75'' by 1.3333 and you get 0.5625'', which is 9/16'', the amount of the setback.

The Ridge Cheek Cut Angle

The ridge cheek cut angle is the alternate interior angle of the 36.87-degree fascia cut angle. It will be made at the same circular saw setting.

In Figure 10-31, you can see that this angle is also the complementary angle of the known angle of 53.13 degrees. That's 36.87 degrees.

The Hip Backing

Find the irregular hip backing by multiplying the tangent of the rafter angle (28.07 degrees) by the amount of each rafter tail setback. Both of these formulas are shown in Figure 10-32.

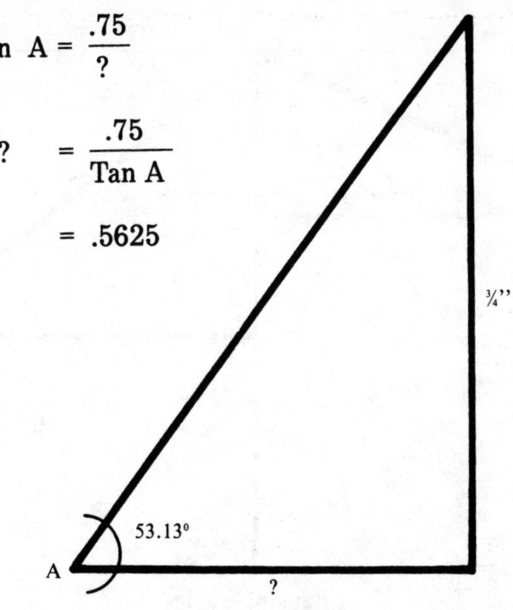

The setback triangle
Figure 10-31

$$\text{Tan B} = \frac{?}{9/16}$$

$$? = 5/16''$$

$$\text{Tan B} = \frac{?}{1''}$$

$$? = 9/16''$$

Irregular hip

Backing

28.07°

B

Actual run

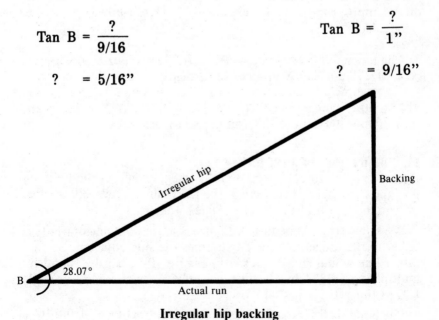

Irregular hip backing
Figure 10-32

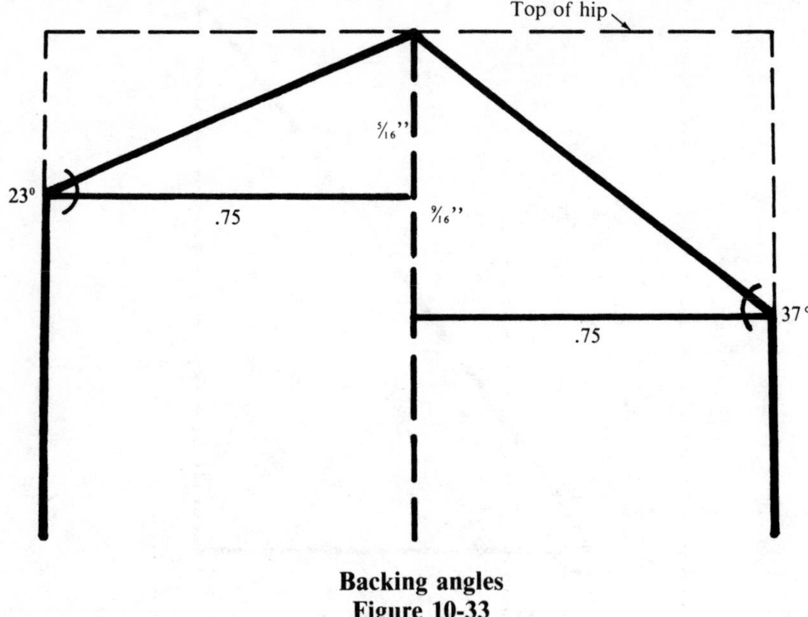

Backing angles
Figure 10-33

The hip backing cuts will be made at 23 degrees and 37 degrees on the appropriate side. See Figure 10-33. Here's how these angles were calculated:

1) The tangent is [5] [÷] [1] [6] [=] [÷] [.] [7] [5] [=] and 0.4166 appears. Now [INV] [tan] and you have 22.62 degrees.

2) The other tangent is [9] [÷] [1] [6] [=] [÷] [.] [7] [5] [=] and 0.75 appears. Now [INV] [tan] and you get 36.87 degrees.

Finding the Run of Jack Rafter 4

Back at Figure 10-24, we related the irregular hip at the building line and the length of that hip at the fascia line. Now we'll use this information to find the length of jack 4.

We know from Appendix B, Figure B-4, that the side opposite is equal to the tangent of the angle times the side adjacent. So the fascia line actual run of jack 4 in Figure 10-34 can be found by multiplying 1.3333 by the distance of 66" minus 54", or 12". 1.3333 times 12" equals 16".

The building line actual run is 6" less. Subtracting 6" from 16", we get 10". This is the figure already entered on our cutting list.

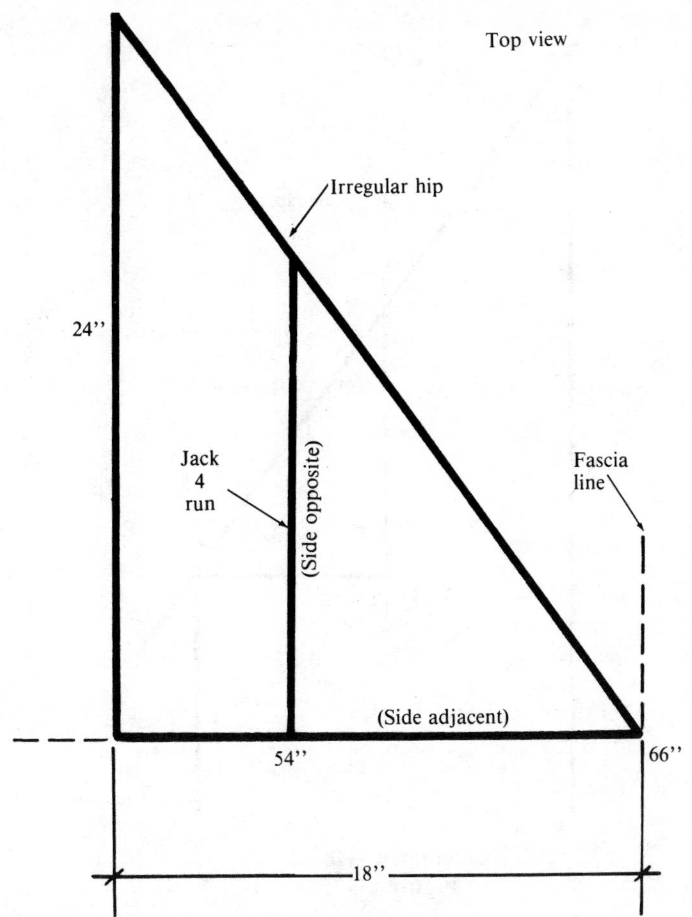

The hip - jack relationship
Figure 10-34

The Jack Rafter 4 Shortening and Setback

Check the setback by the tangent method. Multiply 1.3333 times 0.75'' to get 1''. See Figure 10-35. This checks with the measured distance.

The shortening can also be checked mathematically. To do so, we have to find the triangle hypotenuse rather than the rise. See Figure 10-36.

The cosine is the run divided by the hypotenuse. With 1.3333 showing in your calculator window, punch [INV] [tan] [cos] and you get 0.6. Punch [÷] [.] [7] [5] [=] [1/x] and 1.25'' appears. This is the shortening for jack 4.

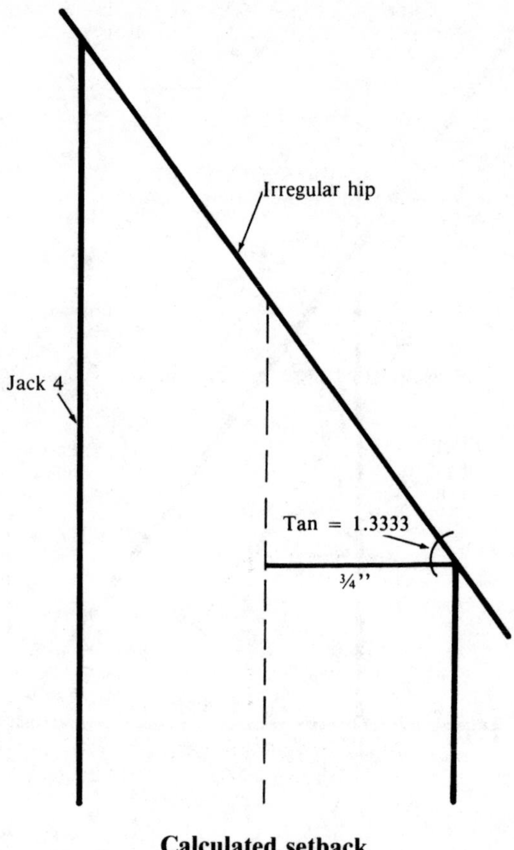

Calculated setback
Figure 10-35

Now multiply the secant of the 8 in 12 roof by the two actual runs to find the mathematical layout measurements of this jack. The answers should agree with the lengths already entered in your cutting table.

Finding the cheek cut angle for rafter 4 will finish our work on this rafter. We've calculated a cheek cut on a regular 45-degree rafter (A in Figure 10-37) like this: The tangent of 3/4'' divided by 3/4'' is one. With 1.0 in the display, punch [INV] [tan] and you have 45 degrees.

For our irregular rafter (B in Figure 10-37) the procedure is the same. The tangent of 1 divided by 0.75 is 1.3333. Then push [INV] [tan] to get 53.13 degrees.

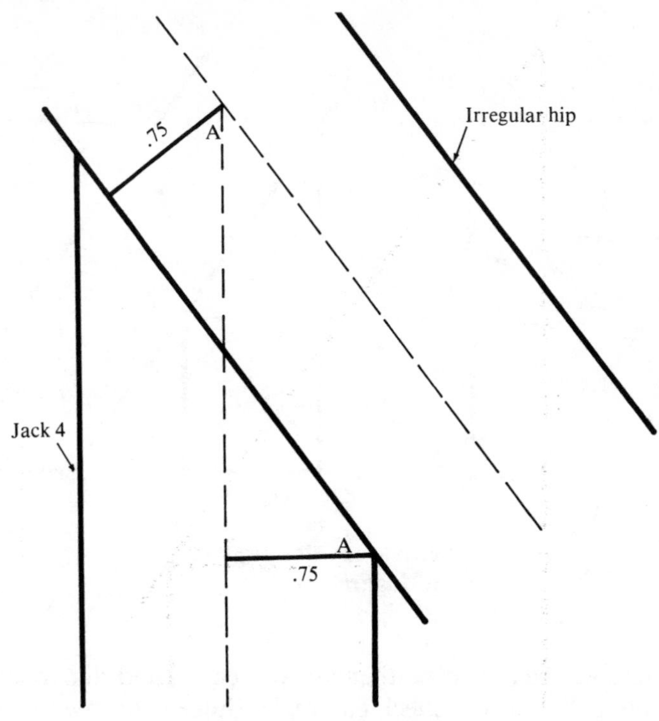

Calculated shortening
Figure 10-36

$$\text{Tan}\ \frac{3/4}{3/4} = 1$$

INV tan = 45°

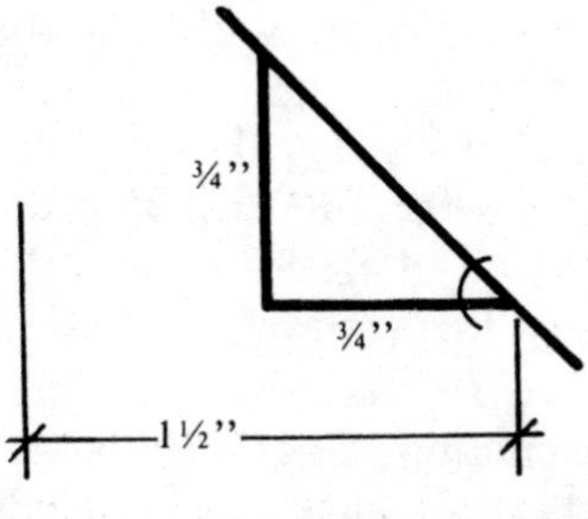

$$\text{Tan}\ \frac{1}{.75}$$

INV tan = 53.13°

A

¾"

¾"

1½"

B

1"

¾"

1½"

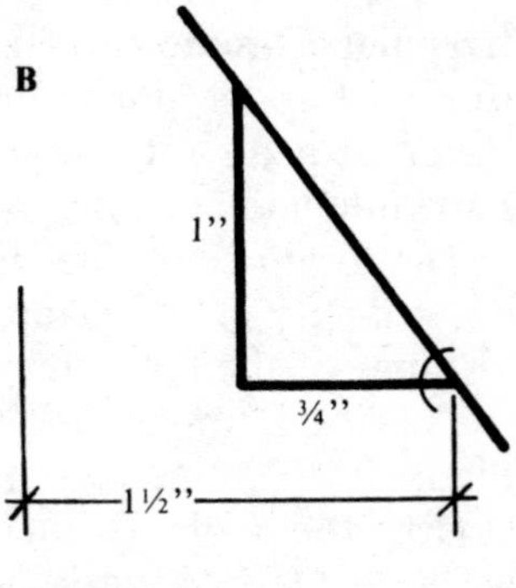

The angle for the cheek cut of rafter 4
Figure 10-37

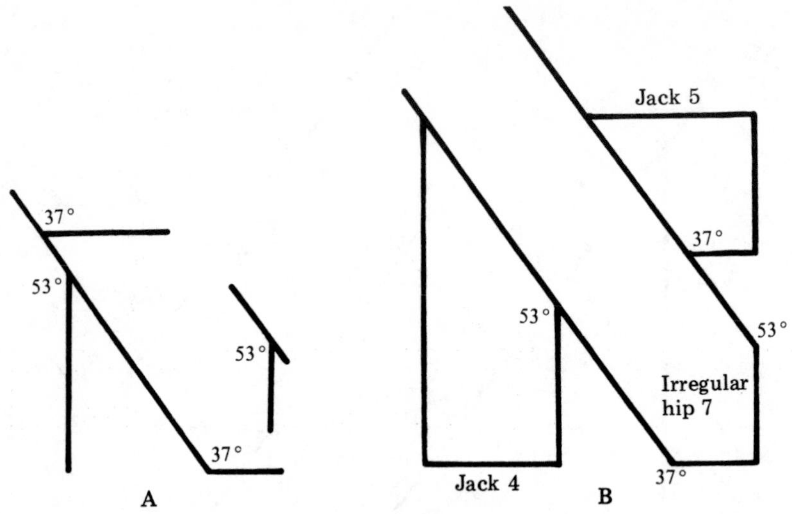

Alternate interior angles
Figure 10-38

The setback has been mathematically calculated above as 1''. Use this to solve for the cheek cut angle. Since your saw probably won't cut at more than 45 degrees, you'll have to lay out 53 degrees and make the cut with a hand saw. See Figure 10-57.

Angular Relationships

The angles of the cheek cuts on jacks 4 and 5 match the angles of the tail cheek cuts of irregular hip 7. This follows from the rule: Opposite interior angles or parallel lines that are cut by a transversal will be equal. This is shown in Figure 10-38

Irregular Common Rafter 6

In Figure 10-39, line (19) shows the pitch of the irregular common rafter. It is also the angle for this side of the roof. We'll use the known lengths of lines (14) and (15) to find the exact angle.

The total rise of line (15) is the same as the rise for the irregular hip triangle. This distance is the 14¾'' rise of line (6) *plus* the distance of the fascia line drop, which is 1¼''. This comes to 16''.

The total run of line (14) is found by inspection, looking at Figure 10-12. One foot plus 6'' is equal to 18''.

Now the angle of line (19) can be calculated. That's the angle with the question mark in Figure 10-39. The tangent of this angle is 16'' divided by 18'', or 0.8888. Convert that to degrees. [INV] [tan]

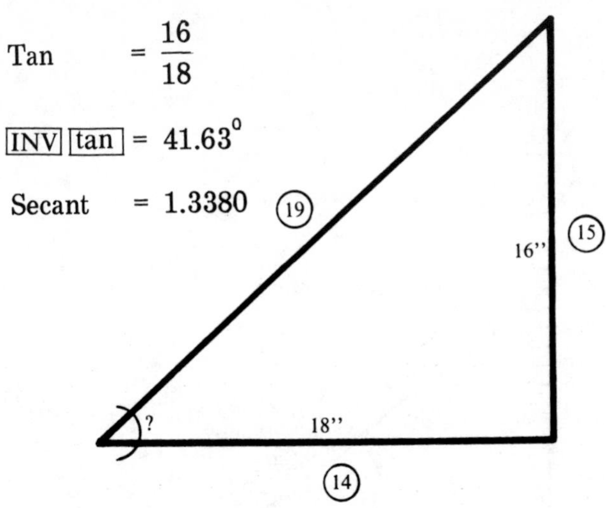

The irregular common rafter angle
Figure 10-39

is 41.63 degrees. Change this to a secant: cos 1/x and you have 1.3380, the secant of this angle. Enter these figures on the cutting lists.

The numbers 18'' and 16'' are hard to use on the framing square. Let's reduce these figures by one-half and use 9'' and 8''.

If, instead, you would like to use the traditional 12'' for the run, then you must find the companion rise. Use the layout method described under Figure 10-29, or the ratio method that follows:

$$\frac{9}{8} : \frac{12}{?} \qquad ? = \frac{8 \times 12}{9} = 10^{11}\!/_{16}''$$

If we set the body stair gauge to 12'', then the tongue would have to be set on 10$^{11}/_{16}$''. Enter the settings on line 5 and 6 of your cutting list.

Jack Rafter 5

Jack rafter 5 has the same angle, secant, HAP and stair gauge settings as the irregular common rafter.

Jack 5 is a proportionate length of the irregular common rafter triangle. Figure 10-40 shows the top view. The side marked with the question mark is jack 5.

The larger triangle has a base of 24'' because the distance from the center of the building to the edge of the fascia is 24''. The rise

$$\frac{18}{?} : \frac{24}{12}$$

$$? = \frac{12 \times 18}{24}$$

$$? = 9''$$

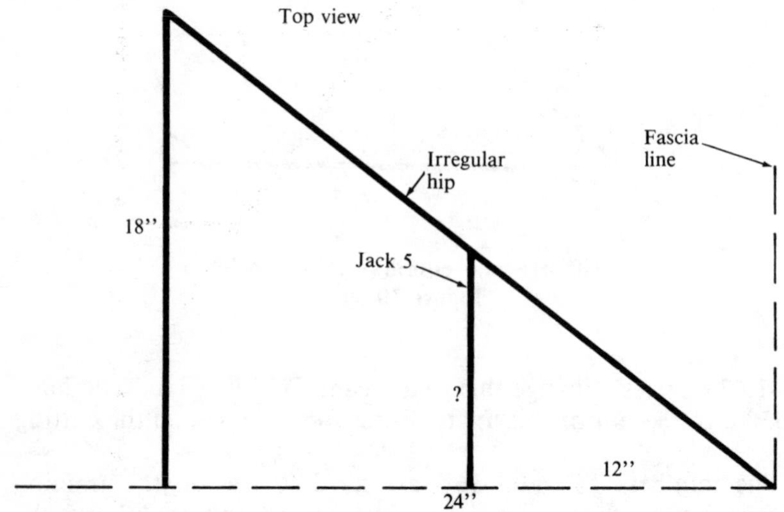

Jack 5 in the larger triangle
Figure 10-40

of the triangle (the distance from the fascia line to the framing point) is 18''.

Use the ratio method to find the length of jack 5. 18'' is to 24'' as jack 5 is to 12''. The formula is shown in Figure 10-40.

The answer is 9'', which is the actual run of jack 5 from its hip framing point to the fascia line. But the building line length is 9'' minus 6'', or 3''. Multiply these two figures by the secant to find the mathematical lengths for layout purposes.

Figure 10-41 shows equal interior angles for the shortening and setback. While the shortening is the hypotenuse of one triangle, the setback is the side adjacent of the other triangle.

We know that the tangent of the angle is 1.3333. For the setback, use the formula shown in Figure 10-42. Remember to multiply by 16 to find sixteenths.

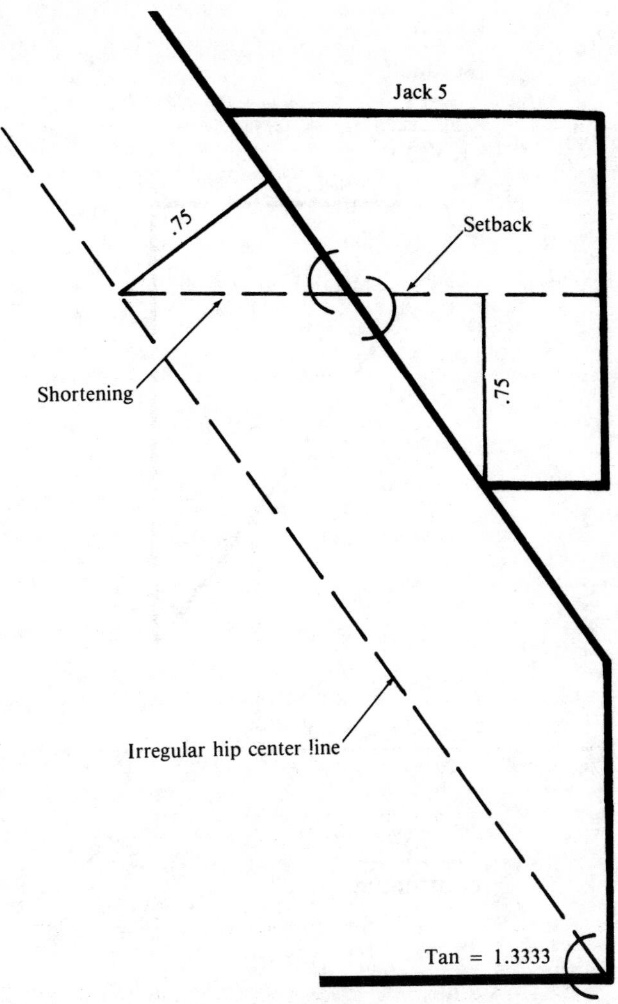

Jack 5 setback and shortening
Figure 10-41

The saw setting for this cheek cut will be the complementary angle of this triangle. To find that angle, put 1.3333 in the display, then punch [INV] [tan] [–] [9] [0] [=] [+/-] and 36.86 degrees appears. These calculations confirm our previous measurements.

Look at Figure 10-43. It shows calculations for the shortening of hip jack 5. With 1.3333 showing, punch: [INV] [tan] [sin] [STO] [·] [7] [5] [÷] [RCL] [=] and 0.9375 appears. Punch [x] [1] [6] [=] and 15/16 appears. This is the shortening for hip jack 5.

$$1.3333 = \frac{.75}{\text{setback}}$$

$$\text{Setback} = \frac{.75}{1.3333} = 9/16"$$

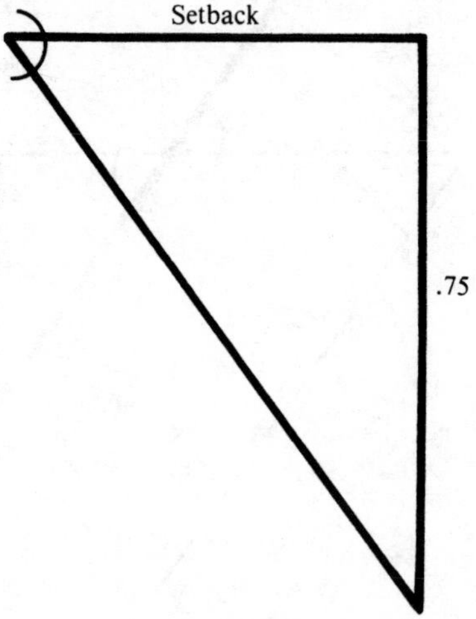

Setback

.75

Jack 5 setback
Figure 10-42

$$\text{Sine} = \frac{.75}{\text{shortening}}$$

$$\text{Shortening} = \frac{.75}{\text{sine}} = 15/16"$$

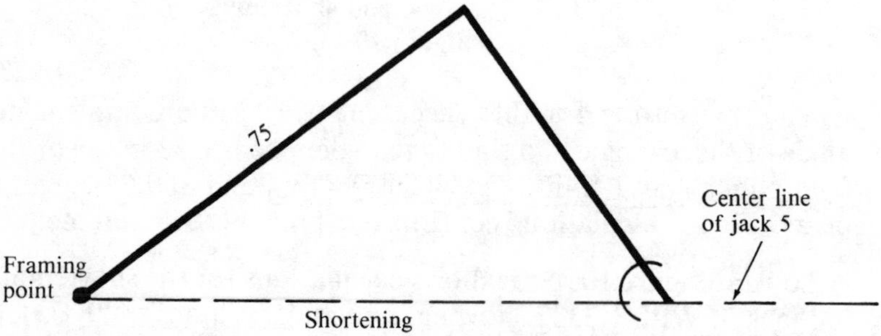

.75

Center line of jack 5

Framing point

Shortening

Jack 5 shortening
Figure 10-43

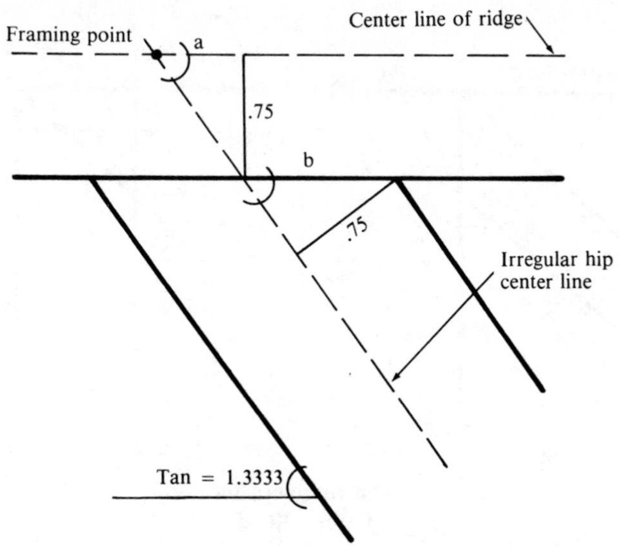

The irregular end ridge growth
Figure 10-44

The Actual Length of The Ridge
The ridge framing point distance was laid out in Figure 10-2 as 2½'. This is the distance between the framing points. Now we have to find the ridge growth at each end to get the actual length of the ridge.

Back in Chapter 3, we found that the growth at the regular hip is one-half the thickness of the common rafter. That's 3/4''.

Ridge growth at the irregular end is exactly the same as the length of the shortening of the irregular rafters. This distance has been broken into different parts of two triangles. See Figure 10-44.

Again, we continue to rely on the information from Figure 10-24 and locate other alternate interior angles.

First, let's find the length of side (a) in Figure 10-44. This growth/shortening dimension is found by the tangent, since the angle and side opposite are known and the side adjacent is the part required. Therefore:

$$\text{Tan} = \frac{.75}{a} \; ; \; a = \frac{.75}{\tan} = \frac{.75}{1.3333} = .5625 = 9/16''$$

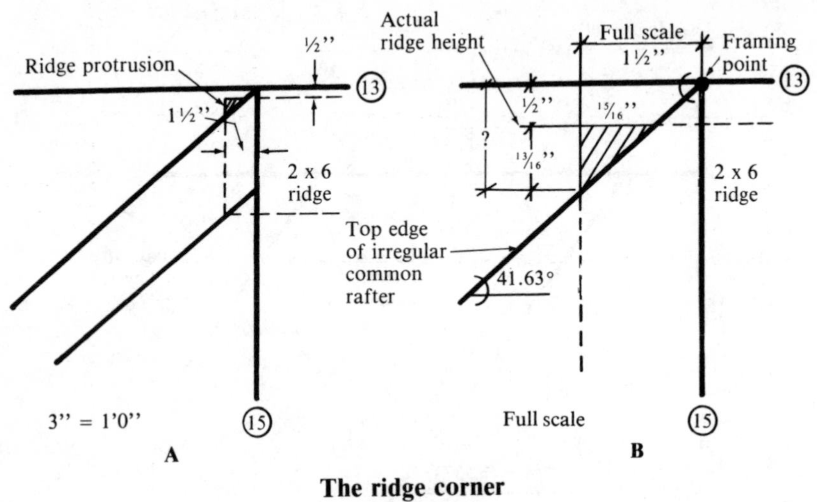

The ridge corner
Figure 10-45

Length (b) is half the distance of the diagonal of the irregular hip cheek cut. In this triangle, the angle and the side opposite are known. It's the hypotenuse, length (b), that's needed:

$$\text{Sine} = \frac{.75}{b} \;;\; b = \frac{.75}{\text{sine}} = .9375 \quad = \quad 15/16''$$

Punch [.] [7] [5] [÷] [1] [.] [3] [3] [3] [3] [INV] [tan] [sin] [=] and 0.9375 shows up in the display. That's 15/16''.

When lengths (a) and (b) are added together, the total is 1½''. This is both the irregular common rafter shortening and the amount of ridge growth on the irregular end.

The framing point length of the ridge was 30''. The growth at the ends was 3/4'' and 1½''. Adding 30'', 0.75'' and 1.50'' gives us the actual ridge length of 32¼''.

Clipping Back the End of the Ridge

If we cut the ridge end off square, the top corner at the irregular hip end would stick out above the roof surface on the irregular side. Figure 10-45 A shows the ridge protrusion. In B, a line parallel to line (15) has been constructed a full 1½'' (the amount of ridge growth) away from the framing point on line (15). Now the vertical distance of the ridge cut can be measured directly in full scale. Remember to make the drawing at the actual ridge height.

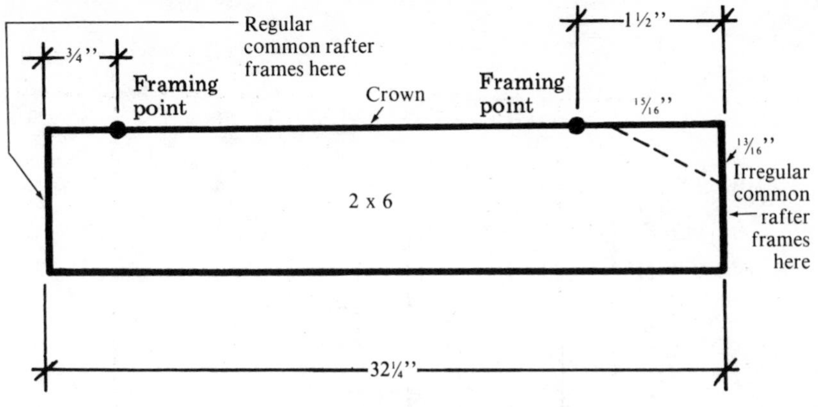

Cutting out the ridge
Figure 10-46

This distance can also be calculated. The angle at the upper right corner at the framing point of 10-45 B is 41.63 degrees because it's an alternate interior angle of a known 41.63-degree angle. The tangent of that angle is equal to the length of the opposite side divided by the length of the adjacent side. We know the adjacent side is 1.5''. We want to find the length of the opposite side, marked with a question mark in Figure 10-45 B.

$$\text{Tan } 41.63° = \frac{?}{1.5''} \qquad ? = (\tan 41.63°)\ (1.5'')$$

Punch [4] [1] [.] [6] [3] [tan] [×] [1] [.] [5] [=] and 1.3332 appears. Subtract one-half inch for the loss of ridge height and you have 0.8332, or 13/16''. This is the rise of the protrusion.

The run along the top edge of the ridge is found by setting up another proportion:

$$\frac{1.3332}{.8332} : \frac{1.5}{?} \qquad ? = .9383 = 15/16''$$

Cutting the Ridge

Cut a 2 x 6 to 32¼'' long. Crown it, and turn the crown up. Measure over 15/16'' and make a mark. Then measure down 13/16'' and make a mark. Cut along the dotted line shown in Figure 10-46.

The Cutting Lists

Correct answers for your cutting lists are at the back of the book. Check yours against the answers there.

#	Kind	Q*	Degrees	Secant	Building line actual run	Math length building line	Fascia line actual run	Math length fascia line
1.								
2.								
3.								
4.								
5.								
6.								
7.								

*Quantity

Mathematical length cutting list
Table 10-47

#	Ridge Cut			HAP	Plancher	Backing		Stair gauges	
	Shortening	Set-back	Cut Degrees			Amount	Degrees	Body	Tongue
1.									
2.									
3.									
4.									
5.									
6.									
*7.									

*Tail setback and angle for each cut ____ at ____ ° and ____ at ____ °

The shortening and setback cutting list
Table10-48

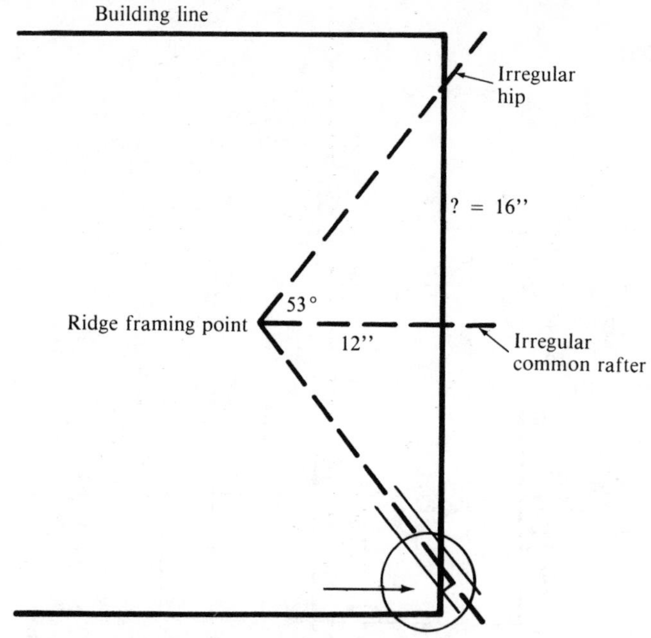

The framing point distance
Figure 10-49

The Raised Plate

It's good practice to install one continuous raised plate from the outside of one irregular hip to the outside of the other. A small piece of 2 x 4 could be placed under each member, but that would interfere with the freeze blocking that will come later.

Finding the Framing Point Distance

Once the plate is installed, you'll have to find the framing points on the plate for the irregular hips. See Figure 10-49. Where do these hips fall? It's important to run the irregular hips to exactly the right point. All of your calculations on the irregular side assume that the hips hit the correct points on the plate.

You could measure the distance on Figure 10-12. But for this part of the chapter, we're *calculating* lengths rather than scaling them off our drawings.

From Figure 10-24, we know the angle at the ridge framing point is 53.13 degrees. The tangent of 53.13 degrees is 1.3333.

$$\text{Tan } 53.13^\circ = \frac{?}{12''} \qquad ? = (1.3333)\ (12'') = 16''$$

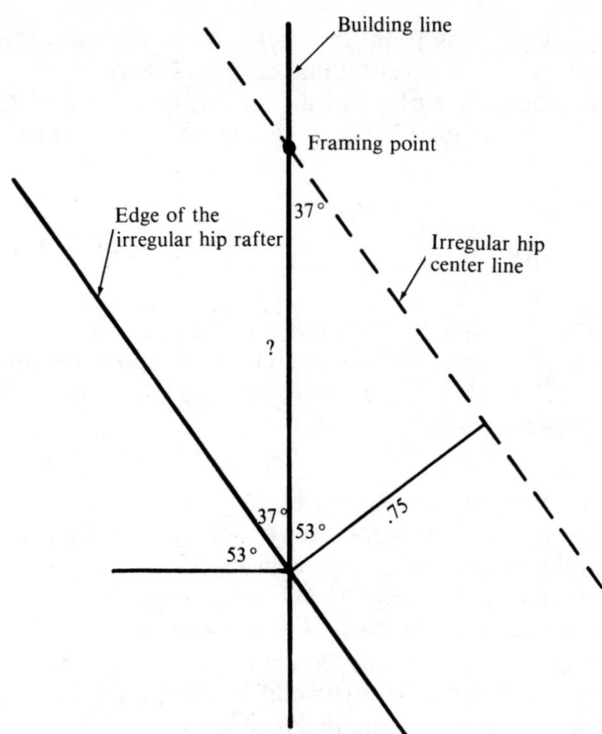

Half the 53° distance
The circled portion of Figure 10-49
Figure 10-50

The distance from the center of the irregular common rafter to the irregular hip framing point is 16''.

Length of the Raised Plate

Twice 16'' gives us 32'' between the framing points of the two irregular hips. But this doesn't provide full bearing for these hips.

How much longer should the raised plate be to provide complete bearing for these hips?

Figure 10-50 shows the circled portion of Figure 10-49. Half the angular thickness of the hip must be added to each end of the plate if we want full bearing under each irregular hip. This distance could be measured directly on your full-size section of the completed plan view, Figure 10-12. But the distance can also be calculated, of course.

You see the framing point in Figure 10-50. We want to find the length marked with a question mark. As in Figure 10-25, we'll use the cosine. The cosine of 53.13 degrees will be equal to the adjacent side, 0.75", divided by the hypotenuse. We want to find the hypotenuse.

$$\text{Cos } 53.13^\circ = \frac{.75}{?} \qquad ? = \frac{.75}{.6} = 1.249 = 1\tfrac{1}{4}"$$

This shows that it's 1¼" from the framing point to the outside edge of the rafter. Add 1¼" to each end of the plate length already calculated: 32" plus 1¼" plus 1¼" equals 34½". That's the long-point to long-point measurement.

Finding the Angle of the Raised Plate

We've calculated the long-point length of the raised plate. But the hip crosses the plate at an angle. We want the end of the plate to be angled so it fits the hip exactly. We could trim the plate after the hip is laid in place, but you'll find that it's easier to cut the plate at the right angle before it's nailed down.

What angle do we use on the end of the plate? Look back at Figure 10-49. Notice that the distance from the framing point at the ridge to the framing point at the plate is 12". The framing point to framing point distance along the plate is 16". This gives us a triangle with a ratio of 16" along the building line to 12" across. Let's set up this same angle on the end of the plate. Remember that a 2 x 4 is only 3½" wide. Here's the proportion:

$$\frac{12}{16} : \frac{3.5}{?} \qquad ? = 4.6667" = 4\tfrac{11}{16}"$$

Laying Out the Raised Plate

Look at Figure 10-51. The 4¹¹⁄₁₆" is laid out at the top side of the 2 x 4. Mark the plate as shown and cut it out. Frame the raised plate to the center of the 3' side on your model.

Layout for the Irregular Common

Crown a piece of 2 x 4 and lay the crown away from you. Set your stair gauges at 12" and 10¹¹⁄₁₆". See Figure 10-52.

The HAP for this rafter is 2⅝". The shortening is 1½". Otherwise, this rafter cuts like any other common rafter. Frame this rafter on the raised plate and against the ridge. It should fit perfectly against the ridge bevel cut.

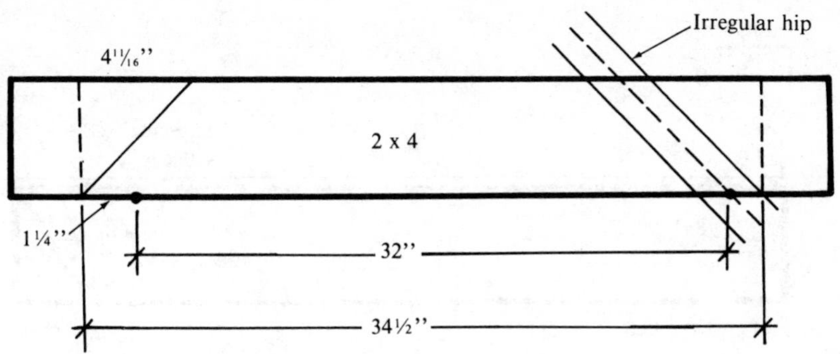

Laying out the raised plate
Figure 10-51

Layout for the Irregular Hip
Crown a piece of 2 x 4 and lay the crown away from you. Set the stair gauges to 16.97" and 9⁄₁₆".

Refer to Figure 10-53 and proceed with the hip layout as usual. Lay out the shortening and setback, and mark the ridge cheek cut line for 37 degrees, as indicated by the cutting list. Lay out the building line and the fascia line and transfer the fascia line to the other side.

Add the plancher cut at 2" and then mark a set forward of 1" on this side. See Figure 10-54. Draw the cheek cut line. Since the angle here is 53 degrees, a construction line on top is necessary. Mark a dot in the center of the fascia line on top of the rafter. From the corner of the cheek cut line, draw line (a). Guide a hand saw along line (a) and the cheek cut line.

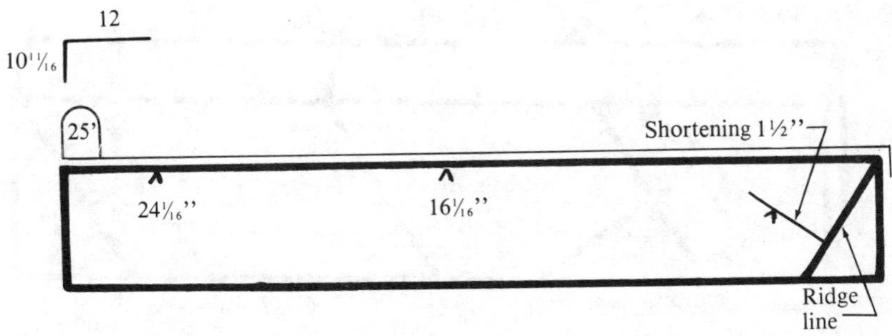

The layout for the irregular common rafter
Figure 10-52

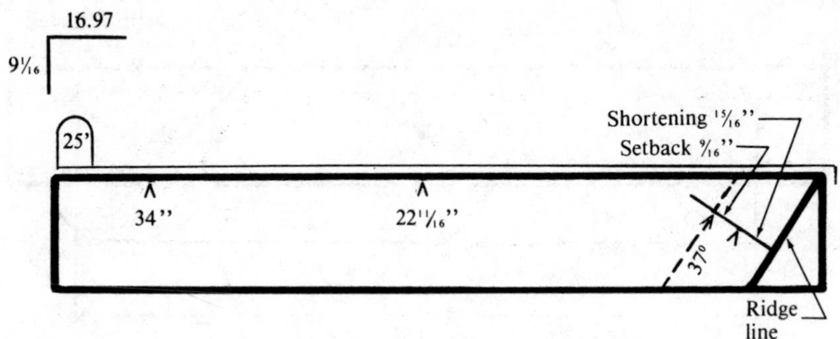

The layout for the irregular hip rafter
Figure 10-53

On the other side at the fascia line, add a set forward of 9/16''. This measurement is from the cutting list, Figure 10-48, under the asterisk on line 7. Mark this line to be cut at 37 degrees.

The Birdsmouth

The birdsmouth for this rafter will be very similar to the birdsmouth of the Tudor hip in Chapter 7, Figure 7-28.

Draw the building line and transfer it to the other side. Mark the HAP of 2¾'' on each side.

Refer to Figure 10-50. The side opposite the 53-degree angle shows the amount of setback.

$$\text{Tan } 53.13 = \frac{?}{.75} \qquad ? = (1.3333)\ (.75) = 1''$$

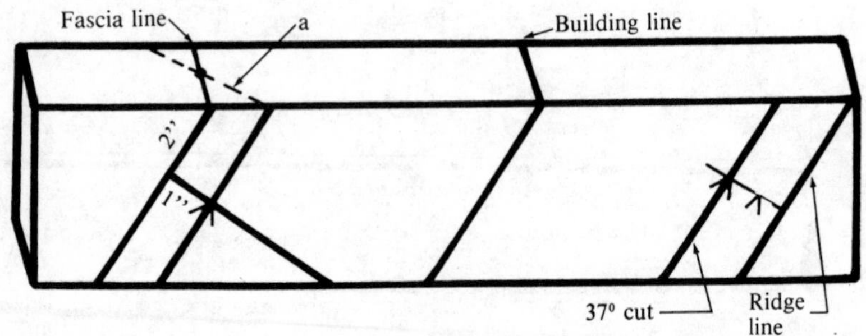

Constructing the cheek cut line
Figure 10-54

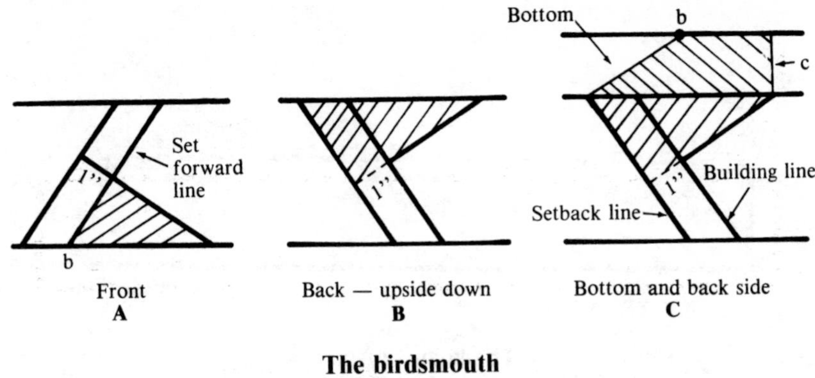

The birdsmouth
Figure 10-55

Mark a 1" set forward on this side of the rafter, and a 1" setback line on the other side. Extend the seat cut line on the back side (the dotted line in Figure 10-55 B) until it reaches the cheek cut line.

The shaded parts of Figure 10-55 show the birdsmouth portion. Look at 10-55 C. Draw a line across the bottom from the lower corner of the setback line to the lower corner of the front side set-forward line at (b). Square line (c) across the rafter.

Laying Out the Backing

Look back to Chapter 4. Figure 4-54 shows how we constructed backing using the building line. Since this rafter is irregular, this construction at the building line won't work. Make the construction using the tail setback lines as shown on the sectional view in Figure 10-13.

Draw line (9e) of Figure 4-54 down 9/16" on the front and 5/16" on the back side. Then construct line (DEF) of Figure 4-54. Draw a center line along the top edge of the rafter.

Cutting Out the Irregular Hip

Make the 37-degree ridge cut and tail cut, and the square plancher cut. Hand saw the 53-degree tail cut. Here's how to handle the birdsmouth. First use a circular saw to cut along the seat cut line, then finish the birdsmouth with the hand saw. For the backing cuts, it's easier to use a table saw.

To cut out the opposite irregular hip rafter, lay out the ridge line on the left side and work toward the right. Follow the same procedure as we used above and the rafter will come out right. Put the two hips side by side and check that all cuts are proportionate.

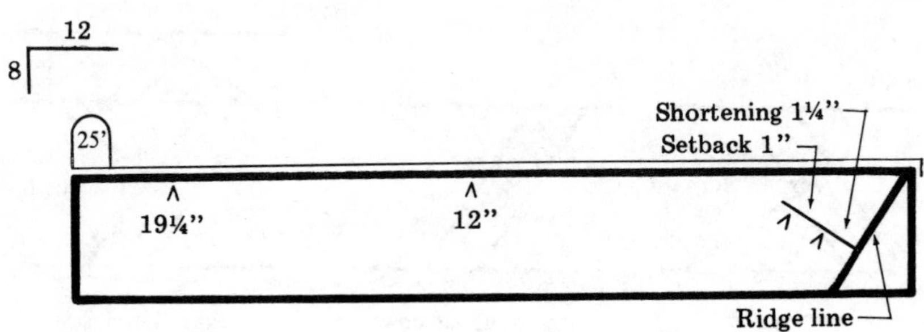

The layout of jack 4
Figure 10-56

Laying Out and Cutting Hip Jack 4

Crown a piece of 2 x 4 and lay the crown away from you. Set the stair gauges to 12'' and 8''.

Refer to the cutting list, and mark 12'' and 19¼'' as shown in Figure 10-56. Draw these plumb lines and then lay out the HAP and plancher lines. Mark the shortening and setback according to the cutting list.

Notice that this cheek cut angle is 53.13 degrees. You can't cut this angle with a circular saw. Instead, draw the construction lines and make the cut with a hand saw.

Square the shortening line across the top and place a dot in the center of this line. See Figure 10-57. From the corner of the cheek cut line, draw the dotted line through the center point. Guide your hand saw along the dotted line and cheek cut line.

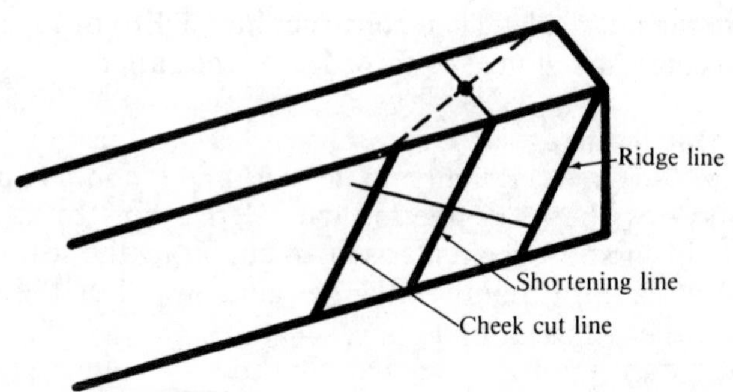

The cheek cut of hip jack 4
Figure 10-57

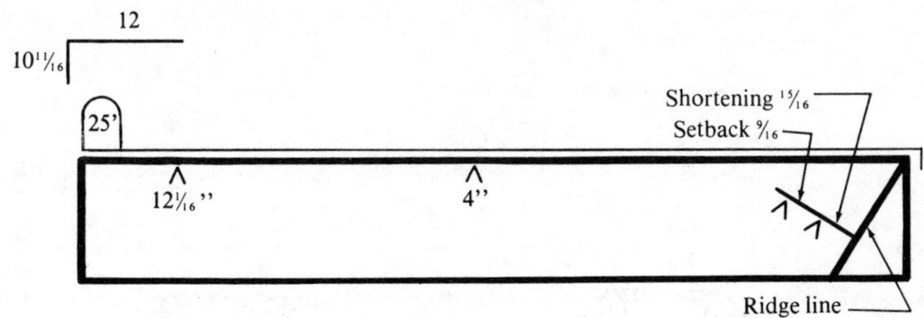

The layout of jack 5
Figure 10-58

Cut another jack rafter with an opposite cheek cut. To find its dotted line, square the cheek cut line across the top and begin the dotted line from the opposite corner. This rafter could also be laid out from left to right, as we did with the irregular hip.

Laying Out and Cutting Hip Jack 5

Crown a piece of 2 x 4 and lay the crown away from you. Set the stair gauges on 12" and 10 11/16". Remember not to look on the twelfths scale.

The cutting list shows mathematical lengths of 4" and 12 1/16". Draw these plumb lines. See Figure 10-58. The HAP for this rafter was measured at 2 5/8". Draw the seat cut accordingly. Add the usual plancher line.

Add the shortening and setback. Since the angle for this cheek cut is under 45 degrees, you don't need a construction line for the cheek cut. The saw setting automatically takes care of it. However, you'll have to transfer a cheek cut line to the other side for the opposite pair rafter. Set the saw at 37 degrees and cut along the cheek cut line.

Nail these jacks on the model. Notice how nicely everything fits together. Hold a 4-foot straightedge against the fascia boards to be sure the tail cuts are all the same length.

Congratulations. You've just framed an irregular hip roof. Most experienced roof cutters couldn't do the work you've just finished.

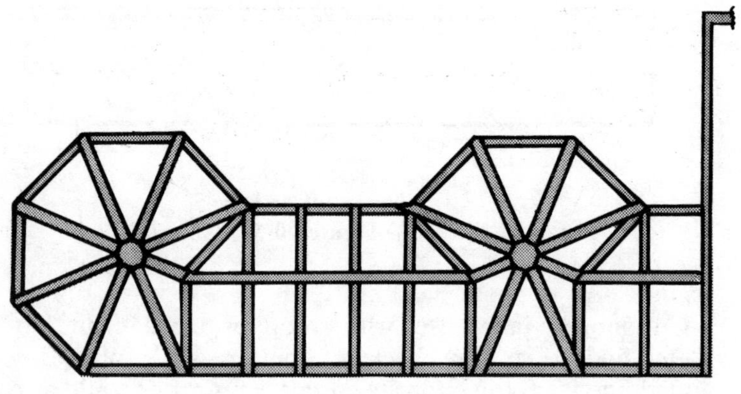

11

A Complex Irregular Roof

The contractor who had the irregular hip job in Chapter 10 was pleased with the work you did on his cabin roof. The word is getting around that you can handle roof framing jobs that others won't touch.

You take on several more irregular hip roofs in the next five months. Then a contractor calls with a remodeling job that's unusual, even for an experienced roof cutter. The owner wants to put a solar greenhouse on one side of his home. The architect drew it as a double octagon connected by a hallway. The roof pitch is 4 in 12. Figure 11-1 shows a plan view of the addition.

The owner wanted the largest floor area possible under the roof. But the city has setback requirements that keep the addition at least five feet from the property line. The architect drew the addition as octagons that give the maximum square footage without intruding on the setback line. That satisfies the zoning ordinance and the owner's need for space. But it makes framing the roof more complex. That's why you've got the job. And that's why you can name your price.

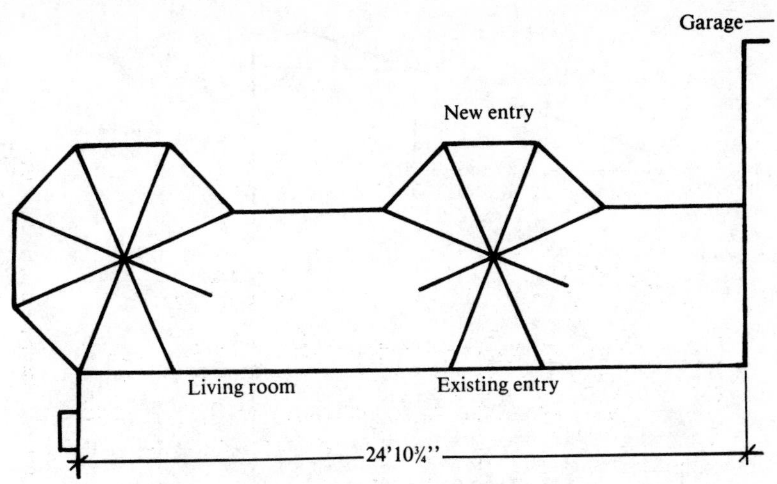

The proposed addition — plan view
Figure 11-1

We're going to need some basic information to do our planning. First, the plans show the circumscribed square as 8'9⁷⁄₁₆'', or 105⁷⁄₁₆''. See Figure 11-2. Knowing the size of the circumscribed square, we can calculate dimensions (b), (c) and (d). Then we can figure the octagon hip actual run, and the mathematical length of a 4 in 12 octagon hip rafter.

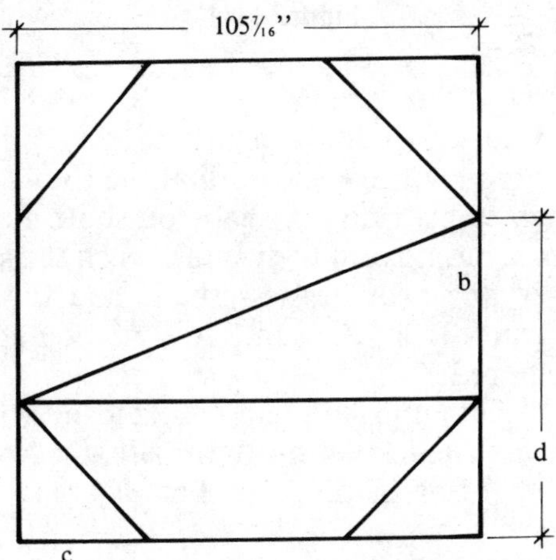

The circumscribed square
Figure 11-2

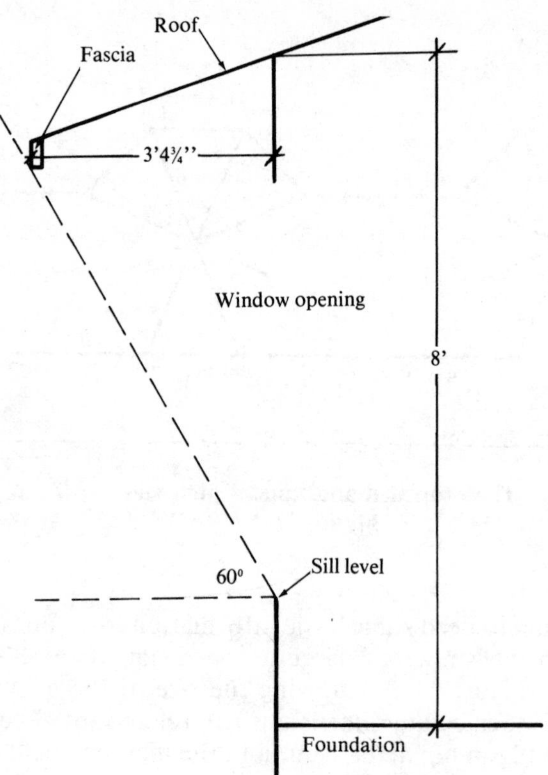

The solar angle
Figure 11-3

We also need to know the run of the overhang. The plans specify a solar overhang. That means that the roof overhang must extend far enough to provide full shade on south-facing windows during the summer months. In the winter, when the sun is lower in the sky, the overhang provides less shade and the windows get more natural warmth from the sun. How are you going to figure the overhang required?

The angle of the sun above the horizon depends on the latitude of the building site. The farther north the site, the lower the sun at noon on the longest day of the year. For this particular site, the angle is 60 degrees.

Look at Figure 11-3. We've laid out a 60-degree angle from the bottom of the window. Now the 4 in 12 roof line is extended until it intersects the angle. This determines the run of the overhang.

With this arrangement, the window will be in full shade when the sun is higher than 60 degrees above the horizon. The run of the overhang for this 4 in 12 roof will be 3'4¾'' to the outside of the fascia board.

The ridge for the addition will run parallel to the existing building. At first glance, you assume that the ridge will be at the center of the addition. That's always worked before. But now, you begin to wonder. Would a centered hall ridge intersect the octagon broken hip on each end, or would the two members miss each other, causing the roof to remain open?

The ridge has to frame exactly against the octagon broken hip rafters on both ends. Suppose the ridge is centered between the two walls of the hall. Will the broken hip be at the right height on both ends?

You can see that the location of the ridge becomes a prime worry on this job. We'll have to draft a plan view and raise sectional views to locate the hall ridge exactly.

Drafting the Plan View

Notice in Figure 11-1 that the existing building measures 24'10¾'' along the line where the addition will be built. Let's make a scale drawing at 3'' to the foot so that the whole project fits on a 4' x 8' sheet of plywood. Use 1/2'' plywood with one good side.

Lay out the original building wall plate within 1½'' of the lower right hand corner of the plywood. See Figure 11-4. In our 3'' scale, 24' will actually be 72''.

According to the plan, a portion of the existing wall will become one side of each of the octagons.

Math for the Octagon Floor Plan

In Figure 11-2, the circumscribed square was given as 105$\frac{7}{16}$''. To find dimension (b), we need to multiply the tangent of 22.5 degrees by 105$\frac{7}{16}$''. The tangent of 22.5 degrees is 0.4142, as you probably remember from Chapter 9. So the length of line (b) is:

$$.4142 \times 105.4375'' = 43.6736'' = 43\tfrac{11}{16}''$$

To find dimension (c), divide (b) by the square root of 2, or 1.414.

$$43.6736'' \div 1.414 = 30.8820'' = 30\tfrac{7}{8}''$$

Dimension (d) is equal to (b) plus (c).

$$43.6736'' + 30.8820'' = 74.5557'' = 74\tfrac{9}{16}''$$

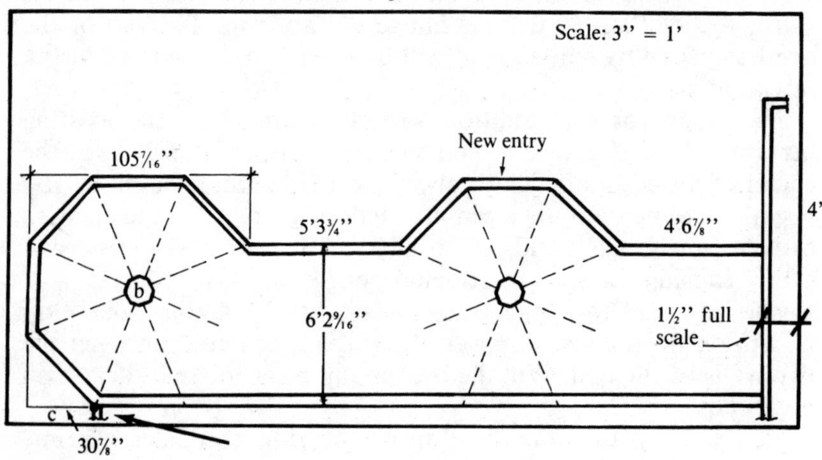

Beginning the plan view
Figure 11-4

Now let's make a template for laying out these octagon plates, as shown in Figure 11-5. First, draw the length of dimension (b), 43¹¹⁄₁₆", then cut the ends back at the necessary 22½-degree angle.

Drawing the Circumscribed Square

The arrow and framing point in the lower left corner of Figure 11-4 indicate one octagon corner. The circumscribed square begins distance (c) from the existing side of the house, and at the inside line of the existing plate. Draw this rectangle carefully on your plywood. Then inscribe the octagon. Draw the plan view of the rafters by connecting the center point with each outer octagon corner. Then add the thickness of the rafters.

The architect shows a combination paddle fan and light hanging from each octagon peak. We should provide a flat surface for mounting these fans. Let's provide a ridge center piece for each oc-

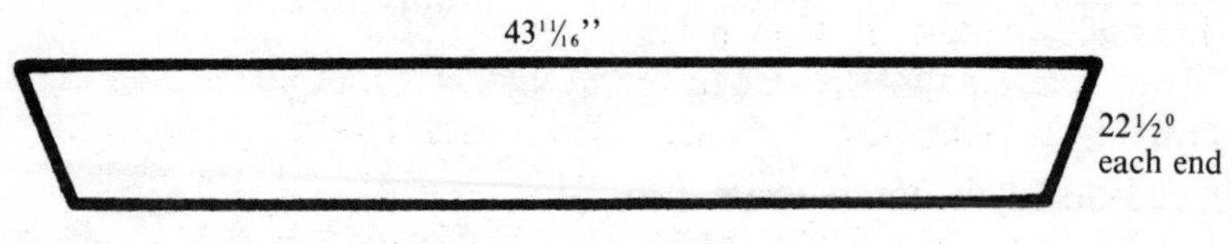

Octagon template
Figure 11-5

tagon. A piece of rough-sawn 4 x 4 will be big enough. We'll cut this stock so that each octagon face of the ridge measures 1½'', the thickness of the rafters. If side (b) is to be 1½'', then the size of the square should be 1.5'' divided by the tangent of 22.5 degrees. Divide 1.5'' by 0.4142 to get 3.6213, or 3⅝''.

Each octagon hip will be shortened by half of 3⅝'', or 1 13/16''. Draw the center octagon ridge piece and then draw in the rafter plate. See Figure 9-41 also.

Drawing the Hall Outline

We calculated length (d) in Figure 11-2 as 6'2 9/16''. This will be the dimension to the outer edge of the hall. It's also the measurement to the corner of the octagons joined by the hall. Make the outside of the hall, which is the distance between the octagons, a distance of 5'3¾''. Then add the thickness of the plate. Refer again to Figure 11-4.

The Second Octagon and Hall

Draw the second octagon as you did the first. Add the hip rafters in scale and the center octagon ridge point. Also add the thickness of the plate. The outside length of the last section of the hallway will be 4'6⅞''.

Locating the Hall Ridge

We can't draw in the rest of the plan view until we know where the hall ridge is located. We suspect that the hall ridge can't be centered, but haven't decided for sure yet. To make this decision, we need to make up some sectional views.

There won't be any common rafters within the octagons. Nor will there be any jack rafters. It is imperative, however, if we are to match the existing 4 in 12 roof, that we establish our sectional views by first laying out a common rafter whose run is equal to half the circumscribed square. This is the correct way to establish the ridge, plate, and fascia lines for the sectional view drawings.

This is not a new concept, since all regular hip rafters fit within their respective common rafter plane.

Establishing the Section Lines

To locate the hall ridge, the sectional view drawing must be "locked" to the plan view drawing. This is established by projecting line (6), the ridge line, and line (2), the building line, as in Figure 11-6.

Establish line (1) 32'' above the bottom edge of the plywood and draw this line along the entire 8' length. This is the plate line.

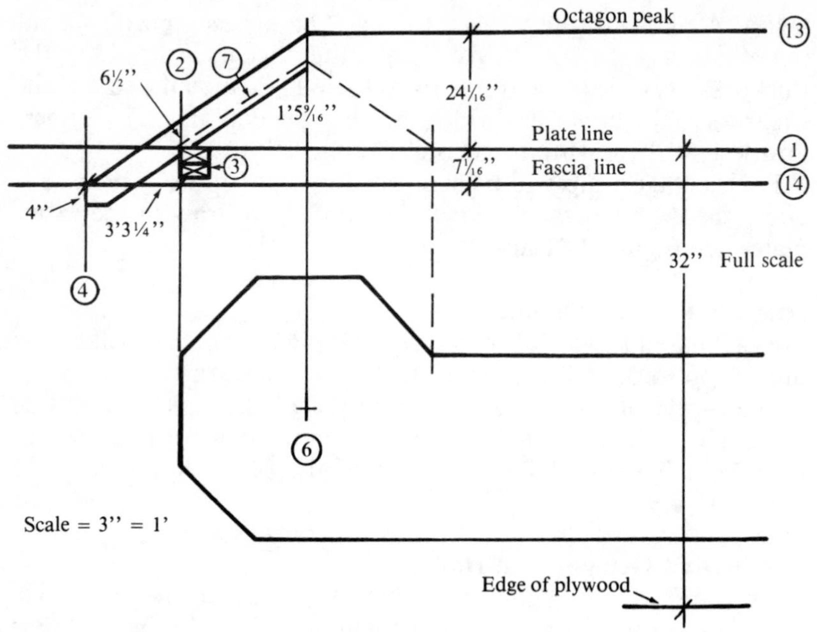

Establishing the section lines
Figure 11-6

Project line (6), the ridge line, from the center of your octagon plan view drawing. Also extend line (2), the building line.

Add the two plates, (3), and measure over 3'3¼", the run of the overhang, to establish line (4).

Laying in the Common Rafter

Now we have to mark the rise on line (6). To do this, multiply the total run in feet (half the distance of the circumscribed square) by the 4" unit rise.

$$105.4375'' \div 2 \div 12 \times 4'' = 17.573'' = 17\tfrac{9}{16}''$$

This is the rise at the octagon peak of the roof. Measure along line (6) for 1'5⁹⁄₁₆" above line (1) to locate the height of the total triangle. That's the dotted line in Figure 11-6. This point is the same as the 12" mark on line (6) of Figures 10-13 and 10-15.

This dotted line (7) closes the total triangle and establishes the slope of the octagon side as 4 in 12. Notice that the dotted line is drawn from the intersection of lines (1) and (2) *only for this first sectional view*. The slope on other sectional views will be drawn to the fascia line (14) at the intersection with the overhang run.

Adding the HAP and Plancher

The HAP for this roof is 6½''. The plancher is 4''. The rafters will be cut from 2 x 8 material that has been surfaced on all four sides (S4S). The rafter material actually measures 1½'' x 7⅜''.

Add the distance of the HAP along lines (2) and (6), then draw the top edge of the rafter through these points to line (4). Add the plancher distance from line (14). Draw the rafter bottom. That's the same as line (11) in Figure 10-13. This line will be 7⅜'' from the top line of the common rafter.

The Octagon Peak and the Fascia Line

Line (13), the octagon peak, will be 24 1/16'' above the plate line (1). This is the rise of 17 9/16'' plus the HAP of 6½''. Line (14), the fascia line, will drop below line (1) by 7 1/16''. Draw line (14) carefully across the plywood.

Projecting to Find the Ridge

Notice the extra dotted lines of Figure 11-6. They weren't needed in Chapter 10. But here they'll help you locate the hall ridge on this project.

Figure 11-7 shows a trial projection for a centered ridge. The hall center line has been extended until it crosses the octagon hip line. This cross point is then projected up to the sectional drawing. Notice that it falls across the dotted line of the roof pitch angle established by the 4 in 12 common rafter. If we measured the scaled distance from the plate line (1) to this dotted intersection, it would be 5⅜''.

This means that if we placed the 2 x 6 ridge at the center of the hall, in order to intersect the octagon broken hip, the ridge would have to be so low that the bottom of it would fall 1/8'' below the plate line. We can't allow that. But if we raised the hall ridge to a better height, the ridge could end up above the broken hip at each end. Some compromise is going to be necessary . . . and a little bit of skillful roof cutting.

Selecting a Framing Point

Look at Figure 11-8. Notice the arrow. That will be a good reference point. The ridge could join the octagon broken hip directly opposite this point on the octagon. Let's make a trial drawing based on this meeting point.

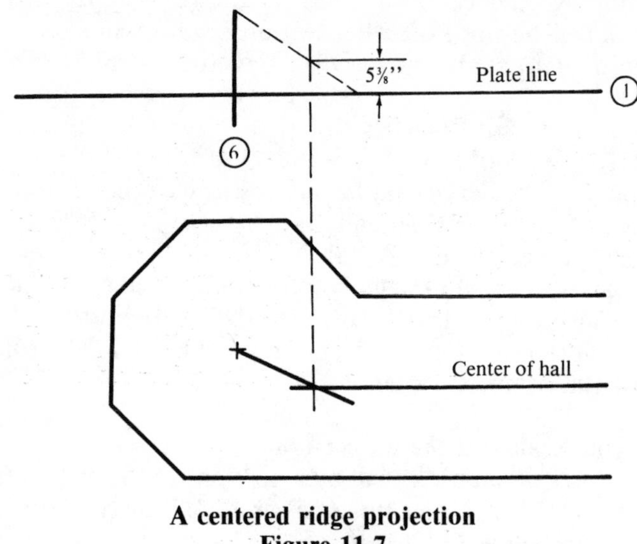

A centered ridge projection
Figure 11-7

Project this point upward until it reaches the dotted roof slope line and downward until it intersects the broken hip. This last point is where you begin drawing the ridge line down the hallway.

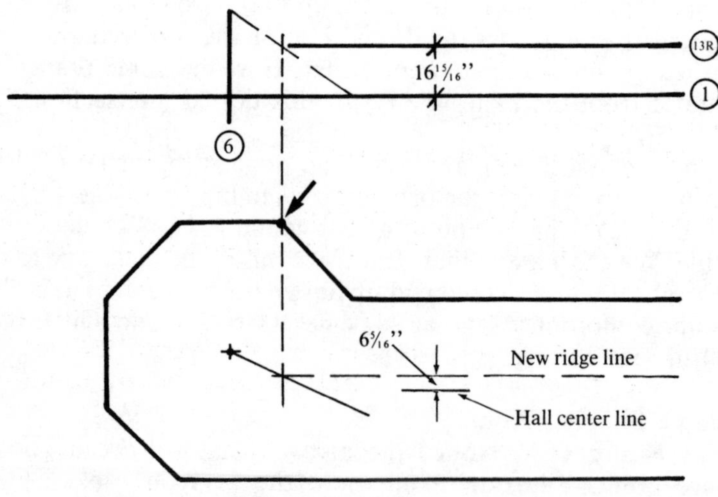

Projecting the framing point
Figure 11-8

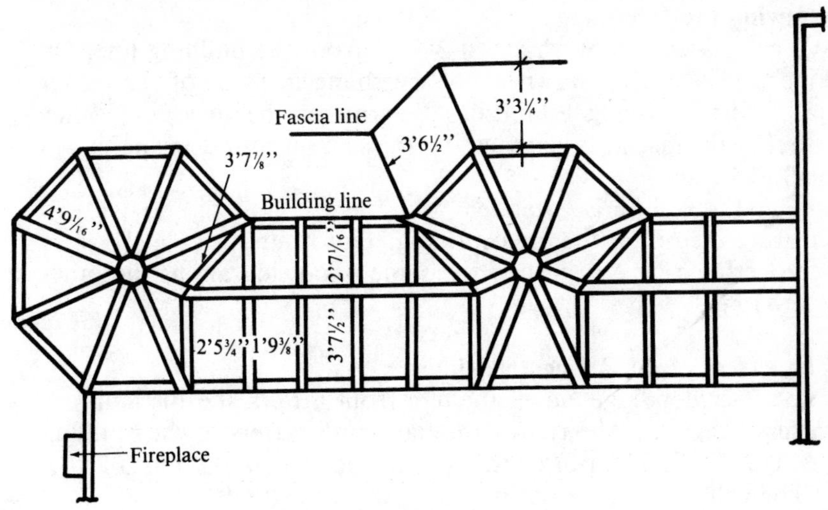

The completed plan view
Figure 11-9

Notice that the ridge line you've just drawn is forward of the center line of the hallway by 6⁵⁄₁₆".

Measure on the sectional view and you'll find that the ridge height at this point is 16¹⁵⁄₁₆". This would give us a closed roof, with a center line that isn't far enough off-center to be noticed by anyone except the architect and the roof cutter. Let's use this ridge location for this job.

Completing the Plan View

Now draw the hall ridge line and connect the broken hips on the plan view. Add both sections of hallway. Add the valley rafters that go from the broken hip intersection to the outside edge of the octagon-front wall intersection. Figure 11-9 shows the completed plan view.

Next, lay out the irregular common rafters of the hall. The octagon-front hall intersection is our starting point. Divide the length of the center hallway into thirds to find a rafter spacing of 1'9⅜".

Spacing for the final rear side irregular common rafter that goes against the broken hip intersect is 2'5¾". This spacing isn't so much more than 1'9⅜" that it will be noticeable on the finished exposed beam ceiling.

Drawing the Overhang

We calculated the overhang at 3'3¼'' from the building line. On Figure 11-9, we've drawn in the overhang in front of the center hall. This overhang establishes the run of the overhang valley rafter to the fascia. If you measure this run on your drawing, it will be 3'6½''.

The fascia line is extended past the octagon hip so we can measure the run of the hip overhang. This run is also 3'6½'', since the overhang here is part of the same parallelogram as the other overhang.

The Effect of an Off-center Ridge

The off-center ridge makes the hall front rafters and the hall rear rafters irregular. The run of the hall front rafters to the building line is 2'7¹⁄₁₆''. The run of the rear rafters is measured as 3'7½''.

The valley rafter is also irregular. We measure its run as 3'7⅞''.

The overhang valley rafter is irregular, too. You've already measured its run as 3'6½''.

We'll have to make sectional views of all these members to determine their mathematical lengths and the amount of bearing at the plate. What you learned in Chapter 10 will come in handy here.

The octagon hip rafters are regular octagon hip rafters, but are not "regular" in terms of rafter manuals or rafter slide rules. You won't find lengths for these members in any rafter table. The rafter tables only contain common rafters, based on 12'', and regular hip rafters, based on 16.97''. Octagon hips are based on 12.98'' and are a different bird altogether. The secant for a 4 in 13 (actually 4 in 12.98) octagon hip is listed in Figure 9-50 as 1.0463. This secant times the hip rafter actual run will give the mathematical length for the octagon hip rafter.

The rafter could also be laid out in a sectional view to find the mathematical lengths. Let's get started making sectional views that show the exact lengths of all members needed to frame this roof.

The Sectional Views

Line (13) in Figure 11-6 represents the height of the octagon peak. It isn't the height of the hall ridge. We have to draw in the hall ridge to establish the irregular members. We'll identify this ridge as line (13R) and project it across the drawing. This line begins at the dotted line intersection in Figure 11-8, 16¹⁵⁄₁₆'' above line (1).

The Overhang Valley Rafter

This rafter runs from the outside wall to the fascia line to support the inside turn of the overhang. See Figure 11-9.

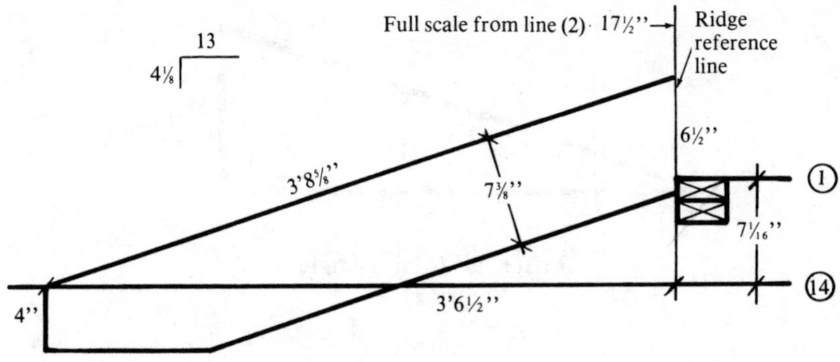

The overhang valley rafter
Figure 11-10

Figure 11-10 shows a building line 17½'' from the common rafter building line, which is line (2) in Figure 11-6. The 17½'' is in full scale and is given here only as a convenience to correctly space your drawing on the 8' sheet of plywood. The other ridge reference lines that follow are for the same purpose. Refer back to Chapter 10, Figure 10-13. Notice how in that drawing, lines (23) and (15) need to be set at the correct distance from line (2) in order to get those drawings in.

Lay out this valley rafter with a run of 3'6½'', as measured on the plan view. Add the HAP of 6½'' to the building line, and draw in the top of the rafter. The length of the sloping line will be 3'8⅝'', the mathematical length for this rafter.

Rafter Tail Mathematics

The rise for this rafter tail will be 13⁹⁄₁₆''. This is the sum of the 6½'' HAP plus the fascia line drop of 7¹⁄₁₆''. Measure the run from the plan view.

To check the mathematical length of this rafter, multiply the run by the secant of 17.70 degrees, as shown in Figure 11-11. The tangent of the angle is 13.5625 divided by 42.5''. That gives you 0.3191. Then punch [INV] [tan] and you have the angle, 17.70 degrees. Push [cos] [1/x] and you get the secant, 1.0497. Multiplying the secant times 42.5'' gives you 44.6116'', or the same 3'8⅝''.

Since this is a valley rafter on an octagon roof, the framing square settings will use a run of 13''. Set up your proportion based on a 13'' run:

$$\frac{42.5}{13.5625} : \frac{13}{?} \quad ? = 4.1485'' = 4\tfrac{1}{8}''$$

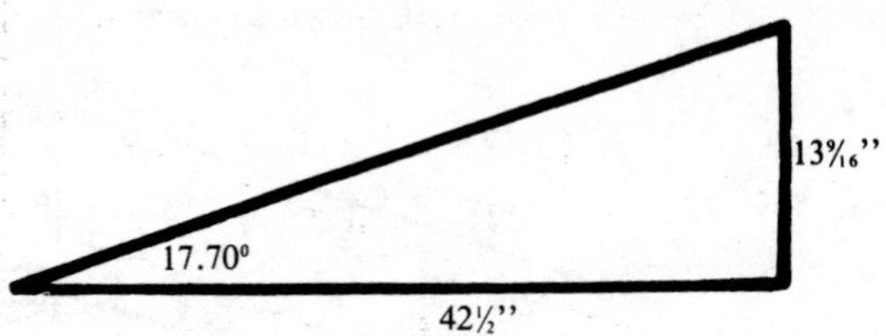

Rafter tail total triangle
Figure 11-11

This rafter will be cut with the stair gauges on 4⅛'' and 13'', or a more convenient proportionate setting, so the framing square will go across the 7⅜'' width of the rafter.

The Octagon Hip Rafter

Draw another ridge reference line 45½ full-scale inches to the right of line (2). See the top of Figure 11-12. On the plan view, the measured run to the building line for this rafter is 4'9¹⁄₁₆''. Lay off this distance to the left of the reference line, then draw in the two plates, as in Figure 11-12. This distance can also be calculated. Look back to Figure 9-17.

The run of the hip overhang will be greater than the run of the common rafter overhang. We measured this greater run at 3'6½''. This becomes the distance from the building line to the fascia line.

Since the hip rafter goes to the octagon peak, the upper intersection for this hip must be at line (13), not (13R). Draw the top line of the hip from the fascia point to line (13).

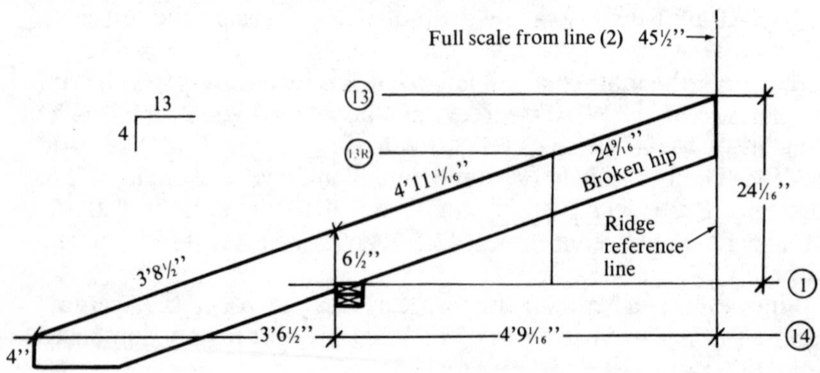

The octagon hip
Figure 11-12

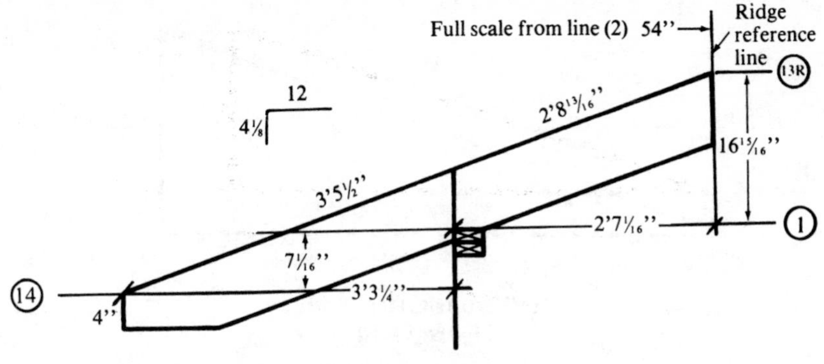

The hall front rafter
Figure 11-13

Now we can measure the mathematical lengths of the hip. They're 4'11 11/16'' and 3'8½'', as in Figure 11-12.

This rafter needs no further calculations. The stair gauges will be set at 4'' and 13''. Notice that the HAP measures 6½''.

The Octagon Broken Hip

Figure 11-12 is a picture of the full octagon hip rafter. Superimposed on it, at the top, is the broken hip rafter. Line (13R) intersects the bottom of the broken hip rafter. The distance from this intersection to the octagon peak is the mathematical length of the broken hip. This distance measures 24 9/16''.

The Hall Front Rafter

Figure 11-13 shows a ridge reference line drawn 54 full-scale inches from line (2), the building line in Figure 11-6. The run of the hall front rafter is 2'7 1/16'' as measured on the plan view. The overhang run would have to be identical with the common rafter run — the distance between lines (2) and (4) in Figure 11-6. This distance is 3'3¼''. Draw the top line of the rafter from the fascia to line (13R). Measure the two mathematical distances. They should be 3'5½'' and 2'8 13/16''.

Hall Front Rafter Math

Look at Figure 11-14. The rise of this triangle is the distance between lines (13R) and (14) in Figure 11-13, which is the sum of the fascia line drop (7 1/16'') and the height of the hall ridge (16 15/16''). This equals 24''. The run is the total run from the ridge to the fascia line. That's 5'10 5/16'', or 70.3125''.

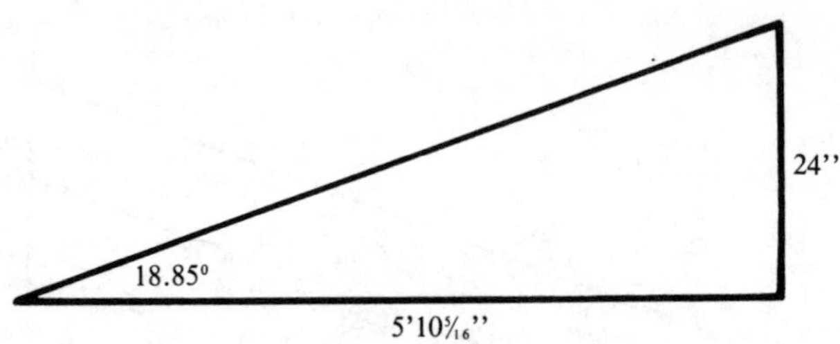

Hall front rafter math
Figure 11-14

To find the length of the rafter, first find the tangent. 24 divided by 70.3125 is 0.3413. Then punch INV tan and you have 18.85 degrees. To find the secant, push cos 1/*x* The display will show 1.0566. You can multiply the run by 1.0566 to get the rafter length, from the ridge to the fascia line.

70.3125'' x 1.0566 = 17.2957'' = 6'2 5/16''

Find the setting for your framing square with the ratio:

$$\frac{70.3125}{24} : \frac{12}{?} \qquad ? = 4.096'' \quad = 4\frac{1}{8}''$$

Your stair gauges will be set at 4⅛'' and 12''.

The Hall Rear Rafter

Draw the ridge reference line 70 full-scale inches from line (2), the building line. We measured the run of this rafter on the plan view, Figure 11-9. It was 3'7½'' to the far side of the existing wall.

The height of plate line (1) has been designed so that it would be just above the *existing wall* frieze blocks at the level of the roof. This is shown in Figure 11-15. Draw the top line of the rear common rafter to a point 6½'' above the inside of the plate line. This distance is the HAP. Then draw in the lower edge of this rafter.

Now you can measure the top and bottom of this common rafter, and use these dimensions for your rafter layout.

Hall Rear Rafter Math

Here's how to calculate the length of the hall rear rafters. The run of the triangle in Figure 11-16 is 3'4''. That's calculated from the 3'7½'' dimension of Figure 11-15, minus 3½'' for the thickness of

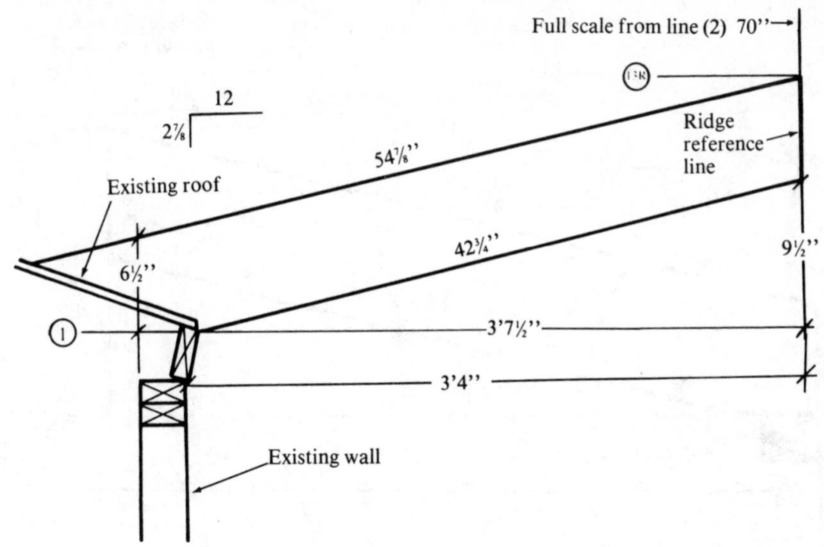

The hall rear rafter
Figure 11-15

the wall. The height of this triangle is 9½''. The tangent of the angle is 9.5 divided by 40'', or 0.2375. Now use your calculator to get the angle, 13.36 degrees. Then punch [cos] [1/x] and you have the secant, 1.0278.

The framing square settings are found by the ratio:

$$\frac{40}{9.5} : \frac{12}{?} \qquad ? = 2.85'' \qquad = 2\tfrac{7}{8}''$$

This rafter will require stair gauge settings at 2⅞'' and 12''.

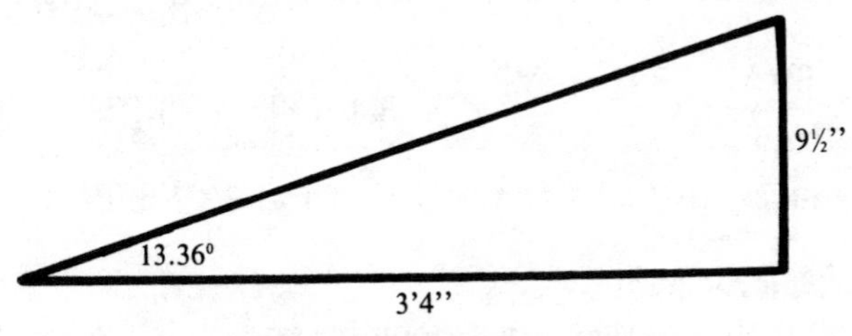

Hall rear rafter math
Figure 11-16

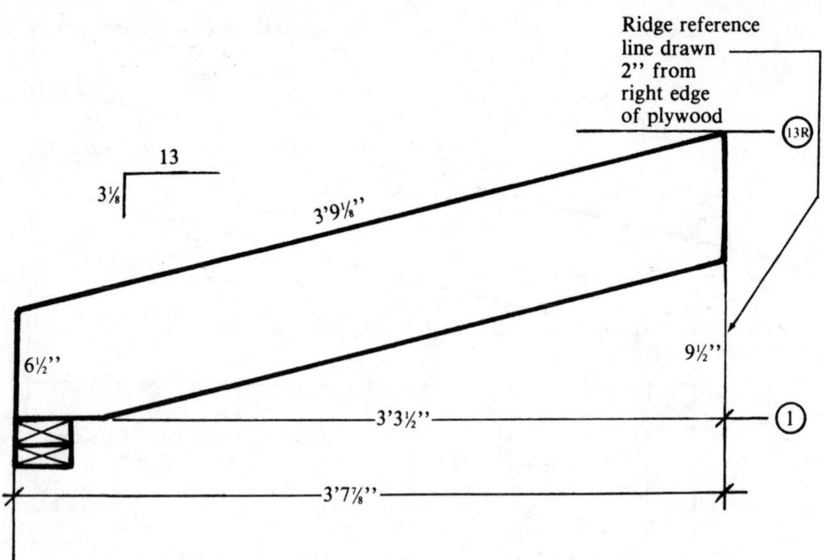

The irregular valley rafter
Figure 11-17

The Irregular Valley Rafter

Start this ridge reference line 2" in from the right edge of the plywood. The plan view run of the irregular valley rafter measures 3'7⅞" (from Figure 11-9). Lay out this distance from the ridge reference line as shown in Figure 11-17.

Lay out the HAP as shown. This becomes one point for the rafter line. The mathematical length is measured as 3'9⅛". Draw in the bottom line of the rafter 7⅜" from the top line.

Irregular Valley Rafter Math

We've already measured the space below this rafter. Use these numbers to find the stair gauge settings. Set up the proportions as follows:

$$\frac{39.5}{9.5} : \frac{13}{?} \qquad ? = 3.1266" = 3\tfrac{1}{8}"$$

The stair gauges for this rafter will be set at 3⅛" and 13".

Plan View Enlargements

We've figured the mathematical lengths and stair gauge settings for all these rafters. Next, we have to find their setbacks and shortenings on the plan view.

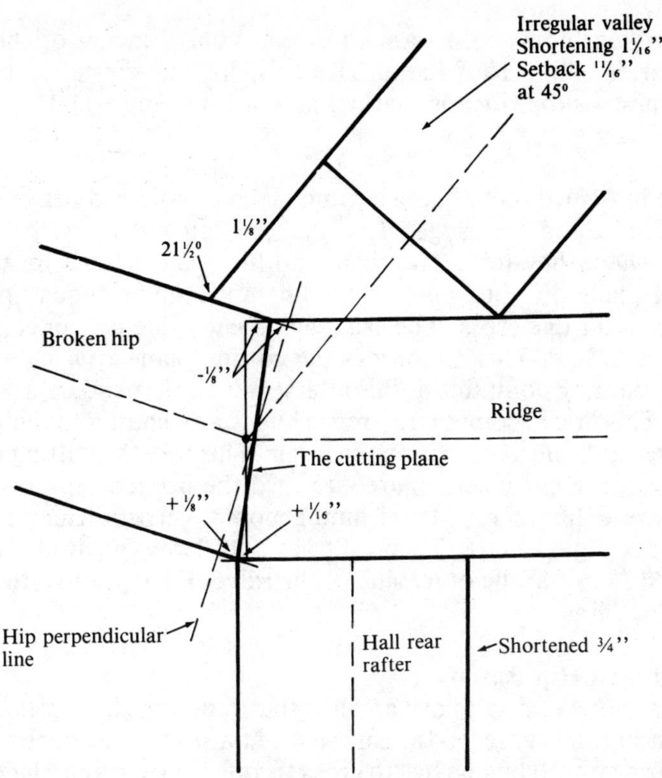

The broken hip intersection
Figure 11-18

First, we'll need a full-size drawing of the broken hip intersection, where the broken hip, ridge, irregular valley rafter and the hall rear rafter meet. That's Figure 11-18.

Another important intersection is the point where the octagon and hall meet. That's where the overhang valley, the irregular valley, the octagon hip, and the irregular hall front common rafter come together. Figure 11-19 shows this point.

These two full-size drawings will give us all the information we need to cut the irregular members of this roof.

The broken hip intersect can be drawn in full scale directly over the same plan view intersect.

Figure 11-18 shows the intersection where the broken hip meets the ridge. Notice that the hall rear rafter is set over from the framing point by 1/16''. This aligns it with the cutting line of the ridge and broken hip. You'll want to extend your drawing of the hall rear rafter to the roof in full size to help find the tail cut. The shortening at the ridge is marked as 3/4" in Figure 11-18.

The Ridge

Notice in Figure 11-18 that a cutting plane is indicated between the broken hip and the ridge. This makes both members the same width where they join. The ridge and hip are extended up to the cutting plane on both sides of the line until the extended lines of the hip and ridge cross. The crossing on each side is connected by the broken line. This establishes the cutting plane.

The framing point line at this intersection is perpendicular to the ridge. This line becomes a reference line for extending the ridge on one edge and shortening it on the other. This way the cutting plane angle will present itself squarely toward the broken hip.

Square a line across the framing point, perpendicular to the ridge. See Figure 11-18. Then subtract 1/8" from the inside corner and add 1/16" to the other side of the ridge. Finally, draw the cutting plane line.

The Broken Hip Rafter

On this roof, we don't have any line that runs at right angles to the broken hip in Figure 11-18. But we need a perpendicular here, so we'll draw one. The line has to cross through the cutting plane and the ridge center line. Growth and shortening will be measured from this line so that the broken hip can present itself squarely toward the ridge.

Subtract 1/8" from the inside corner and add 1/8" to the outside edge of the broken hip. Connect these points to get the cutting plane line for the broken hip.

The Irregular Valley Rafter

The shortening for the irregular valley rafter is measured at 1 1/16" from the framing point. Setback for the cheek cut will be 11/16". A protractor will show this to be a 45-degree cut. The tip of the cheek cut is interrupted by the broken hip. Figure 11-18 shows that the growth side of the cheek cut will be cut off at 1 1/8". The cut is made at 21 1/2 degrees. This rafter is laid out beginning at Figure 11-20.

The Hall-Octagon Intersection

Draw the hall-octagon intersection on your plan view in the space toward fascia line (14).

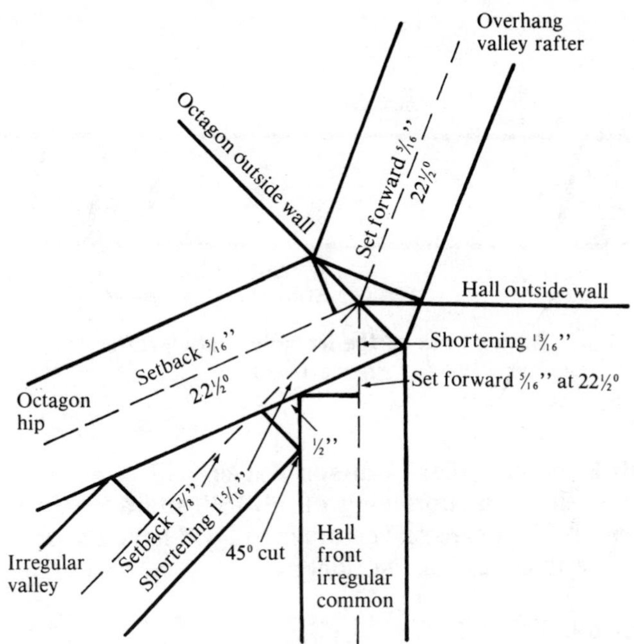

The hall-octagon intersection
Figure 11-19

Now we have to decide which members go through the intersection and which members are interrupted, as we did in Chapter 3, Figure 3-39. In Figure 11-19, we show the octagon hip and overhang valley rafter meeting at the framing point, and the other two rafters being interrupted. That's a good choice, but not the only way to form this intersection.

The Octagon Hip Rafter

The hip will end at the mathematical building line length. The setback will be 5/16'' and the cut will be at 22½ degrees.

The Overhang Valley Rafter

The overhang valley will have a mathematical length of 3'8½''. There will be no shortening, but the set forward will be 5/16''. Again, the cut will be at 22½ degrees. The fascia end will require a valley tail cut with an octagon 5/16'' set forward. See Figure 9-30. Each cheek will be cut at 22½ degrees.

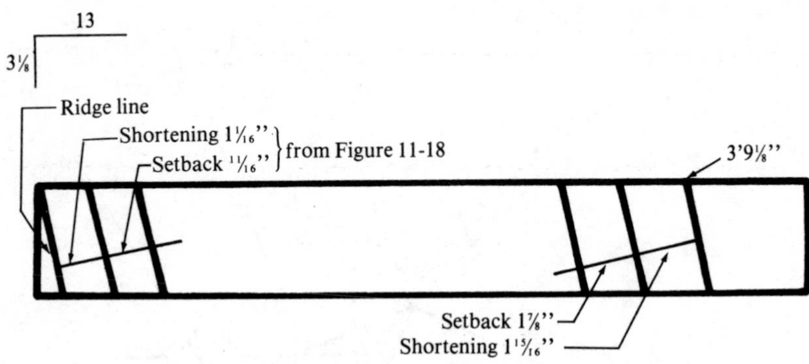

Major lines of the irregular valley rafter
Figure 11-20

The Hall Front Irregular Common Rafter
This rafter shows a shortening of 13/16" and a set forward of 5/16", cut at 22½ degrees. The other end will have a shortening of one-half the thickness of the ridge.

The Irregular Valley Rafter
Lay out this end of the valley rafter very carefully. Part of this cut will have to be made with a hand saw.

Figure 11-20 shows the irregular valley rafter in reversed position. We've drawn it this way so that it fits better with the plan view drawings. Lay out the ridge end first, beginning from the left side.

Draw all the lines as in Figure 11-20. At the tail, square the shortening and cheek cut lines across the top of the rafter, then put a dot in the center of the shortening line. Also square the ridge cheek cut line across the top.

Figure 11-21, which is a top view of the rafter, shows markings for the cheek cuts. Draw dot (a) 1⅛" from the cheek cut line as shown in Figure 11-18. Place a plumb cut line, which passes through point (a), on the far side of the rafter. The plan view shows a 21½-degree cut is to be made along this line.

Next, cut along the cheek cut line at 45 degrees. At the tail end of the irregular valley rafter, draw line (b) from the corner of the cheek cut line through the dot at the center of the shortening line. This gives you the extremely long cheek cut that fits against the octagon hip. See Figure 11-19.

From Figure 11-19, measure off a 1/2" from the framing point along the top edge of the cheek cut line. Draw line (c) of Figure

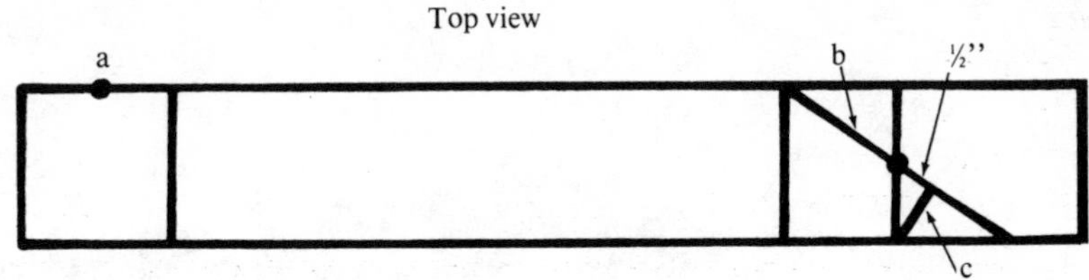

Marking the cheek cuts
Figure 11-21

11-21 from this point back to the shortening line. Line (c) shows the construction, but is otherwise not needed, since the cheek cut that fits against the front irregular common rafter is a 45-degree cut along the shortening line. Now it's easier to hand saw the remaining cut.

More on Irregular Roof Jobs

If you've followed along with the descriptions and made the drawings suggested, you're prepared to take on a fairly complex roof. But don't be concerned if you can't remember everything we've covered in this chapter. Just have this book handy to use as a reference when a complex roof job comes up.

Even though we've covered two irregular roofs, there are a few important points on irregular roof cutting that we haven't touched yet. And that's the subject of the next chapter.

12

Irregular Roof Problems

Irregular roof jobs come in several varieties. We've covered three common types. Non-centered ridges were the subject of Chapter 8. In Chapter 10, we cut a roof with hips that didn't cross the building lines at the corners. Chapter 11 was a more complex job, including both an octagon and an off-center ridge.

There are four other common irregular roofs that you may come across: the unequal pitch/equal ridge roof, the irregular California, the shed roof dormer, and the hip or gable dormer. We'll cover each of these in this chapter. First we'll look at a hip roof that has two ridges at the same height, even though each span has a different pitch.

The Unequal Pitch/Equal Ridge Roof

Get out your model again so you can follow along. We are going to attach a 2' addition at one end of the 3' x 5' rafter plate, as shown in Figure 12-1.

On this job, the owner wants the main building to have an 8 in 12 hip roof. That's the easy part. But he wants the ridge on the ad-

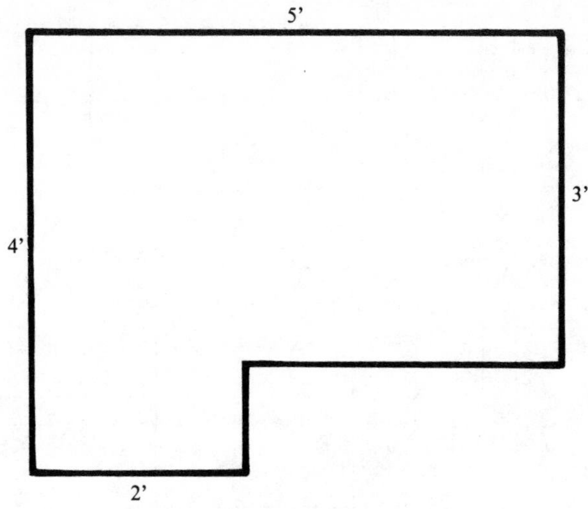

The model plate
Figure 12-1

dition to be at the same height as the main ridge. The pitch of the main roof and the addition roof will have to be different, because the spans are different.

Start by figuring the main span ridge height. The main roof is 8 in 12. Its run will be 1½'. An 8 in 12 roof with a 1½' total run will have a mathematical ridge height of 12" (1.5 times 8" is 12").

The 2' addition ridge also has to be at 12". But the total run is 1'. If the run is 12" and the rise is 12", the addition has to be framed with a 12 in 12 roof. That's the only way to make the two ridge heights equal.

Figure 12-2 shows the plan view of this roof. Notice that the irregular hip and the irregular valley are directly in line, and the two ridges are at 90 degrees. Even though this roof is irregular, it follows the principles we learned back in Chapter 3 for a Rule 4b ridge.

Drafting the Plan View

Draw this roof on a 19" by 24" piece of drafting paper. Use a scale of 3" to the foot. Leave room above the plan view for the sectional view drawing, Figure 12-4.

Draw the building lines as dotted lines and indicate the 6" overhang with a solid line. Make the 8 in 12 hip a double cheek cut hip. The 12 in 12 hip will be a single cheek cut hip.

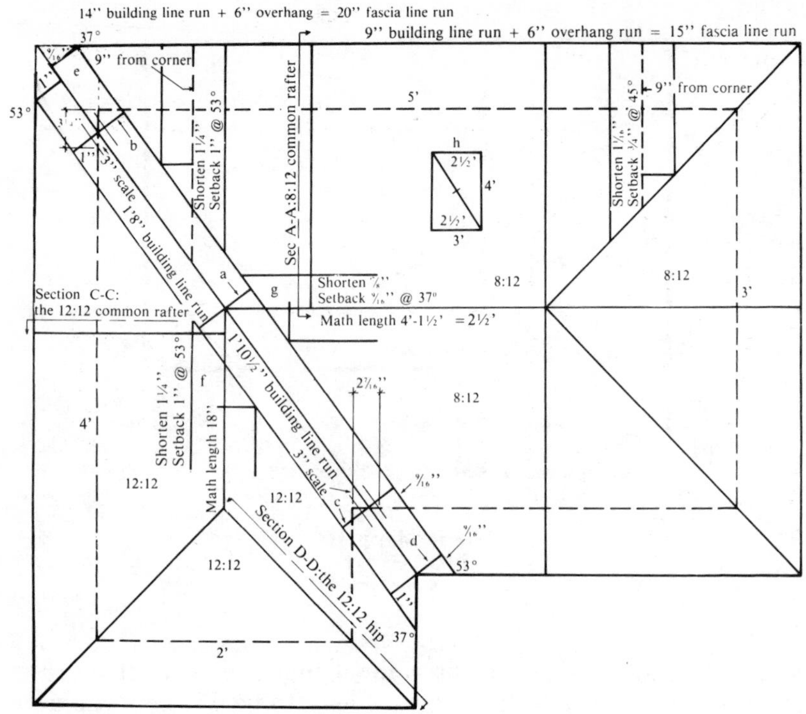

The completed plan view
Figure 12-2

Draw the irregular hip and valley as one continuous line from the outside corner of one fascia line to the inside corner of the opposite fascia line, *not* the building line. Draw the thickness of these irregular rafters in 3'' scale where they cross the building line.

The Full-size Drawing

Now, draw in the 1½'' thickness of the irregular hip and valley rafter in actual size. Draw a line at the point where they meet, line (a). Square each building line across the rafters at lines (b) and (c). At line (d), square the fascia framing-point line across the irregular valley rafter, so you can measure the length of each setback. At line (e), draw the irregular hip set-forward lines so the amount of each set forward can be measured.

At (f) and (g), draw each ridge in full size. Then you can measure the shortening and setback at each ridge.

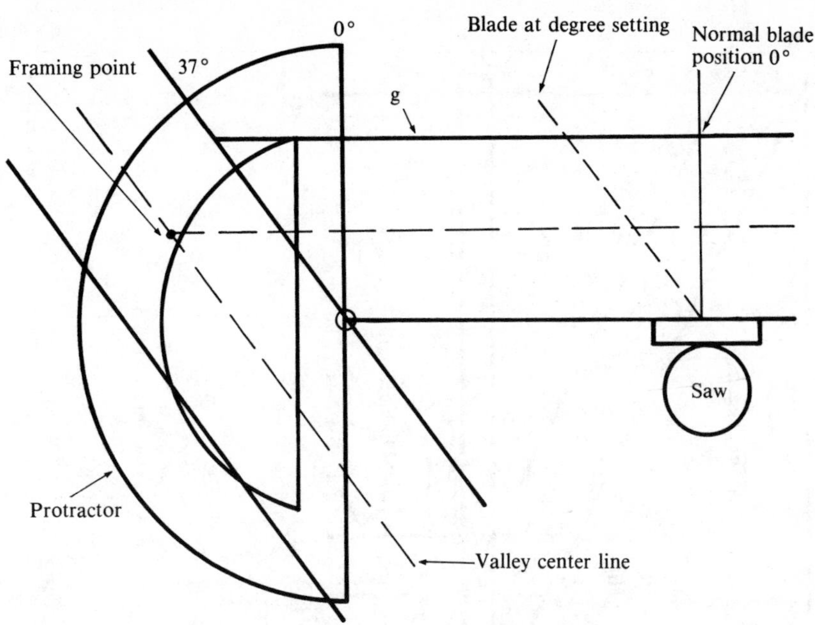

Measuring the angle of the main ridge
Figure 12-3

At (h), a small drawing shows the circumscribed rectangle of the irregular members. From the 3-4-5 triangle, we can see that the fascia line run (the top view, or plan view run) for each member is 2'6".

Carefully measure each building-line distance. The irregular hip has a building-line distance of 1'8". The irregular valley has a building-line distance of 1'10½".

Measuring the Angles

On the full-size drawing, use a protractor to measure the cheek cut angle for each ridge, and the angles of the cuts at (d) and (e). Figure 12-3 shows how to do it.

Drawing the Sectional View

Now we'll draw the sectional view we mentioned earlier. Draw it above the plan view. See Figure 12-4.

Since the addition roof has to conform to the main roof, a main roof common rafter will be used as the beginning point for three major lines: the plate line (1), the ridge line (13), and the fascia line (14).

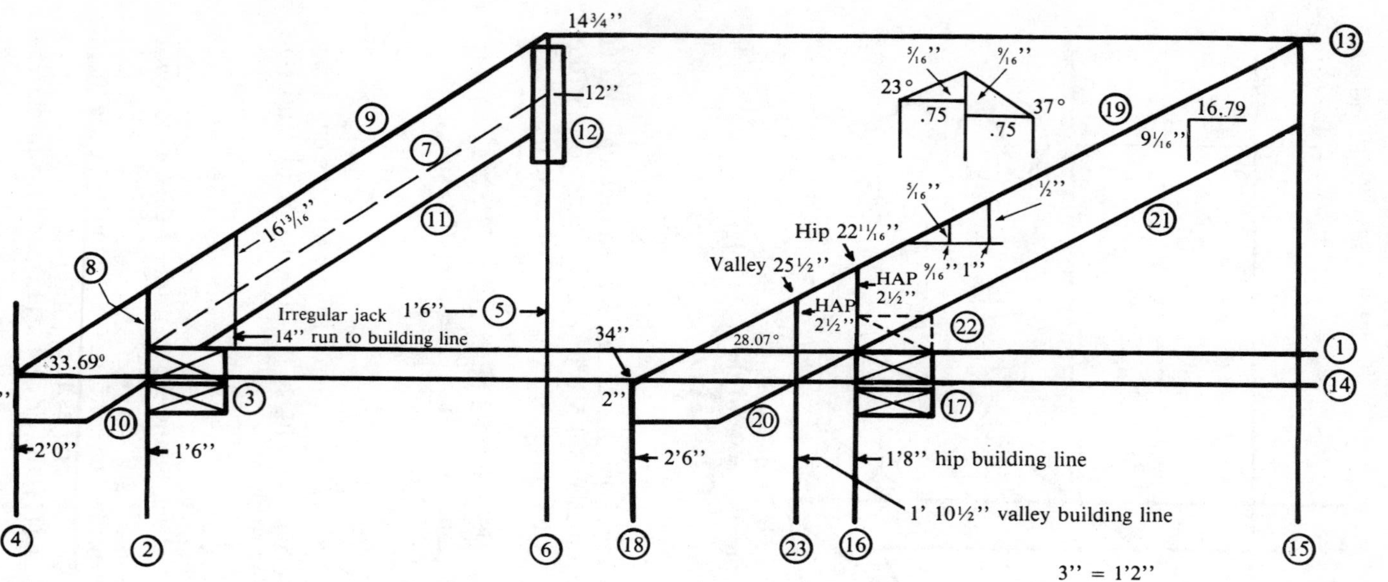

Section A-A:
The 8:12 common rafter

Section B-B:
The irregular hip and valley

The sectional view drawing
Figure 12-4

The left side of Figure 12-4 shows Section A-A from Figure 12-2. Make the same drawing on your paper. You may need to refer back to Chapter 10 for directions on how to establish lines (1), (13) and (14). These directions are given with Figures 10-14 through 10-16.

Finding the Lengths of the Irregular Hip

From line (15), lay out a distance of 1'8'' for the irregular hip building line. Follow Section B-B at the right side of Figure 12-4. Line (18) is 2'6'' from line (15). Draw the top of the rafter, line (19), from the meeting point of lines (13) and (15), to the meeting point of lines (14) and (18). The irregular hip and valley rafters will be made of 2 x 4's, so draw line (21) 3½'' from line (19).

Notice that the plate must be raised 1½''. Follow the directions in Chapter 10, under Figure 10-19. The new HAP measures 2½''.

The method for making the irregular hip backing cuts was explained at Figure 10-22 in Chapter 10. Add these details to the sectional views and compute the angles of the backing cuts, as shown in Figure 10-33.

Finally, measure along line (19) to find the mathematical lengths of the irregular hip rafter. The building line length is $22\frac{11}{16}$'', and the fascia line length is 34''.

Finding the Lengths of the Irregular Valley

Measure over 1'10½'' from line (15). That's the run of the irregular valley rafter to the building line. See line (23) in Figure 12-4.

Find the mathematical length of the irregular valley rafter by measuring along line (19) from the ridge to line (23) at the irregular valley building line. This length is 25½''. The fascia line length is 34''.

The 12 in 12 Common Rafter

One 12 in 12 common rafter is needed for the model. If there were no HAP *and* no overhang, this would be a regular 12 in 12 rafter. But because of the HAP and the overhang, this common rafter is slightly irregular.

For this roof, the common rafter has somewhat shorter mathematical lengths than on a standard 12 in 12 roof. The building line is $16\frac{1}{16}$'' rather than 17''. The fascia line is $24\frac{1}{16}$'' rather than $25\frac{7}{16}$''. The new pitch becomes $10\frac{11}{16}$ in 12. Look at Section C-C of Figure 12-5.

A regular-cut 12 in 12 rafter would be too low on the ridge, and the fascia point would fall far below the fascia line (14).

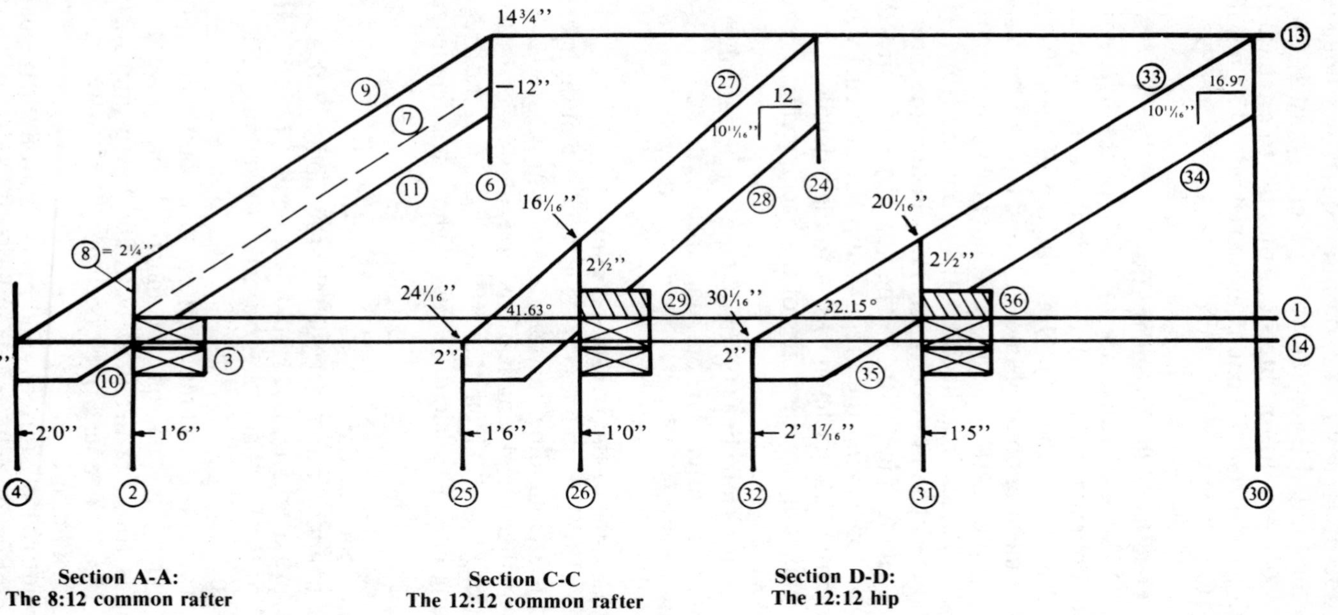

The 12:12 sectional views
Figure 12-5

The 12 in 12 rafter must be laid out like any irregular rafter. Notice in Figure 12-5 that we've used the 8 in 12 common rafter from the main roof to generate the three major lines. The scale should be 3'' to the foot.

Draw line (24) as the ridge center line. Lay out the fascia, line (25), the proper distance away. Add line (26), the building line, and the two plates. Next, draw the common rafter top line (27), between the meeting point of lines (14) and (25) and the meeting point of lines (13) and (24).

Now the rafter will fit properly against the ridge, and the overhang will remain level and at a constant 6'' from the building line.

Perpendicular to line (27), measure down 3½'' and draw line (28). This shows the thickness of the 12 in 12 rafter, and helps you see that a raised plate is needed. The raised plate is labeled (29). Now you can measure the HAP for this rafter.

Finding the 12 in 12 Common Rafter Angle

Refer again to Figure 12-5. The total run for the 12 in 12 rafter is the distance from line (24) to line (25). This is 18''. The total rise is measured from line (14) to line (13). The distance from line (1) to line (13) is known to be 14¾''. Refer back to Chapter 10, Figure 10-27, to see that the distance from line (14) to line (1) is 1¼''. The total rise, therefore, is 16''. The tangent of this rafter is 16'' divided by 18''.

$$\text{Tan 12:12 common rafter} = \frac{16''}{18''} = .8888$$

$$\text{Inverse tangent} = 41.63^\circ$$

$$\text{Secant} = 1.3380$$

The 12 in 12 Common Rafter Mathematical Lengths

Now we can use this secant to calculate the mathematical lengths to the building line and fascia line. If you prefer, scale these dimensions off on the sectional view.

Find the settings for your framing square by setting up the ratios:

$$\frac{16''}{18''} : \frac{?}{12''} \qquad ? = \frac{12'' \times 16''}{18''} = 10.6667'' = 10\tfrac{11}{16}''$$

Set the framing square to 10¹¹⁄₁₆'' and 12''. Avoid the 12ths scale on the back of the framing square. Shorten the rafter one-half the thickness of the ridge.

The 12 in 12 Single Cheek Cut Hip

This hip is also slightly irregular. See Section D-D in Figure 12-5. Line (30) represents the ridge line.

The apparent run of this hip to the building line is 1' and the run to the fascia line is 1'6". But you can't use the apparent run for the sectional view layout. You must find the actual run. There are two ways to do this. First, you could scale off the distance on Figure 12-2. The other way is to multiply by the square root of 2 (1.414) to get the actual run.

The building line	The fascia line
$1' \times \sqrt{2} = 1'5''$	$1.5' \times \sqrt{2} = 2'1\frac{7}{16}''$

Line (31) in Section D-D shows a run of 1'5" to line (31) and 2'1$\frac{7}{16}$" to the fascia line (32).

Now draw in the top of the hip rafter, line (33), from the intersection of (14) and (32) to the intersection of (13) and (30). Add the 3½" thickness of the rafter, line (34).

If the plate for this rafter is raised 1½" with plate (36), the birdsmouth will provide adequate bearing. The new HAP can be measured at 2½".

The 12 in 12 Hip Angle and Mathematical Lengths

The rise for this rafter is the same 16". The run is 2'1$\frac{7}{16}$".

$$\text{Tan 12:12 hip} = \frac{16''}{25.4558''} = .6285$$

$$\text{Inverse tan} = 32.15^\circ$$

$$\text{Secant} = 1.1811$$

Now use the secant to calculate mathematical lengths to the building line and fascia line on Section D-D, Figure 12-5. Another way is to measure the drawing.

Cutting the 12 in 12 Hip

Set up a ratio to find settings for your framing square.

$$\frac{16''}{25.4558''} : \frac{?}{16.97''} \qquad ? = \frac{16'' \times 16.97''}{25.4558''} = 10.66'' = 10\frac{11}{16}''$$

Set the framing square to 10$\frac{11}{16}$" and 16.97". Notice that the 12 in 12 common rafter was set to 10$\frac{11}{16}$" and 12". Because of the HAP and overhang, the 12 in 12 roof is shifted to a 10$\frac{11}{16}$ in 12 roof. This maintains a level fascia line and constant 6" overhang.

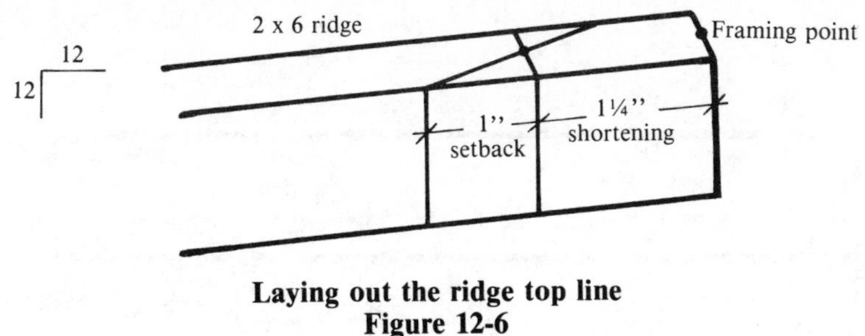

Laying out the ridge top line
Figure 12-6

Cut out two opposite cheek cut rafters. Either drop or back these hips. The angle for the backing cut is the same angle as the hip, 32.15 degrees.

The 12 in 12 Ridge

The mathematical length of this ridge is 18", as shown on the plan view. Add ridge growth for a single cheek cut hip to one side, and subtract shortening and setback from the other side. Don't try to do the subtraction first and then cut the ridge length. That wouldn't leave you with enough material to finish the cheek cut.

On the plan view, the shortening measures 1¼". The setback measures 1". See Figure 12-6. Since the cut is 53 degrees, you'll need a hand saw. Lay out the top line as a guide for the hand saw. Refer back to Chapter 10, Figure 10-57.

Remember to cut back the upper corner of the ridge at the hip end.

The 8 in 12 Ridge

The mathematical length of the 8 in 12 ridge is 2'6". Since it frames to a double cheek cut hip, the 3/4" growth must be added. That brings the ridge total to 2'6¾". The shortening and setback, as measured in the plan view, are 7/8" and 9/16". The cheek cut is at 37 degrees.

Cut out these ridges and frame them to the model. The extra common rafters at the irregular end will hold these ridges in place. Notice that the 12 in 12 side ridge will be lower than the 8 in 12 side ridge by 3/16", but the roof sheathing will be at the same level. (Look back to Figure 7-23 in Chapter 7.)

The Angle of the Irregular Hip and Valley

The run and rise of these irregular members just happen to be identical to those in Chapter 10. See Figures 10-27, 10-28 and 10-29.

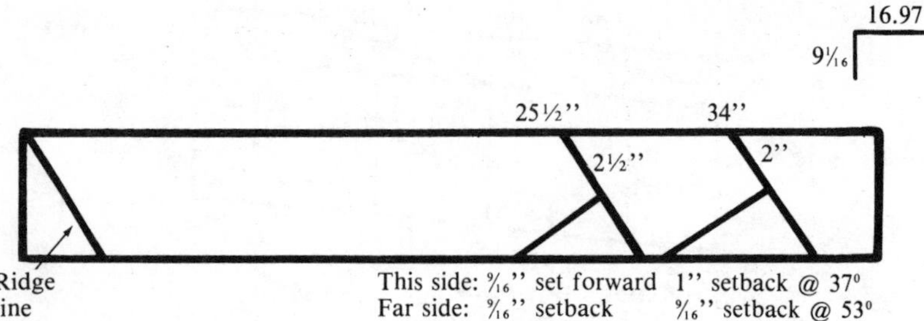

Beginning the valley rafter layout
Figure 12-7

The Irregular Hip
This irregular hip is almost identical to the one in Chapter 10, except that it had a single side cut. This hip is cut on the ridge line without shortening or setback.

The Irregular Valley
Begin the valley rafter layout from the left side. Crown a piece of 2 x 4 and place the crown away from you. Set the stair gauges on 15" and 8", or 16.97" and 9⁹⁄₁₆".

The building line for this rafter is at 25½". See Figure 12-7. The fascia line is the same 34". Add the plancher line. Without a raised plate, the valley HAP measures 2½". Square the bottom of the seat cut line across the bottom of the rafter. Square the top of the building line and fascia line measurements across the top of the rafter.

Add the set forward and setback on each side. If you vee this rafter, the seat cut must be raised.

Framing the Irregular Hip and Valley
Start by measuring, on the plan view, the distance each rafter frames away from the corner. Note on Figure 12-2 that the edge of the hip rafter is 3¼" in from the corner. Note also that the valley rafter edge is 2⁷⁄₁₆" from the corner. Add the raised plate under the hip rafter and frame the rafters to your model.

The 8 in 12 Jack Rafters
Both jacks are to have their framing point 9" from the corner of the building. The regular-side jack rafter will have a run of 9" to the building line, a shortening of 1¹⁄₁₆", and a setback of 3/4" at 45 degrees. See Figure 12-8.

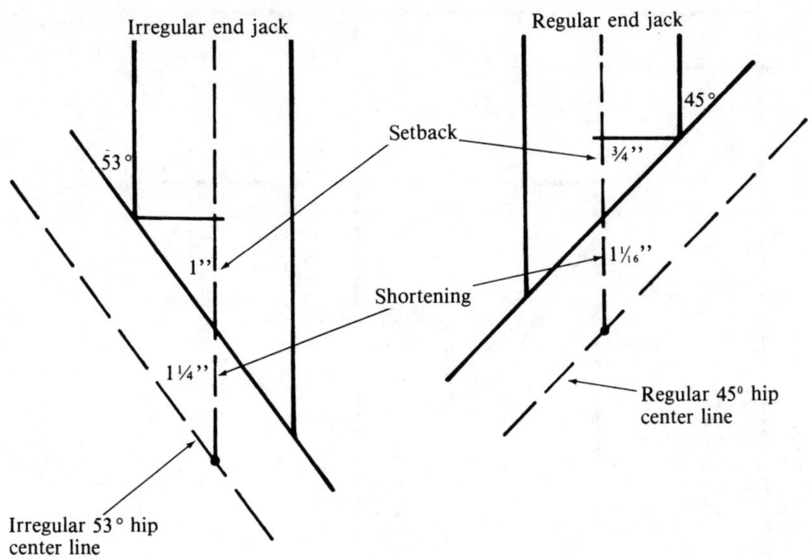

Comparing the 8:12 jacks
Figure 12-8

Finding the Run of the Irregular Jack
We need the run of the irregular-side jack. There are three ways to find it. First, you can measure your plan view to find the length, 14". Then measure 14" from line (6) on the sectional view, and project upward to the roof plane to find the mathematical length of 16¹³⁄₁₆". See Figure 12-4, Section A-A.

You could also find the jack length using the secant for an 8 in 12 roof: 1.2018 times 14" equals 16.8259", or 16¹³⁄₁₆".

The third way is to find the run of the jack mathematically. But the run of this jack can't be calculated the usual way. Here, you have to combine the building line run of 9" with the run of the 6" overhang, shown in Figure 12-9. Then subtract the rafter's own overhang to find the mathematical run to the building line. You must use the irregular rectangle (in this case the 3' x 4' rectangle). The 9" building line run plus the 6" overhang makes 15" of run. Now, set up a proportion:

$$\frac{3''}{15''} : \frac{4''}{?}$$

$$? = \frac{15'' \times 4''}{3''} = 20''$$

6"
9"
Building line
Run of the jack
Irregular hip
Fascia line

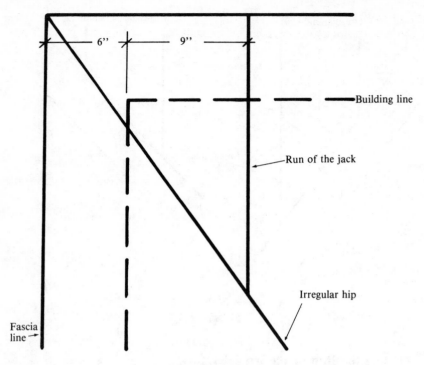

The mathematical run
Figure 12-9

Subtract the 6" overhang from the 20" run to the fascia line to find the 14" run to the building line.

Laying Out the Irregular End Jack

Crown a piece of 2 x 4 and lay the crown away from you. Set the stair gauges to 8" and 12". Lay out the building line and fascia

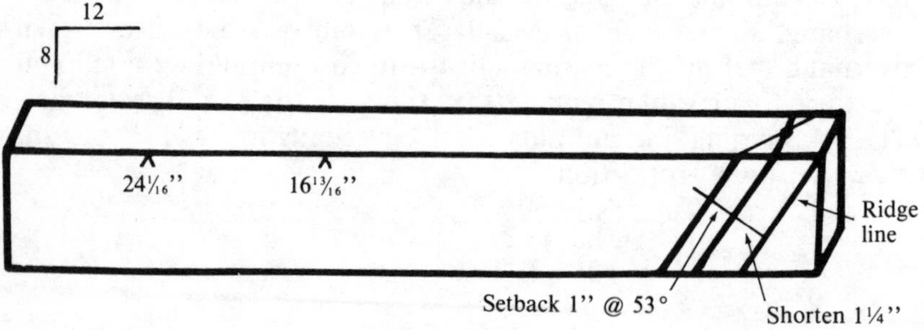

Laying out the jack
Figure 12-10

line as in Figure 12-10. Add the HAP and plancher, then lay out the shortening and setback as measured on the plan view. The cheek cut will be 53 degrees. Construct the top line and saw the cheek cut by hand. Then mark the rafter plate at 9¾'' and nail this rafter to the model.

Congratulations. You've mastered another type of irregular roof!

The Irregular California

Figure 12-11 shows a different type of irregular hip roof. Notice that the 3' addition is centered along the 5' side of the main building. The main roof pitch is 8 in 12 and the addition pitch is 12 in 12. The pitch is unequal and the addition ridge is higher than the ridge on the main roof.

Some parts of this roof present problems we haven't explored yet. Let's look at these differences.

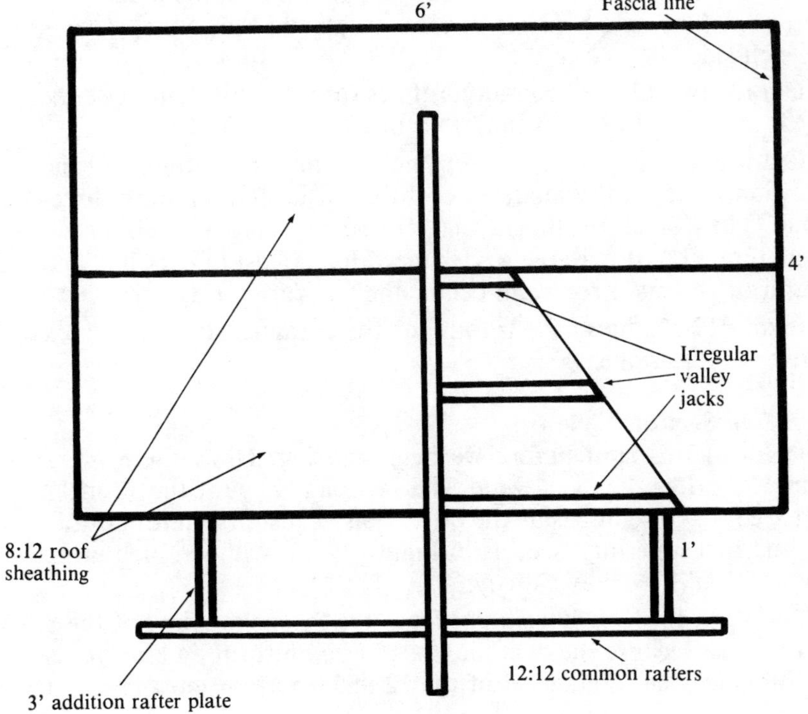

The irregular California model
Figure 12-11

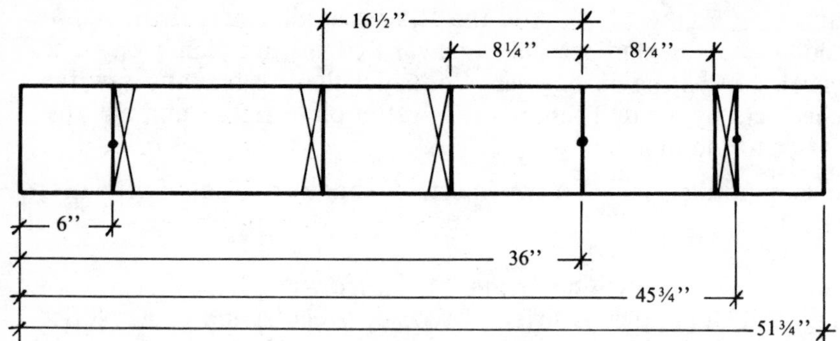

Laying out the ridge
Figure 12-12

Use your 3' by 5' plate. With a 6" overhang on all sides, the roof will be 4' by 6' to the fascia line. Add a 3' addition to the center of the model frame, as in Figure 12-11. Nail the main ridge and 8 in 12 common rafters in place, then add the roof sheathing to each side.

Cut the two 12 in 12 common rafters for a 1½' total run. Use the HAP and raised plate as indicated in Figure 12-13.

Cut a 2 x 6 ridge to 51¾". Frame the common rafters and the addition ridge on the model. Level the rafter plate in both directions. Then block up the 12 in 12 ridge so that it's level. You'll need about 1" of material under the ridge. Measure over half the length of the main roof and center the California ridge.

Figure 12-12 shows the layout for the framing points and jack rafters on the ridge.

The Two Sectional Views

Let's draft this roof before we begin cutting. Draw the 8 in 12 common rafter the usual way. The sectional view of the 12 in 12 rafter could go right beside the rafter you've just drawn. But drawing the two superimposed, as in Figure 12-13, will be much more useful. You'll see why shortly.

The 12 in 12 slope starts right from the fascia level line of the 8 in 12 roof. Because the overhang has to be uniform on both parts of this roof, the fascia level of the 12 in 12 portion must begin at this point.

The 12 in 12 common rafter distance from line (4) to line (6) is 2'; therefore, the rise must be set at 2 times 12", or 24".

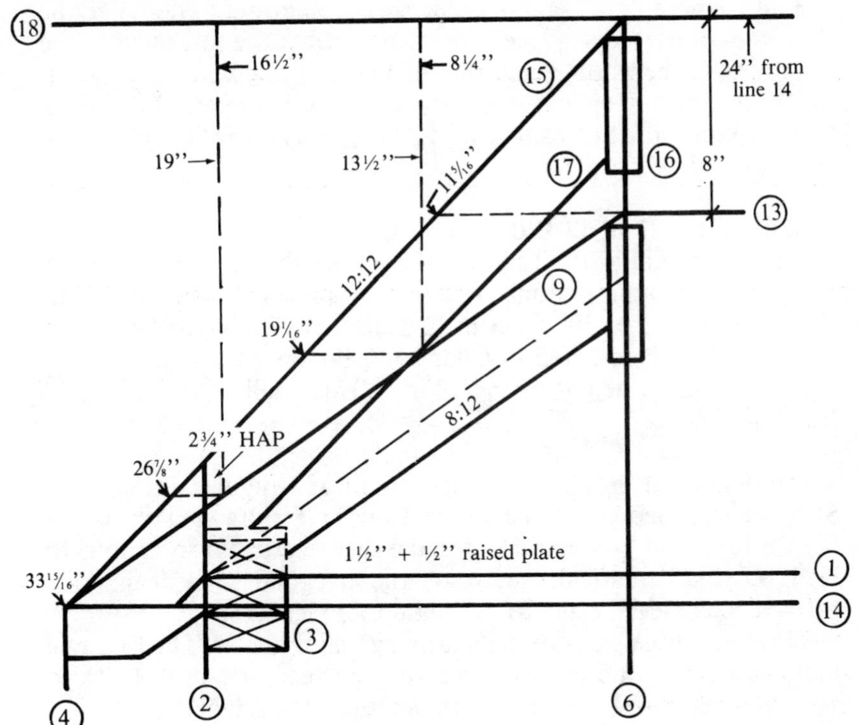

The superimposed sectional views
Figure 12-13

From line (14), measure up 24" along line (6). This establishes the slope of the 12 in 12 rafter. Draw line (15), the top of the rafter. Draw in the California ridge (16) as a 2 x 6 ridge. Draw line (17) 3½" from line (15). Notice that the plate must be raised for the 12 in 12 common rafter. Use a 2" raising for the plate. The HAP will be 2¾".

Draw line (9) and line (14) in red. Add line (18), the California ridge. On line (18), lay out 8¼" and 16½". These are the short point measurements of the jack rafters.

Square down from the 12 in 12 ridge to the 8 in 12 pitch line, as shown by the dotted lines. From the intersections, project a level line to the left until it intersects the 12 in 12 line. These intersections on the 12 in 12 line give us the mathematical lengths for the jack rafters that end at each position. Measure down along line (15) from the intersection of lines (15) and (18).

From these mathematical lengths, subtract one-half the thickness of the ridge at the top. Also subtract the thickness of the sheathing at the bottom and the thickness of any valley strips that are used. The bottom cut is a compound seat cut made at the angle of the main roof. Set your saw to 34 degrees and follow along the seat cut line.

Height of the 12 in 12 Ridge

The actual height of the 12 in 12 ridge above the rafter plate will be 24", minus two dimensions. You have to subtract both the 1¼" of the fascia line drop on the 8 in 12 sectional view, and the loss of ridge height as presented in Chapter 7, Figure 7-23.

So, the actual height of the 12 in 12 ridge will be 24" less 1¼" less 3/4", or 22".

A Mathematical Solution for the Length of Irregular Jacks

Suppose you had not done the sectional view drawing shown in Figure 12-13, so you could not make the necessary projections to find the mathematical length of the irregular California jack rafters. You would have to calculate these lengths.

There are three parts to the mathematical solution. The first two parts combined find the total rise of the irregular jack. The third part uses this total rise to find the mathematical length.

Part 1— The first part of the rise is uniform for each jack. It is the difference between the two *mathematical* ridge heights, which is the distance between line (18) and line (13) in Figures 12-13 and 12-14.

You just calculated the actual height of the 12 in 12 ridge from the rafter plate. But here we need the *mathematical* height, which is simply the 2' total run to the fascia line times the unit rise of 12". Therefore, the 12 in 12 ridge has a mathematical rise of 24".

Two things must be remembered if we are to correctly determine the *mathematical* ridge height of the main roof rafter. This ridge height must be figured from the *fascia* line (14). Therefore we must take into consideration the fascia line drop described under Figure 10-27. This is true because all other irregular hips, valleys, and common rafters are generated from the edge of the fascia line. You must also add in the main roof HAP.

To find the total rise of the main roof common rafter, multiply the unit rise by the total run to the *building* line. Then add the amount of the fascia line drop and the HAP.

For this roof, the run is 1.5' times 8". or 12", plus 1¼" of fascia line drop plus the HAP of 2¾". This totals 16".

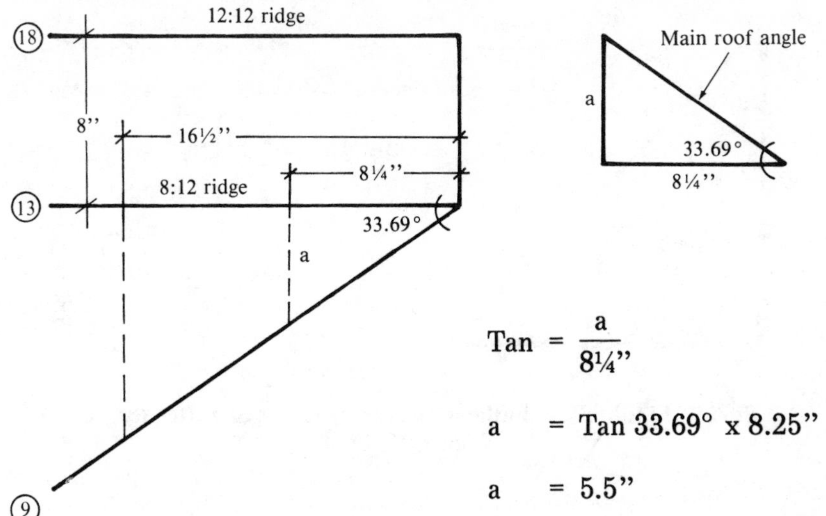

Finding part 2 of the run
Figure 12-14

For this irregular California, then, the difference in ridge heights is 24" minus 16", or 8". This 8" is the same for all of the irregular California jacks, and it is entered all the way across in line 1, Table 12-16.

Part 2— The second part of the total rise for each jack in dependent upon its short-point distance away from the main ridge center line. This is (a) in Figure 12-14. The formula given in this figure uses the tangent of the main roof rise. To find the tangent, divide the unit rise by the unit run. Eight divided by 12 is 0.6667. Multiply this by the short point distance from the ridge center line, 8¼". The answer is 5½". Enter the 5½" on line 2 of Table 12-16, under the 8¼" distance from the ridge.

Part 3— Now that we know the total rise, we can find the mathematical length of the irregular California jack rafter. Look at Figure 12-15. Divide the sine of the angle of the addition roof into the total rise. Enter the answer, 19 1/16", on line 4 of Table 12-16, in the column under 8¼" distance from the ridge.

Laying Out and Cutting the Jacks

Crown a piece of 2 x 4 and lay the crown away from you. Set the stair gauges to 12" and 12" (or 9" and 9" if you like). Shorten the ridge line by half the thickness of the ridge, and shorten the seat

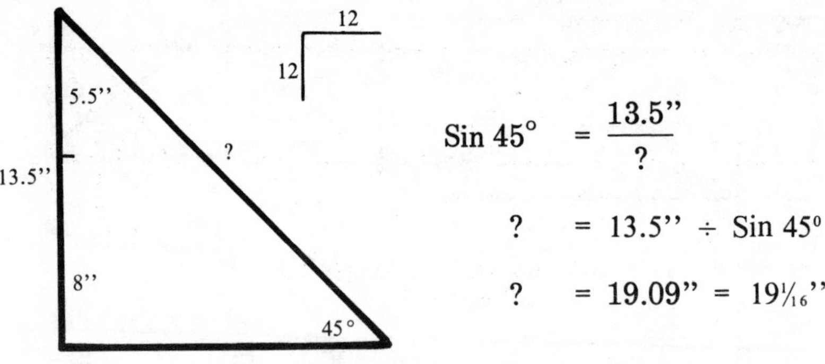

$$\sin 45^\circ = \frac{13.5''}{?}$$

$$? = 13.5'' \div \sin 45^\circ$$

$$? = 19.09'' = 19\tfrac{1}{16}''$$

Part 3: Finding the mathematical irregular jack rafter length
Figure 12-15

	Distance from ridge	16½"	8¼"	0"
1.	Part 1: The ridge height difference	8"	8"	8"
2.	Part 2: Part of rise Figure 12-14	11"	5½"	0"
3.	Equals: Total rise of addition jack	19"	13½"	8"
4.	Part 3: Math rafter length Figure 12-15	26⅞"	19 1/16"	11 5/16"
5.	Common difference	7¾"	7¾"	

The irregular jack rafter lengths
Table 12-16

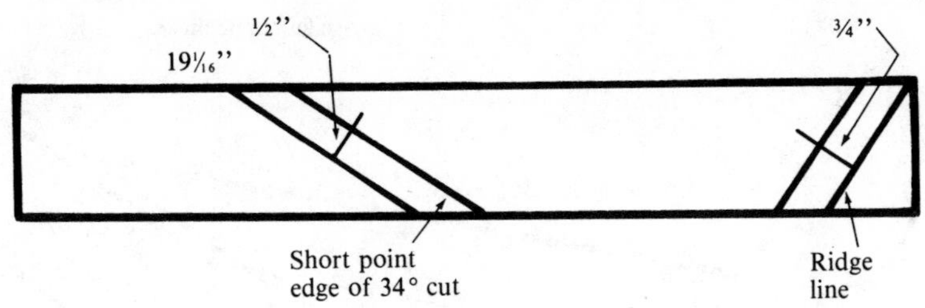

The lines of the jack rafter
Figure 12-17

cut by the thickness of the main roof sheathing, shingles, or the valley strip if one will be used. See Figure 12-17. Cut the short point line on the angle of the main roof common rafter.

Cut one pair of each rafter, plus one extra pair of the center rafter to be used for the California side. Frame these to the model and check the common difference.

Notice that the valley is not at 45 degrees. You'll have to cut a dummy valley tail rafter for the inside corner of the fascia board. This rafter will be the same type as the dummy valley tail rafter in Chapter 11.

You've just finished another type of irregular roof!

The Shed Roof Dormer

There are two other irregular roof problems that you may see from time to time. Both involve irregular dormers: the shed dormer, and the hip or gable dormer. We'll start with the shed dormer because it's more difficult. The hip or gable dormer is simpler, and uses principles that we have covered already.

Figure 12-18 shows a typical problem. To frame this shed roof dormer, you need to know three things: the run of the dormer rafter (if we know this run, we can use the total run and the unit rise to find the mathematical length of the rafter), the seat cut at the upper end of the rafter where it joins the main roof (this is an irregular cut and must be calculated separately), and the common difference in stud length for the end rafters.

The Run of the Dormer Rafter

On any roof where you know the total rise and the pitch, you can find the total run in feet by dividing the total rise in inches by the number of inches per foot of rise.

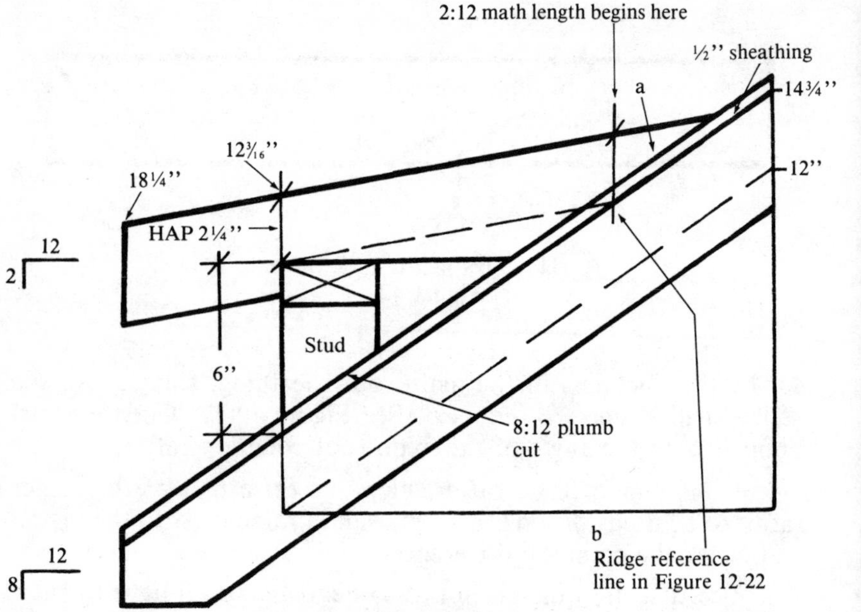

2:12 math length begins here
½'' sheathing
a
14¾''
12''
12³⁄₁₆''
18¼''
HAP 2¼''
12
2
6''
Stud
8:12 plumb cut
12
8
b
Ridge reference line in Figure 12-22

The model shed roof dormer
Figure 12-18

Look at Figure 12-19. You can see that the total run will be 1.5 feet.

$$\text{Total run'} = \frac{\text{Total Rise''}}{\text{Unit Rise''}} \qquad \frac{12''}{8''} = 1.5'$$

12
8
Total rise = 12''
Pitch angle
Total run = ?

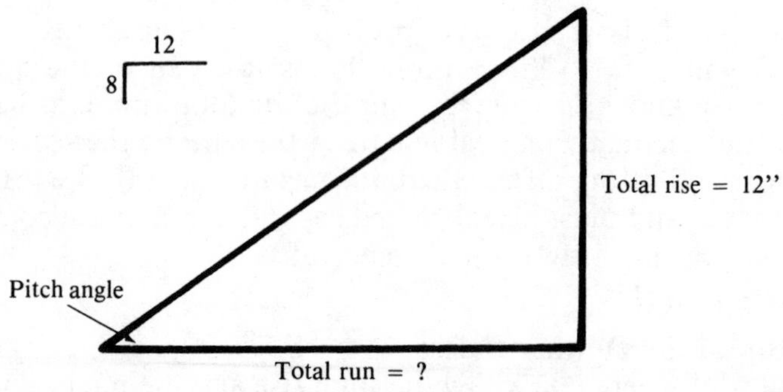

The basic relationship
Figure 12-19

We use this same principle to compute the rafter length for the shed dormer. But here, one other element is involved: The main roof rises at a certain rate, while the dormer roof rises at a lesser rate. The difference between these two rates is the effective rise of the shed dormer roof.

For these calculations, "total rise" means the rise of the dormer front wall. Here is how it would be expressed in mathematical language:

$$\frac{\text{Dormer Rafter}}{\text{Total run in feet}} = \frac{\text{Total rise in inches}}{\text{Main roof unit rise - Dormer unit rise}}$$

In Figure 12-18, the dormer total rise is 6". The main roof has an 8" unit rise, while the dormer roof has a 2" unit rise.

$$\text{Total run} = \frac{6''}{8'' - 2''} = 1'$$

This shed dormer rafter has a total run of 1' and a unit rise of 2". Now the mathematical length can be calculated. Punch [2] [÷] [1] [2] [=] [INV] [tan] and you have the angle, 9.46 degrees. Then, to find the secant, punch [cos] [1/x]. 1.0138 shows up in the display. This is the secant of a 2 in 12 roof, as given in Figure 1-14 in Chapter 1.

Multiply 1.0138 by 1' to get 1.0138', or 12³⁄₁₆". This is the building line measurement for the shed dormer rafter.

If this shed rafter is to have a 6" overhang, the mathematical length for the fascia line would be 1.0138 times 1.5', which is 1.5207", or 18¼".

Now begin the layout of the shed dormer rafter. Crown a piece of 2 x 4 and lay the crown away from you. Set the stair gauges on 2" and 12" and make a ridge reference line 6" in from the end of the material. See Figure 12-20. Mark the building line at 12³⁄₁₆" and the fascia line at 18¼". Draw the 2 in 12 plumb cuts at the two mathematical lengths. At the building line, use a 2¼" HAP so you have two-thirds of the material remaining above the birdsmouth cut.

The Upper Seat Cut

The rafter seat cut, (a) in Figure 12-18, is irregular because the upper end of the shed rafter does not fit against a perpendicular ridge like a plumb cut. Instead, it fits against the slant of the main roof and is more like a seat cut.

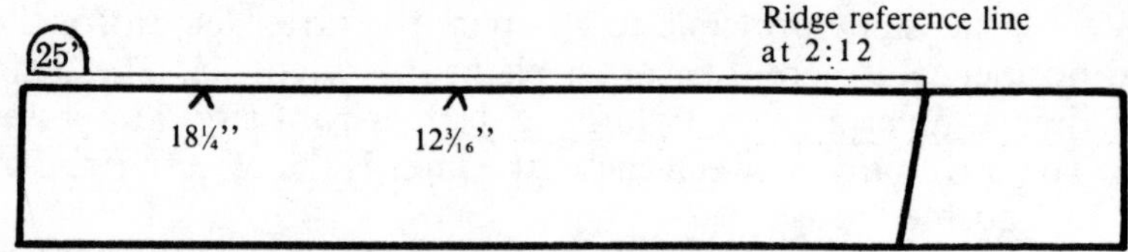

Beginning the rafter layout
Figure 12-20

Here's how to figure this seat cut. Subtract the angle of the dormer roof from the angle of the main roof, as shown in Figure 12-21. This gives you a resultant complementary angle, which is used to find the seat cut. The angular measurements must be used for these calculations.

$$\frac{8}{12} = .6667 \boxed{\text{INV}}\boxed{\text{tan}} = 33.69^\circ$$

$$\frac{2}{12} = .1667 \boxed{\text{INV}}\boxed{\text{tan}} = 9.46^\circ \qquad \begin{array}{r} 33.69^\circ \\ -\ 9.46^\circ \\ \hline =\ 24.23^\circ \end{array}$$

The complementary angle is 24.23 degrees. Use this to calculate the proper framing square setting.

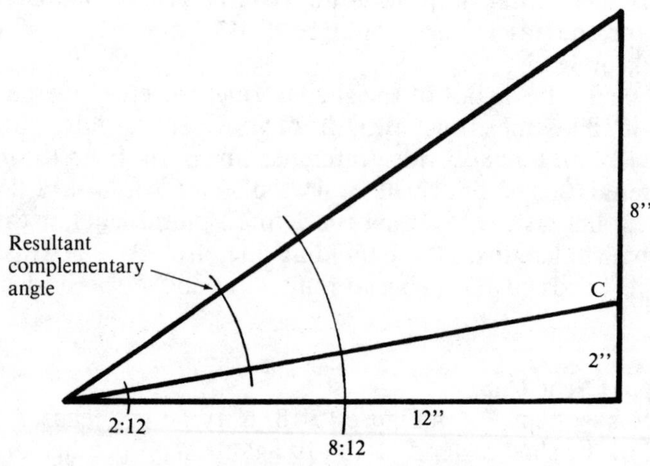

Finding the proper angle
Figure 12-21

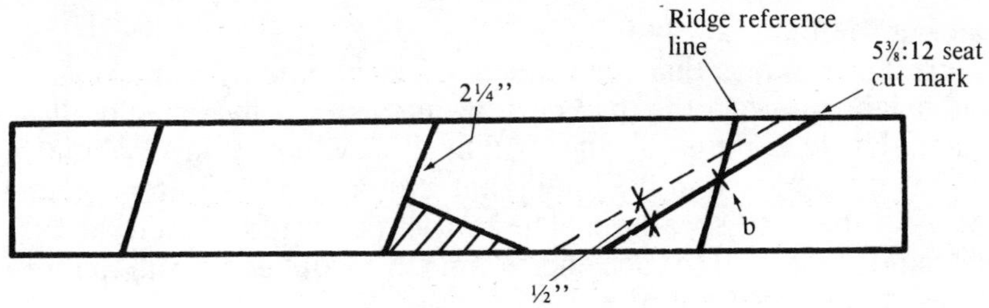

The completed layout
Figure 12-22

The framing square setting will be 12" times the tangent of the complementary angle, in this case 24.23 degrees. Twelve inches times the tangent of 24.23 is 5.4". So our framing square setting will be 5.4", or 5⅜".

Set the framing square to 5⅜" and 12". Draw the upper seat cut on the 12" side with the square reversed so that it passes through the HAP distance of 2¼", point (b) in Figure 12-22.

Don't try to subtract the two pitches from each other. Subtracting 2" from 8" and then using a 6 and 12 setting on the framing square doesn't work. Neither does subtracting one tangent from the other tangent. Look back at angle C in Figure 12-21 to see why. Angle C isn't a right angle. It's an oblique angle.

To complete the layout of this rafter, mark the 1/2" shortening for the roof sheathing and cut on the dotted line. See Figure 12-22. Cut out two of these shed dormer rafters.

Finding the Common Difference in Stud Length

The effective rise of the dormer roof is the difference between the main roof unit rise and the dormer roof unit rise. In our case, the effective rise for the model is 8" minus 2", or 6" for every 12" of unit run. If the end wall studs are to be 16" on center, the common difference is found by:

$$\text{Common difference"} = \frac{\text{Spacing"}}{12"} \times \text{Effective rise"}$$

$$\text{Common difference"} = \frac{16"}{12"} \times 6"$$

$$= 8"$$

Cutting the Longest Stud

Figure 12-18 shows that the total wall height is 6''. From the 6'' wall height, subtract the thickness of one plate, as indicated in the figure. The long point distance will be 4½'' with an 8 in 12 plumb cut line drawn through the long point. Mark a 1/2'' shortening to allow for the 1/2'' sheathing. Then cut out two studs. Cut a rafter plate and nail a stud to each end. Frame this front gable wall to the model, then add the shed rafters.

On a larger dormer, the upper end of the side wall studs will have a 2 in 12 plumb cut. The lower cut is the plumb cut of the main roof. Look back to Figure 12-18.

You'll probably want to double the main roof rafters at each side wall of the dormer, so they can carry the extra weight.

Congratulations on another fine roof!

The Hip or Gable Dormer

Frame the roof of a hip or gable dormer just like the roof for a blind valley. If the pitch of the dormer roof is the same as the main roof pitch, the valley jack rafters will be regular valley jacks. If the pitch of each roof is different, the valley jacks will be irregular. In that case, follow the procedure explained for irregular California jacks.

The side wall studs of the dormer are basically the same as upside-down gable studs, covered in Chapter 7.

Conclusion

Congratulations once more. You've made it to the end of the last chapter. You've followed all the calculations and cuts, answered the problems, checked your answers in the back of the book and found them correct, you've drawn the diagrams and you've built the models. Now you should consider yourself a master roof cutter, even if just a beginning one.

If you find there's something that still baffles you, go through the exercises again. I've tried to put down every step you'll go through. The answer should be in the chapter.

At first, take on the simpler jobs. Take care to check all your calculations before you do any cutting, and take this book with you.

The appendix that follows provides additional information to help round out your understanding and add to your skills. So do check it out.

Appendix A

The Framing Square Scales

Even if you learn to use the power of a modern hand-held calculator, your framing square is still a useful tool. Some parts of the framing square you use on every roof job. Those parts of the square will be very familiar to you. But some of the scales are not fully understood by some carpenters and roof cutters. This section is intended to acquaint you with information on the square that you may not be accustomed to using.

Refer to your framing square, or Figure 2-10, as we discuss each of the scales on the square.

The Regular Brace Table

On the back of the tongue is the regular brace table. At the end of the regular brace table, near the heel, is a set of three numbers: 18, 24, and 30. You'll recognize these numbers as being six times the lengths of the basic 3-4-5 right triangle. A triangle with sides 18, 24 and 30 units long will be a right triangle. But it will not be a triangle with angles of 90, 45 and 45 degrees. All other numbers in the brace table apply to a 45-degree right triangle.

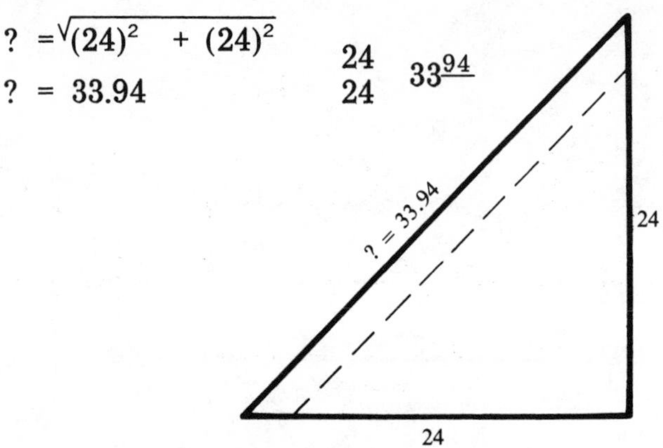

The 45° triangle
Figure A-1

Figure A-1 shows a 45-degree right triangle. If the two sides adjacent to the right angle are 24 inches, then, by the Pythagorean Theorem, the hypotenuse is 33.94 inches.

The brace table depicts 13 triangles beginning with a 24'' triangle and increasing three inches for each of thirteen steps. The largest triangle has sides of 60 inches.

The word *regular* on this brace table indicates that two sides are equal and two interior angles are 45 degrees. The hypotenuse figure on the table shows the long-point to long-point measurement for a brace installed diagonally at 45 degrees.

The brace table is handy for the sizes given. But you don't need the table at all if you can remember the square root of 2, or have your calculator handy. The square root of 2 is 1.414. Multiply the adjacent side by 1.414 to find the brace length long point. If the adjacent side is 24'', the brace length would be 1.414 times 24, or 33.94 inches.

Remember 1.414 and you can calculate the length of any 45-degree brace.

The Hundredths Scale

The hundredths scale is located at the end of the regular brace table on the heel. Find it just above the sixteenths scale. The hundredths scale shows 100 small lines. Each intermediate line represents 5 hundredths; each large line represents 25 hundredths.

You'll use the hundredths scale to convert decimal fractions of an inch to sixteenths of an inch. That's why the hundredths scale is just above the sixteenths scale.

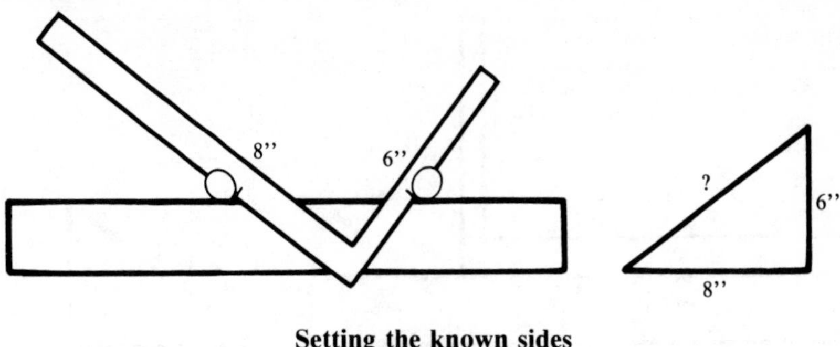

Setting the known sides
Figure A-2

Look under the 3'' mark of the twelfths scale. You'll note that the hypotenuse of a 60'' triangle is given as 84 and 85 hundredths. Locate 85 hundredths on the hundredths scale and read directly down to convert this to sixteenths. The answer is just under 14 sixteenths.

The Irregular Brace Table — The Twelfths Scale

On the outside edge of the back of most framing squares is a twelfths scale. This scale breaks each inch into 12 divisions, or twelfths. Figure 2-10 shows this scale.

Your framing square forms a right angle, so either the sixteenths scale on the front or the twelfths scale on the back can be used as a Pythagorean Theorem Calculator. Let's look at a problem that uses the framing square as a calculator.

Using the Twelfths Scale as a Calculator

It's late afternoon on a Friday. You're the finish carpenter on a commercial job. An interior wall is being covered with decorative 1 x 8 rough-sawn stock. The wall is finished except for one diagonal that has to be installed to complete the design.

You're ready to cut this diagonal but discover that only one 10' piece of 1 x 8 remains on the job. There's no material to make a practice cut and no room for error.

The wall is 80'' long and the decorative finish is 60'' high. What's the correct length of the diagonal from long point to long point, and what are the correct angles of the cuts?

We'll use the square as a calculator to figure this problem. We can't use the full 60'' and 80'', so we'll divide each by 10 to get 6'' and 8'' as the first two sides. See Figure A-2. Now, what's the length of the diagonal?

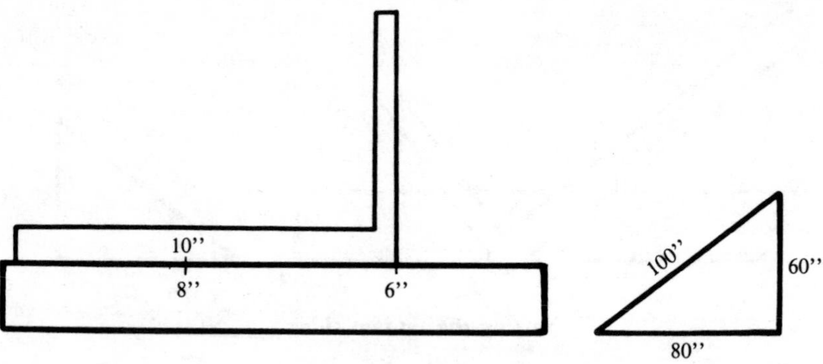

Finding the hypotenuse
Figure A-3

Set the framing square to 8'' and 6'' along any straight edge. Mark each edge at those points with a sharp pencil. Now turn the framing square as in Figure A-3. Move the square until the 6'' mark is directly under the zero point of the square. Then read the distance on the scale that's right above the 8'' mark on the straight edge. You'll find that the 10'' point is directly above the 8'' mark. Multiply this by 10 to convert back to actual size and you have 100''. This is the third side of the triangle.

Lay out 100'' on the brace material and set the stair gauges as in Figure A-4. This gives you the level cut and plumb cut for this brace. Refer back to Figure 2-12.

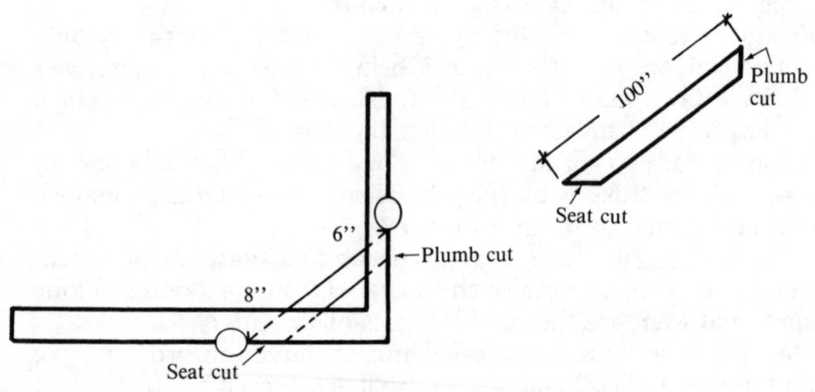

The finished brace
Figure A-4

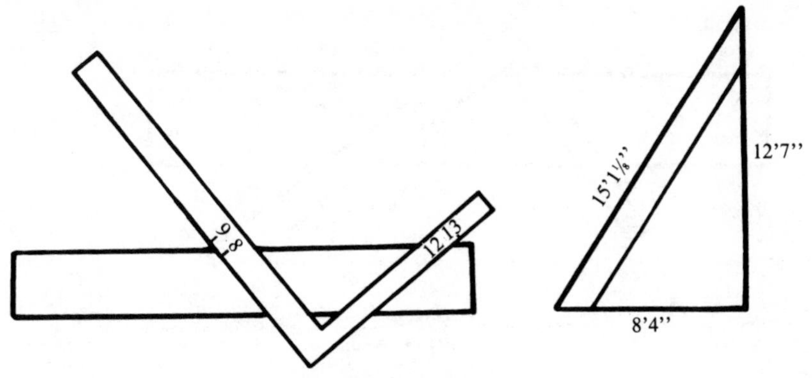

Finding the unknown side
Figure A-5

Using the Twelfths Scale

The twelfths scale is very useful because it lets you make calculations in the scale of 1" to the foot. Each inch on the framing square becomes one foot, and each of the 12 divisions on the twelfths scale becomes one inch.

Look at Figure A-5. The body of the framing square is set to 8" plus 4 marks, while the tongue is set to 12" plus 7 marks. Measuring across, as in Figure A-3, gives a long-point measurement for the brace of 15'1⅛". The stair gauges and framing square give the level and plumb cuts as usual.

Of course, you could have used your calculator and the Pythagorean Theorem to find the same length, but that wouldn't give you the cuts.

The Board Foot Scale

Some framing squares are no longer printed with a board foot scale. In its place, use this simple formula:

$$\frac{\text{Length'} \times \text{Height''} \times \text{Width''}}{12} = \text{Board feet}$$

This formula uses the length in feet, since lumber is usually measured in whole foot lengths. Height and width are in inches.

A twelve-foot length of 2 x 6 material would be 12 board feet. Eight feet of 2 x 6 material would be eight board feet.

The Octagon Scale

The octagon scale was explained in the second part of Chapter 9.

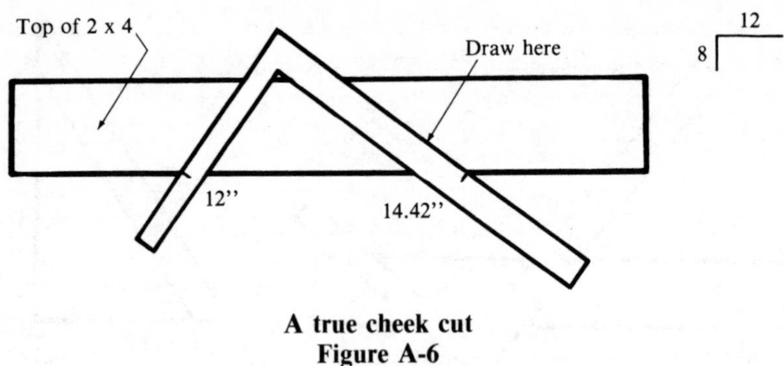

A true cheek cut
Figure A-6

The Rafter Scale

The first two lines of the rafter scale were used in Chapters 2 and 4. Refer back to those chapters if necessary.

The two center lines on the rafter scale show the common difference in length between jack rafters at a certain spacing. This was discussed in Chapter 6.

The last two lines of the rafter scale pertain to the cheek cuts for jacks or hip and valley rafters. We never referred to these two lines in this book, even though we cut and framed fancy regular and irregular roofs. Therefore, the first thing you'll observe about these lines is that we don't need them.

The second thing you'll observe is that all the numbers are less than 12. These numbers are mathematical rearrangements of the truth, as we'll explain later.

Making Cheek Cuts the True Way

You will really find cheek cuts by using the rafter unit run (such as 12, 13 or 17) in combination with the unit length for that particular rise. The unit length side, which is always the bigger number, will be the cut side.

Figure A-6 shows an 8 in 12 jack rafter. Set the framing square to 12, the unit run, and 14.42, the unit length. Draw along the unit length side. This is the angle on top of the rafter, but you already know that you don't need to draw this line. It comes out automatically when you set the saw to the necessary 45 degrees and cut along the 8 in 12 cheek cut line.

For any regular pitch roof, the saw must be set to 45 degrees. But for every pitch the angle of the line on top changes, and therefore all of these numbers are given on the last two lines of the rafter scale.

Yes, the angle on top does change, but *don't* change your saw setting from 45 degrees while cutting the cheek cut line or the rafter won't come out right. You don't need to know the top angle for regular rafters, and it's a waste of time to do a balancing act with the framing square on the top edge of 16' of 2 x 6.

Making Side Cuts the Framing Square Way

The designers of the rafter scale were wrong when they thought roof cutters needed settings for side cuts. But they correctly saw that the larger unit lengths would limit the number of side cuts that could be set on the framing square. Since 16.97'' can't be set on the tongue, there's no way to set the unit length for hip and valley rafters.

Setting the 12'' unit run on the tongue lets you set all 18 unit lengths on the body. It's a little clumsy, but side cuts for jacks can be laid out this way. Side cuts for hips and valleys are impossible, however.

The designers of the square came up with a way to avoid this problem and also to make it possible to lay out all side cuts on the same number. They established 12 as the cut side for jacks, *and* hips and valleys. Then they found the mathematical equivalents, some number less than 12. It's easy to find the changing side with a ratio.

The framing square scales show that the side cut for an 8 in 12 jack rafter would use 12 as the cut side (this is understood) and 10 as the floating side.

$$\frac{12}{?} : \frac{14.42}{\underset{\text{(unit length)}}{12}} \qquad ? = \frac{12 \times 12}{14.42} = 9.986 = 10''$$

Rather than cutting on the 14.42 side, they elected to convert that to the standard of 12. Therefore, the made-up number 12 is to the real number, the unit length, as some number will be to the unit length. The answer is 10, so 10 is printed on the side marked ''side cuts for jack use'' and under column 8.

A side cut for a 10 in 12 hip rafter would be:

$$\frac{12}{?} : \frac{19.70}{16.97} = 10.337 = 10\tfrac{5}{16}''$$

The framing square erroneously prints 10⅜ as the number.

Appendix B

A Roof Cutter's Geometry and Trigonometry

Roof cutting is based on geometric and trigonometric principles that are thousands of years old. But fortunately, it doesn't take a lot of mathematical knowledge to learn roof cutting. This section summarizes about all the math you need to know.

Complementary Angles

One geometric theorem states that the sum of the angles of a triangle is equal to 180 degrees. Every right triangle has one 90-degree angle. Therefore, the two remaining angles must add up to 90 degrees.

Any two angles that add up to 90 degrees are called *complementary angles.* Angles A and B in both of the drawings in Figure B-1 are complementary angles.

Alternate Interior Angles

Another geometric theorem states that if two parallel lines are cut by a transversal, the alternate interior angles are equal. In Figure B-2, angles A and A1 are alternate interior angles and are equal. Angles B and B1 are also alternate interior angles and are equal in degrees.

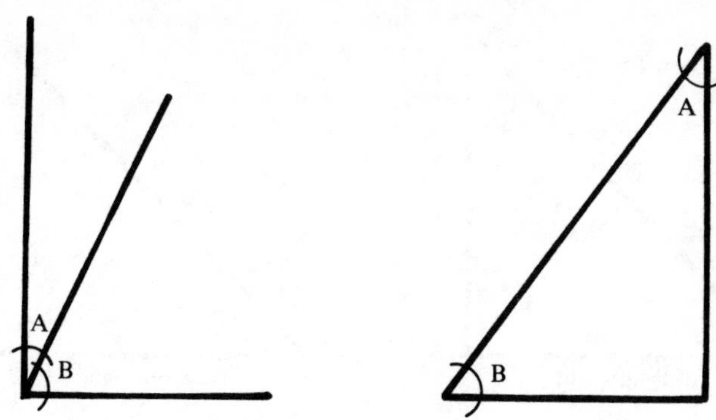

Complementary angles
Figure B-1

The Six Trigonometric Functions

As we saw under Figure 2-5 in Chapter 2, certain mathematical relationships are true of all right triangles. There are three main relationships and three inverse relationships, for a total of six. These six are referred to as trigonometric functions.

Look at the left drawing in Figure B-3. A carpenter would say that the side opposite is the rise, and the side adjacent is the run. The rafter would be the hypotenuse.

But it's sometimes useful to use the words *side opposite* rather than run. Side opposite clearly identifies the side opposite angle A.

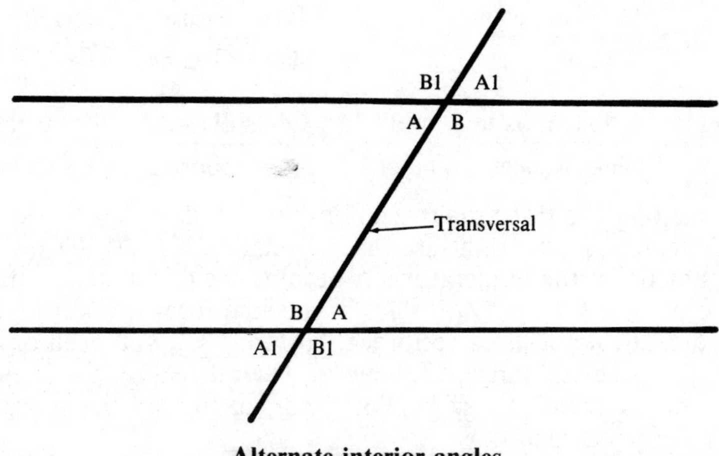

Alternate interior angles
Figure B-2

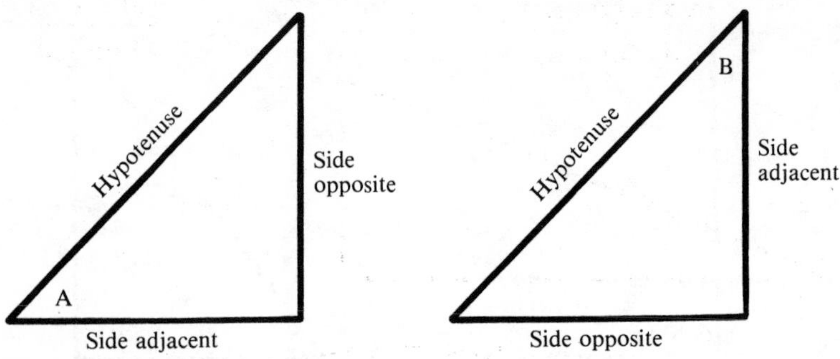

Names of the sides
Figure B-3

If we look at the triangle from angle B, as in the right-hand drawing in Figure B-3, side opposite is no longer the rise. It's the run. To avoid confusion, mathematicians identify the angle they're talking about and then say either "side opposite" or "side adjacent" when talking about the two shorter sides of a right triangle.

Here are the names of the six trig functions and the relationships they describe.

$$\text{Sine (sin)} = \frac{\text{Side Opposite}}{\text{Hypotenuse}} \quad \frac{1}{X} \quad \frac{\text{Hypotenuse}}{\text{Side Opposite}} = \text{Cosecant (csc)}$$

$$\text{Cosine (cos)} = \frac{\text{Side Adjacent}}{\text{Hypotenuse}} \quad \frac{1}{X} \quad \frac{\text{Hypotenuse}}{\text{Side Adjacent}} = \text{Secant (sec)}$$

$$\text{Tangent (tan)} = \frac{\text{Side Opposite}}{\text{Side Adjacent}} \quad \frac{1}{X} \quad \frac{\text{Side Adjacent}}{\text{Side Opposite}} = \text{Cotangent (cot)}$$

Notice that the three main relationships are listed on the left. The functions on the right are inverse relationships of the three main functions: the numerator has become the denominator and visa versa. The symbol *1/X* indicates the reciprocal relationship.

If your calculator has a reciprocal button [1/*x*], you need only the sine, cosine and tangent functions. The other three functions are found by punching [1/*x*]. For example, to find the secant, punch [cos] and then [1/*x*].

Let's review the two functions you use most, the tangent and the secant.

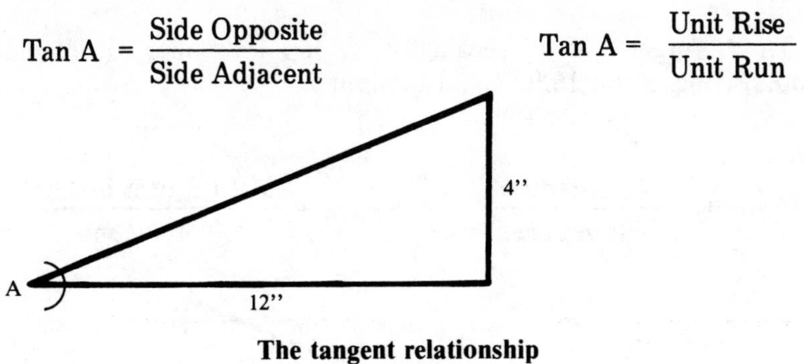

The tangent relationship
Figure B-4

The Tangent Function

The angle of the roof is most conveniently found with the tangent function, since the unit rise and unit run are usually known. Look at Figure B-4.

The unit rise divided by the unit run is the same as the length of the side opposite divided by the length of the side adjacent. That's the tangent. Divide 4 by 12 and you have 0.3333. Change that to degrees either with a trig table or by punching [INV] [tan]. The answer is 18.43 degrees, the angle of a 4 in 12 roof.

The Secant Function

The secant tells you the length of the rafter for each unit of run at a certain pitch. That's very useful information to a roof cutter. Once the roof angle is found with the tangent function, it's a simple matter to find the secant of the angle. It can be done either by looking up the angle in a trig table, or by using a calculator.

Look at Figure B-5. With 18.43 degrees showing on the calculator, punch [cos] (to convert the degrees to the cosine equivalent) and then [1/*x*] to get the the reciprocal of the cosine, the secant. The display will show 1.0540926. Multiply that number by any run in feet for a 4 in 12 roof and you have the mathematical rafter length for the common rafter.

The secant formula states:

Mathematical Rafter Length = Secant x Run

For the problem in Figure B-5, the total run is 12'. Therefore, the mathematical rafter length will be 1.0541 times 12, or 12.6491'. This converts to 12'7$\frac{13}{16}$''.

Don't forget that the secant for the hip and valley (on regular roofs) is based on 16.97'' and the unit rise.

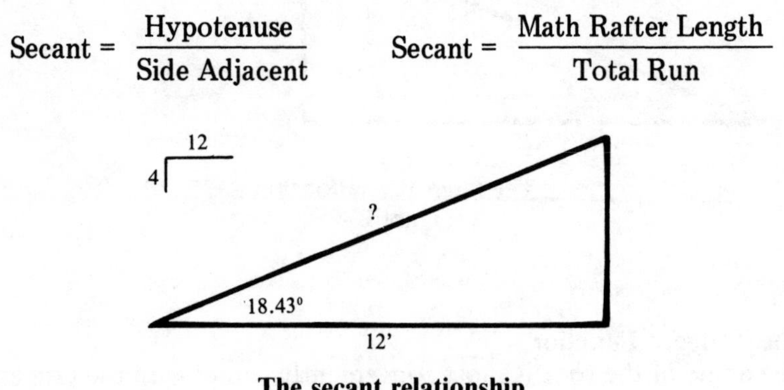

The secant relationship
Figure B-5

Appendix C

Pitch Wall Framing

A pitch wall is usually required when an open beam ceiling is included in a design. The pitch wall uses roof framing principles and will be easy if you have mastered the work explained in this manual.

Figure C-1 shows a typical pitch wall plan. This end wall has been framed with window and sliding glass door openings. Notice also the three beam pockets that will receive rough-sawn purlins and a ridge beam.

There are two parts to this framing job. The first is cutting the plates and marking the stud layout. The second is cutting the studs correctly for framing.

Laying Out the Plates

The mudsill plate, which is treated to resist termites, is laid out along the wall location snap line. The mudsill doesn't have to run completely through the door openings because it will be cut back to create an opening later. See Figure C-2.

Lay the mudsill on edge according to the end lines of the wall and cut it to the correct length.

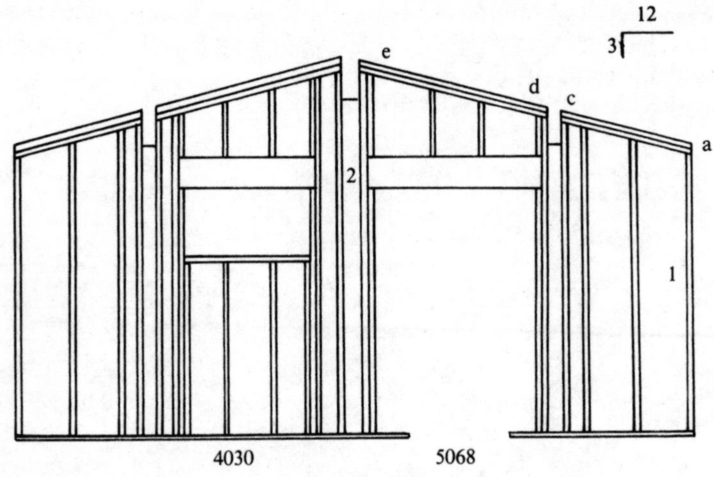

A pitch wall ready for beams
Figure C-1

If the mudsill is to be held down with foundation "J" bolts, mark and drill these holes. Lay the mudsill over the holes to ensure a proper fit. Be sure not to turn the mudsill the wrong way for layout when it is removed from the "J" bolts and set on edge again for layout. If the plate is turned the wrong way for layout, you'll find the holes in the wrong position after you've framed the

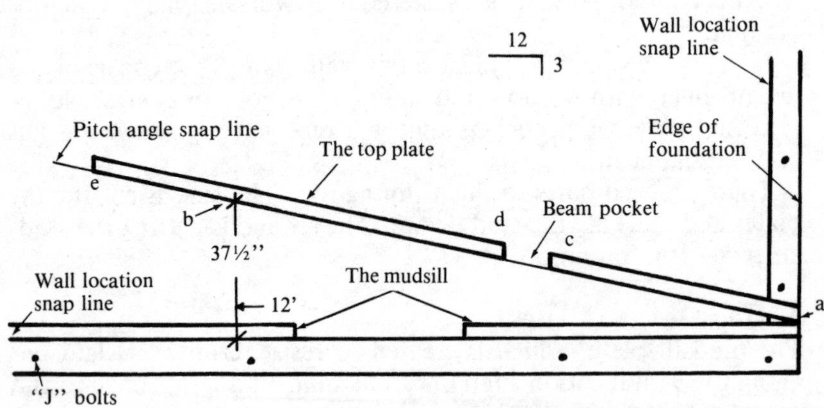

Laying out the plates
Figure C-2

wall and tried to stand it up to rest within the snap lines. If this happens, you'll have to hold up the wall and place a block under the mudsill while drilling new holes.

After drilling the mudsill, lift it off the "J" bolts and rotate it correctly on the inside edge to receive the stud layout. Again, place it exactly to the end lines of the wall.

Cut a piece of top plate to the roof pitch angle. This is a 3 in 12 roof, so set the saw to 14 degrees for the cut at (a) in Figure C-2.

Lay out some convenient number of feet from the end of the mudsill and then measure up the total rise for that amount of footage. In Figure C-2, we have measured 12' from point (a). At the 12' point on this 3 in 12 roof, the rise will be 36" (12 times 3" equals 36"). Add 1½" for the thickness of the mudsill, as indicated in Figure C-2.

Snap a line that extends from (a) to beyond (b). Lay the top plate above this line as indicated.

Points (c), (d) and (e) identify ends of the beam pockets. Make the same 14-degree roof angle cut on the plate at these points.

Begin the layout by marking the location of the sliding door and beam pockets. Then return to point (a), begin the stud layout along the mudsill, and mark both plates.

When the layout is complete, cut the rafter plate.

Cutting the Studs

Find the short-point length of the shortest and longest stud and make a 14-degree cut on each one. These are studs (1) and (2) in Figures C-1 and C-3.

Use the chalk lines to make sure the plates are square and straight. If they aren't, scab some pieces of 2 x 4 to the floor so the plates are held straight until the studs are marked in place.

Lay in the other studs over the layout marks. Use a combination square to mark the short point of each stud for the 14-degree cut. See point (f) in Figure C-4.

Frame these studs to the plates. Add the framing for door and window openings and the correct length 4 x 4's (or 2 x 4's doubled) for the beam pockets.

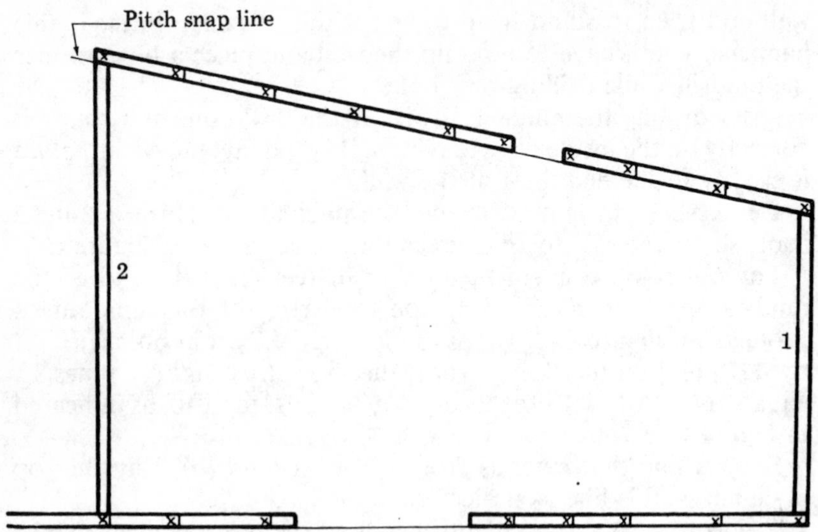

Laying out the frame
Figure C-3

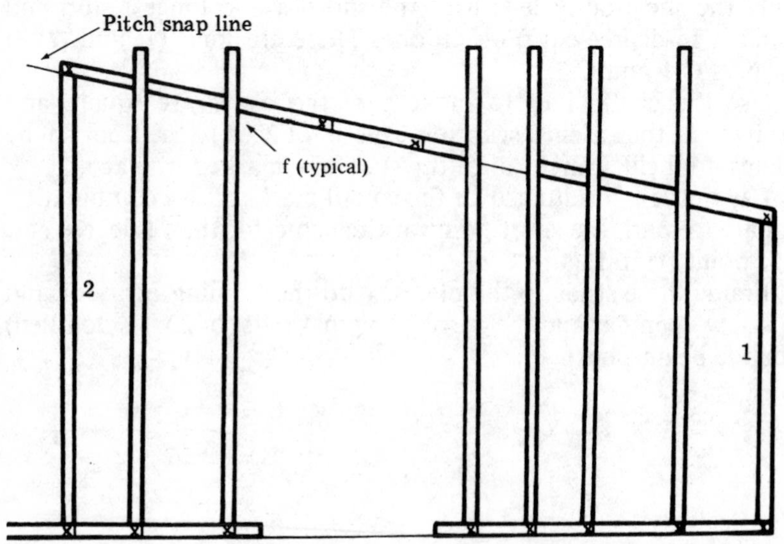

Placing the studs for marking
Figure C-4

Appendix D

Crickets and Hog Valleys

The Cricket

A cricket is a diverter built on a roof to keep water from collecting in certain areas.

Figure D-1 shows a cricket properly installed behind a chimney. It provides a sloped surface to either side. In essence, it's a small roof framed after the main roof sheathing is in place, done in the manner of a blind valley. A small cricket can be made simply from a few pieces of 2 x 4 and plywood.

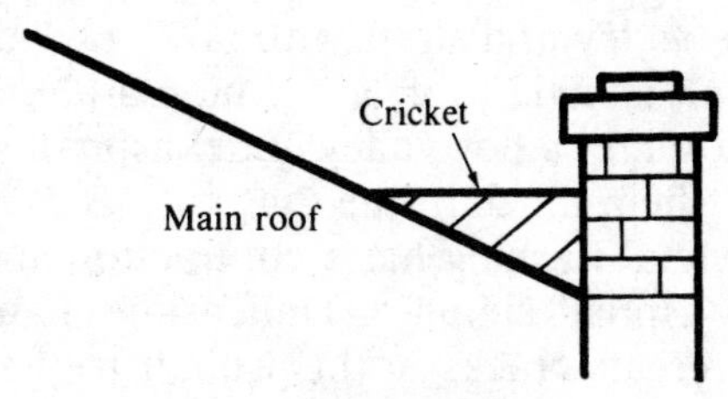

The cricket
Figure D-1

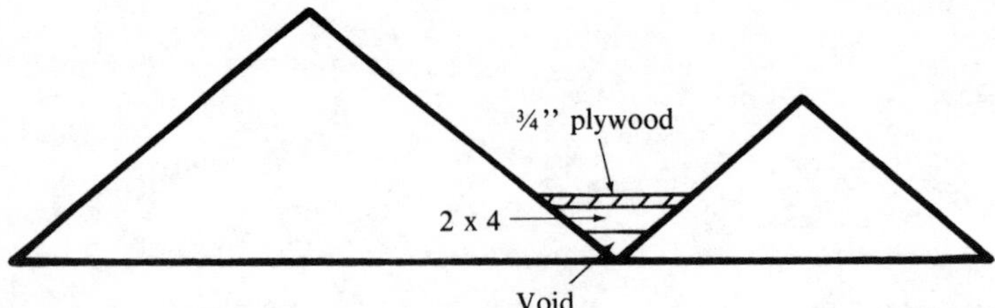

Section view showing the need for a hog valley
Figure D-2

The Hog Valley

Sometimes two sections of roof have to be framed as in Figure D-2. Usually this is the result of a room addition. In this case, a hog valley must be built to divert water toward a downspout.

The slope of the hog valley depends on some judgments:

1) Larger roof areas need a steeper hog valley to carry the heavy flow away quickly. The larger the roof area draining into the valley, the greater the accumulation likely in a heavy rain.

2) If the pitch of the adjacent roof is low, make the valley steeper. A low pitch roof tends to reduce the velocity of the runoff in the valley.

3) Trees overhanging the roof will drop leaves into the hog valley. There they obstruct runoff . . . unless water flowing quickly down the valley flushes them off the roof.

4) The last factor is the diameter of the downspout and the size of the accumulation box at the end of the valley. If heavy rains are common, provide a larger downspout and larger box to handle the heavy rain.

To frame a hog valley, place successively larger 2 x 4 supports every 2' up the valley and sheet with 3/4'' plywood.

For an asphalt shingle roof, a hot mop and valley metal aren't required. To hot mop a hog valley under asphalt shingles is expensive, and won't help much in the long run. But don't lay flashing in the valley. Metal flashing has a contraction and expansion rate different from asphalt shingles. Dimension changes in the metal will eventually create cracks in the asphalt roof surface.

Appendix E

Calculating Roof Area for Different Pitches

At some time in your roof cutting career, someone will ask you to estimate the surface area of the roof you're framing. You could always measure the actual roof. But it's easier to convert the plan view (horizontal) area to roof surface area with a conversion factor. Every master roof cutter should know how it's done.

Let's calculate the surface area of the model gable roof in Chapter 2. The plan view area is 6' by 4', or 24 square feet. See Figure E-1.

Here's how to find the surface area. Multiply the plan view area by a conversion factor, which is the common rafter unit length divided by 12. Figure E-2 shows the unit triangle for an 8 in 12 roof. The unit length is 14.42.

$$\frac{\text{Unit Length}''}{12''} = \frac{14.42''}{12''} = 1.2018$$

1.2018 x 24 square feet = 28.84 square feet of roof surface

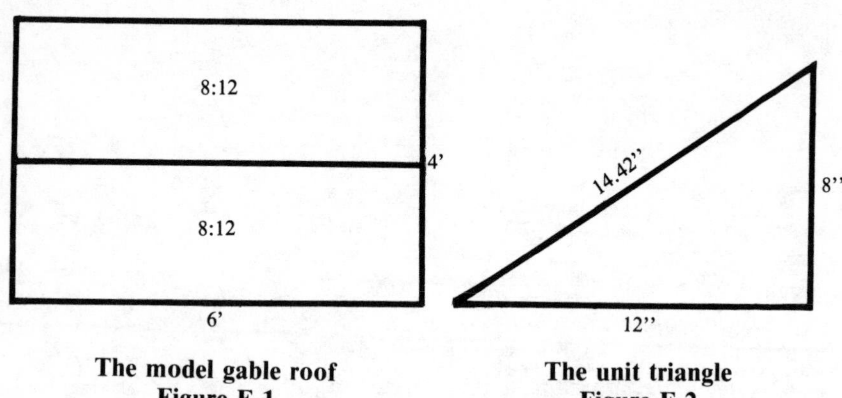

The model gable roof
Figure E-1

The unit triangle
Figure E-2

Let's confirm this figure. Multiply the length of the roof (in feet) by twice the mathematical length to the fascia line (in feet). Under Figure 2-17, in Chapter 2, the fascia line was figured at 2.4037'. Two times 2.4037' is 4.8074'. Multiply this figure by the 6' length of the roof and you have 28.84 square feet.

Table E-3 is a handy conversion table for the common roof slopes.

Rise"	Factor	Rise"	Factor
1	1.0035	7	1.1577
2	1.0138	8	1.2018
3	1.0308	9	1.2500
4	1.0541	10	1.3017
5	1.0833	11	1.3566
6	1.1180	12	1.4142

Area conversion factors
Table E-3

Appendix F

Cutting Roof Sheathing

It takes a little knowledge to make sheathing fit perfectly at the hip or valley. But with your mastery of roof cutting, you'll have no trouble finding the correct cut angles. Figure F-1 shows the layout method. You may find it easier to find the angles with your calculator.

Layout Method

Suppose you're sheathing an 8 in 12 roof. Start by drawing a line at right angles to the edge of the plywood. See Figure F-1. Mark point (c) 12" from (a). Find point (b) by laying your square on the sheathing so that the 12" and 8" marks fall right over the line. This establishes point (b), the distance of the common rafter unit length from (a). In this case the distance is 14.42 inches. Draw a line between points (b) and (c).

Line (c-b) is the angle that will fit exactly on the hip or in the valley. The position shown is for the valley. Turn the sheet upside down for the hip. Use this sheet as a template for cutting all sheets that go on the hip or valley.

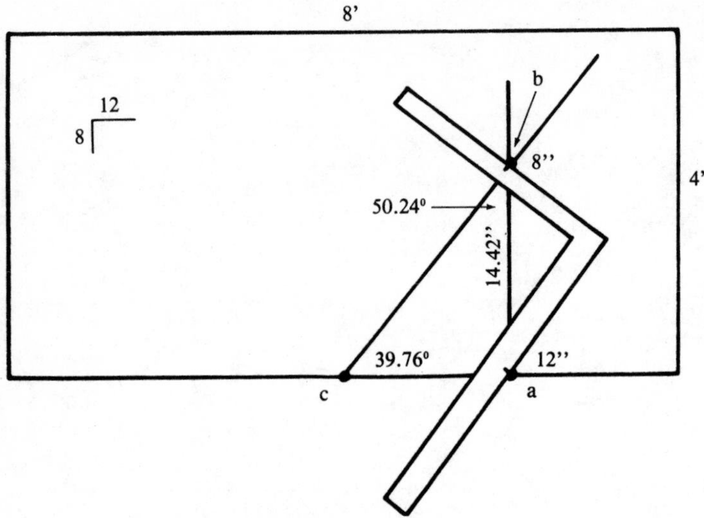

Layout on the roof sheathing
Figure F-1

Math Method

Find the main roof secant and put that number in the display of your calculator. Press [INV] [tan] and 50.24 degrees appears in the display. This is the angle of the cut at (b).

For the angle at (a), put the secant in the display. Press [1/x] [INV] [tan] and you have 39.76 degrees.

Appendix G

Ellipses for Roof Framing

Any time a round pipe passes through a sloping roof, the hole has to be an ellipse. And if the hole is more than a few inches in diameter, or if it won't be covered by some other finish material later, the ellipse must be drawn accurately. This appendix explains how.

Figure G-1 shows the parts of an ellipse. Every ellipse has a major axis, a minor axis and two foci from which the ellipse is drawn. Notice that the distance from either (a) or (b) to either (f1) or (f2) is equal to half the length of the major axis.

The length of the minor axis is equal to the diameter of the pipe. The length of the major axis is determined by the slope of the roof. Let's look at an example.

Horizontal Pipe

Figure G-2 shows how to figure the length of the major axis. A 7" diameter pipe is passing horizontally through an 8 in 12 roof. Start by drawing a right triangle with an 8" rise and a 12" run. Measure up the rise of the triangle for the diameter of the pipe, 7" in this case. Project that point horizontally to the slope that represents

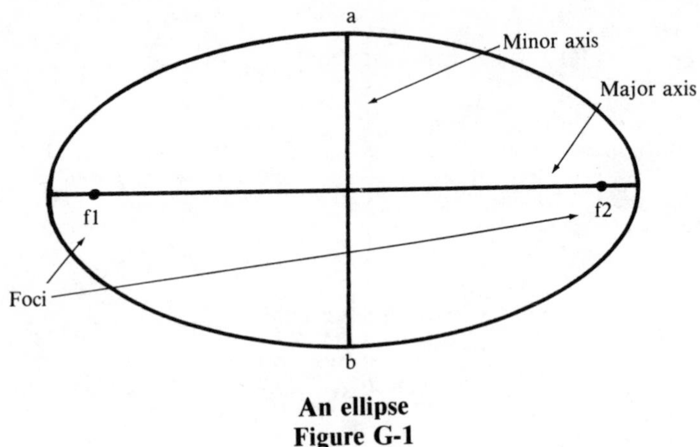

An ellipse
Figure G-1

the roof surface. Call this intersection (x). Next, measure the distance along the slope from point (y) to point (x). That's the length of the major axis. In the example, it's 12⅝''.

Notice that the major axis runs horizontally for horizontal pipe. It runs vertically for vertical pipe.

Layout of the Foci

The final step is to lay out the ellipse. Draw the major and minor axes at right angles to each other on the surface to be cut. Make sure the major and minor axes intersect at the exact center where you want the ellipse.

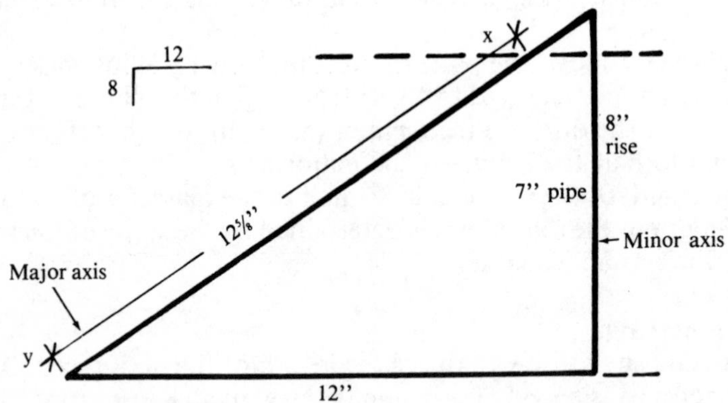

A horizontal 7'' pipe through an 8 in 12 roof
Figure G-2

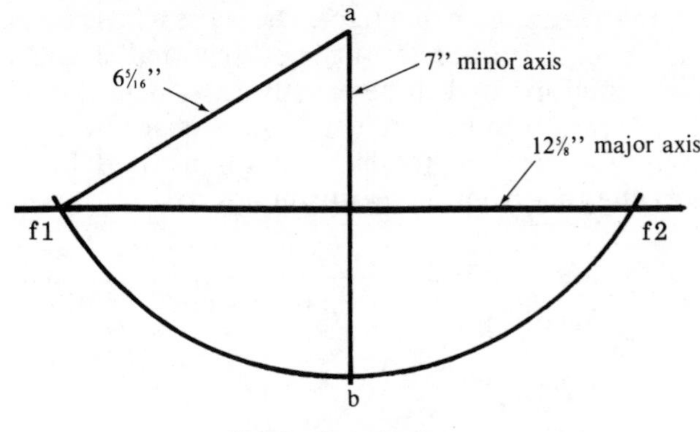

Beginning the layout
Figure G-3

Look at Figure G-3. Measure from the center intersection along the minor axis for a distance equal to one-half the minor axis. In this case, that's 3½". Call that point (a). With a compass or string and pencil, swing an arc from (a), whose radius is one-half the major axis, 6$\frac{5}{16}$" in this case. Points (f1) and (f2) are located where the arc crosses the major axis.

Drive small nails at (f1) and (f2) and tie a string to the two nails. Adjust the length of the string so that it can be extended just to point (a) when the ends are tied to the two foci. This distance is the same as the length of the major axis.

Place a pencil against the string, as shown in Figure G-4. Draw the ellipse as far as the string will let your pencil go from the two

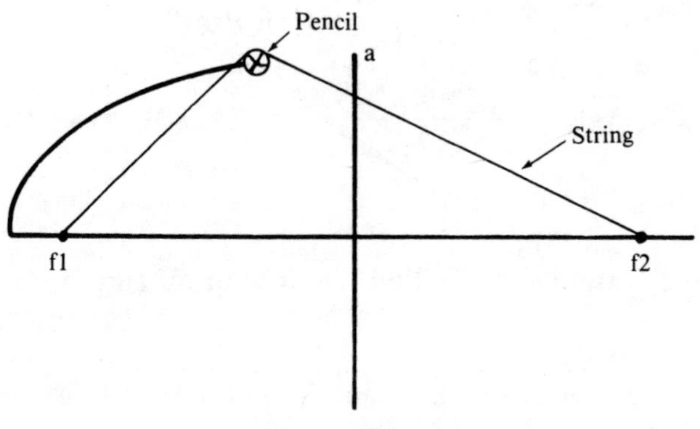

Drawing the ellipse
Figure G-4

foci. Be sure to keep the string tight. Draw the ellipse on both sides of the major axis. If your layout is correct and if you work accurately, the major axis will be exactly 12⅝'' and the minor axis will be 7''. Measure to be sure your figure is accurate.

You may also use the following math method to locate the distance of the foci from the major-minor axis intersect:

$$\text{Distance} = \frac{\tfrac{1}{2}\text{ Minor Axis}}{\tfrac{1}{2}\text{ Major Axis}} \boxed{=} \boxed{\text{INV}} \boxed{\cos} \boxed{\sin} \boxed{\text{x}} \ \tfrac{1}{2}\text{ Major Axis} \boxed{=}$$

Mathematical Method for Horizontal Pipe

As you may have guessed, it's easy to calculate the length of the major axis. For any pipe or any rise, the major axis is:

$$\text{Major Axis} = \frac{\text{Pipe Diameter'' x Roof Unit Length''}}{\text{Unit Rise''}}$$

$$\text{Major Axis} = \frac{7\text{'' x } 14.42\text{''}}{8\text{''}} = 12.6194\text{''} = 12\tfrac{5}{8}\text{''}$$

Mathematical Method for Vertical Pipe

The major axis is equal to the diameter of the pipe (in inches) times the unit length for the roof pitch (in inches), divided by the unit run (in inches).

$$\text{Major Axis} = \frac{\text{Pipe Diameter'' x Roof Unit Length''}}{\text{Unit Run''}}$$

$$\text{Major Axis} = \frac{7\text{'' x } 14.42\text{''}}{12\text{''}} = 8.4130\text{''} = 8\tfrac{7}{16}\text{''}$$

The Area of an Ellipse

Here's the formula for finding the area of an ellipse:

$$\text{Area} = .7854 \text{ x Major Axis x Minor Axis}$$

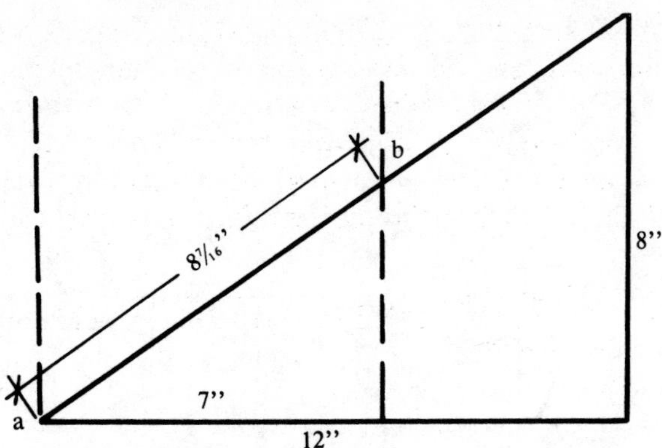

A vertical 7'' pipe through an 8 in 12 roof
Figure G-5

Appendix H

Steps to Calculate an Irregular Roof

This appendix offers a concise, systematic procedure for framing an irregular roof. If you've mastered the information in Chapters 8 through 12, you won't find anything new here. But you may find that following the steps recommended in this section saves valuable time and prevents an error when actually doing the work.

We'll start with a typical plan supplied by an architect. Figure H-1 shows the only information you're likely to have as you begin work.

Step 1: Solving for Non-centered Ridges
The first thing to find is the total run for each side of the main ridge. Figure 8-5 and following illustrations from Chapter 8 depict a roof with a non-centered ridge.

Decide at the outset whether you're going to draw a layout on four squares to an inch graph paper and measure the dimensions, or calculate all lengths as explained in Chapter 8. Figure H-2 shows the layout triangle. Here are the steps needed to calculate the run of each pitch and the rise:

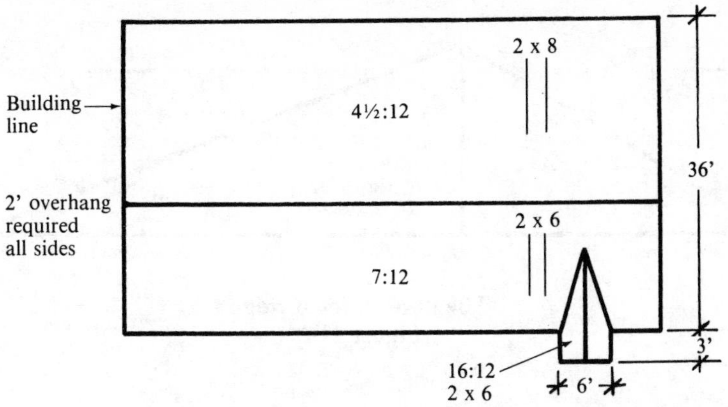

An irregular roof plan
Figure H-1

$$AB = \frac{\text{Tan C (Span)}}{\text{Tan A + Tan C}} \qquad BC = \frac{\text{Tan A (Span)}}{\text{Tan A + Tan C}}$$

$$\text{Tan A} = \frac{7}{12} = .5833$$

$$\text{Tan C} = \frac{4.5}{12} = .375$$

$$\text{Tan A} + \text{Tan C} = .9583$$

$$ABC = 36'$$

$$AB = 14'1\tfrac{1}{16}'' \qquad BC = 21'10\tfrac{15}{16}''$$

$$\text{Total Rise} = 14.0874' \times 7'' = 8'2\tfrac{5}{8}''$$

$$\text{or:} = 21.9125' \times 4.5'' = 98.6065'' = 8.2172'$$

$$= 8'2\tfrac{5}{8}''$$

Selecting a Common Rafter— Select one common rafter from the main roof to generate the sectional view drawings. The rafter selected establishes the HAP for the entire roof. If the 7 in 12

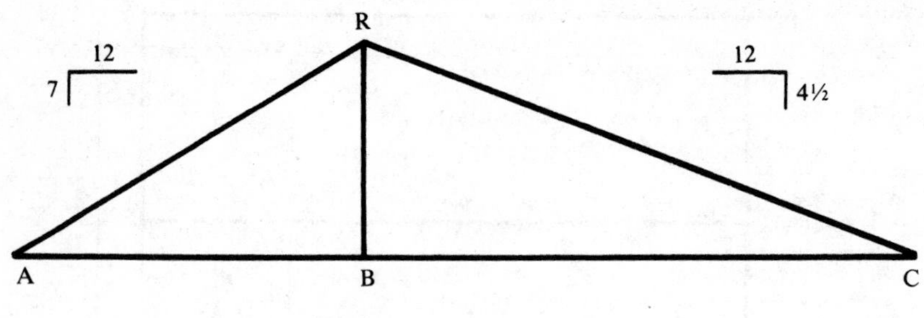

The non-centered ridge
Figure H-2

rafter is selected, you'll discover that the 4½ in 12 rafter will have only a one-half inch HAP. That's why it's best to select a rafter from the lower pitch side as the controlling rafter.

Beginning the Layout— Figure H-3 shows the beginning layout for this rafter. If the scale of 3" per foot is used, the length of Figure H-3 will be 66". If the scale were reduced to 1" per foot, the rafters would be too small and the space needed for the drawing would still be large.

For a large roof, make the drawing in a slightly different way. Use a scale of 3" per foot. But rather than draw in line (6), draw the rafter center line (7) at the common rafter pitch of 4½ and 12.

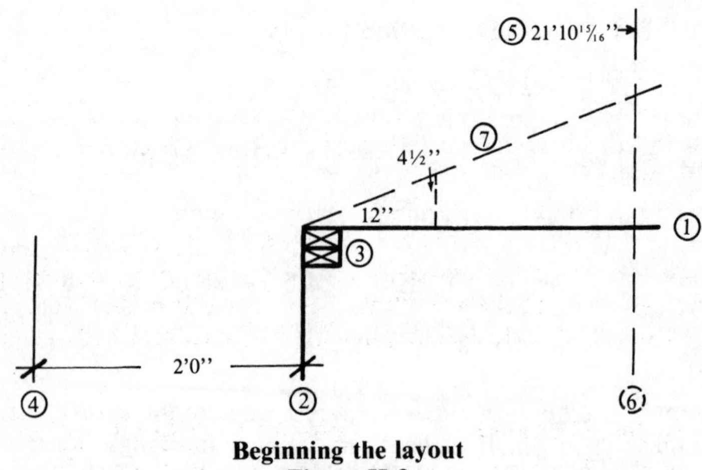

Beginning the layout
Figure H-3

Step 2: Establishing the HAP

A HAP must now be selected. Back at Figure 2-22 in Chapter 2, we recommended a HAP equal to two-thirds the thickness of the rafter.

Figure 7-51 and 7-52 show the layout on graph paper for a plumb cut. If we draw a line two-thirds of the way from the top edge of the rafter, its intersect with the plumb cut line will determine the ideal HAP.

Using this two-thirds rule, the formula back in Figure 7-52 could be rewritten:

$$\text{HAP} = \frac{\text{2/3 Thickness of Rafter}}{\text{Cosine of Common Rafter Roof Angle}}$$

This rafter is 7½'' thick. Multiply 2/3 by 7.5'' to get 5''. The slope of this rafter is 4½ in 12. To find the angle on your calculator, punch: [4] [·] [5] [÷] [1] [2] [=] [INV] [tan]. The display will show 20.56 degrees. Use the cosine of this angle to find the proper HAP.

$$\text{HAP} = \frac{5''}{\text{Cos } 20.57°} = \frac{5''}{.9363} = 5.34'' = 5\tfrac{5}{16}''$$

The right HAP for this roof is $5\tfrac{5}{16}$''.

Adding the HAP— Add the HAP (8) to line (2). See Figure H-4. Through this point draw in the top of the rafter, line (9). Extend this line until it touches line (4), the fascia length line. From this intersection generate line (14), the fascia level line. Add the thickness of the rafter in 3'' scale.

Step 3: The Plancher Distance

Figure 7-60, in Chapter 7, shows correct treatment for a fascia board. But this is not always the best procedure. Figure H-5 shows two methods.

The First Plancher Method— Here's how to find the plancher distance by the first method in Figure H-5. First, calculate the angular thickness of the sheathing. Refer back to Figure 7-60. The first formula for calculating plancher distance says (in general terms):

$$\begin{array}{l}\text{Angular Thickness}\\ \text{of the Sheathing}\end{array} = \frac{\text{Sheathing Thickness}}{\text{Cos of Common Rafter Angle}}$$

9
7
8
5⁵⁄₁₆"
7½"
3
1
3¹¹⁄₁₆"
14
4½"
10
2'0"
4
2

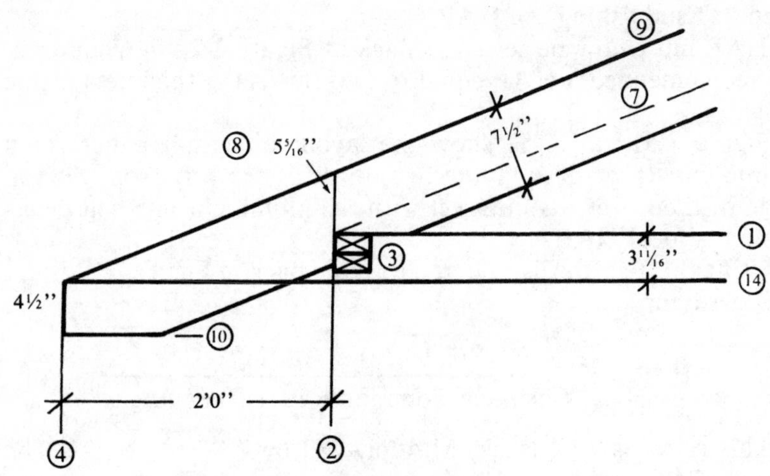

Finishing the partial layout
Figure H-4

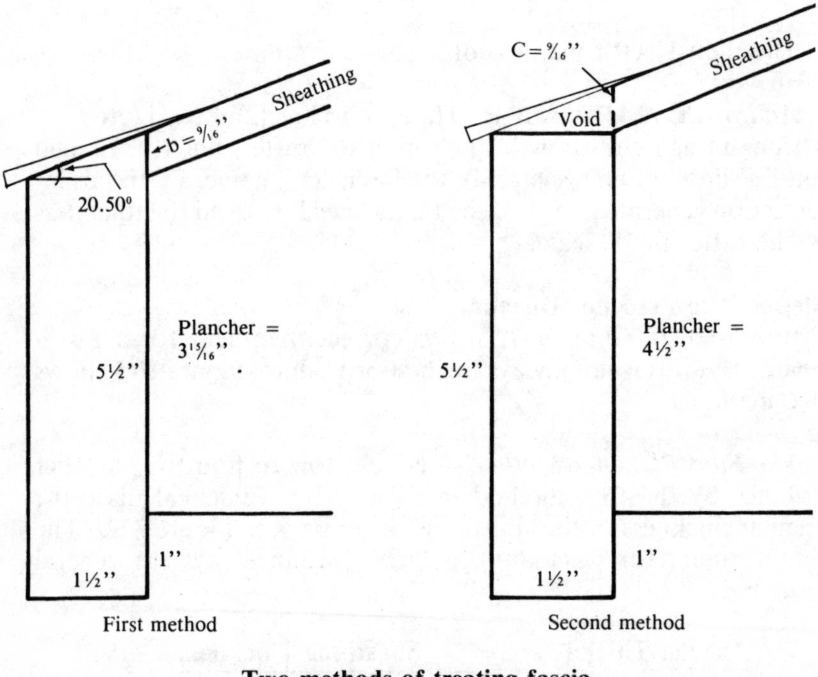

Sheathing
b = ⁹⁄₁₆"
20.50°
Plancher =
3¹⁵⁄₁₆"
5½"
1"
1½"
First method
C = ⁹⁄₁₆"
Sheathing
Void
Plancher =
4½"
5½"
1"
1½"
Second method

Two methods of treating fascia
Figure H-5

The irregular roof we're framing will be covered with 1/2" plywood sheathing.

$$\text{Angular Thickness of Sheathing} = \frac{.5}{\text{Cos } 20.56^\circ} = .5340 = 9/16"$$

The fascia will be 2 x 6. Since an inch of material is usually required below the plancher cut, the cut depth will be equal to the width of the fascia minus the angular thickness of the sheathing and minus 1":

Plancher Cut = Width of Fascia - b - 1"

Plancher Cut = 5.5" -.5340" - 1" = 3.966" = $3\frac{15}{16}$"

The top cut of the fascia is made at the common rafter angle, in this case 20.56 degrees.

The Second Plancher Method— For this method, apply both formulas from Figure 7-60. The first formula length for this roof was found to be 9/16".

The second formula, in general terms, states that the loss of fascia width equals the tangent of the common rafter angle times the thickness of the fascia:

Loss of Fascia Width = (Tan Com. Rafter Angle) (Thickness of Fascia)

Loss of Fascia Width = (Tan 20.56°) (1.5")

= .5625 = 9/16"

Now the plancher cut distance can be found:

Plancher Cut = Width of Fascia + C - B - 1"

Plancher Cut = 5.5" + .5625" - .5340" - 1" = 4.5285" = 4½"

Let's use the plancher length developed in the second method. Add 4½" on line (4) and draw the plancher line (10). This concludes the common rafter layout.

Beginning the 7 in 12 Layout— We can't draw the 7 in 12 rafter from the ridge line (13) to the fascia line intersection, as we did at Figure 10-18 in Chapter 10, because there is no ridge line (13) in Figure H-4. However, the angle of inclination for this rafter can be calculated and drawn.

Figure 10-28 shows the total run and the total rise of an irregular rafter. Find the total run on the plan view. For the 7 in 12 rafter, the total run was distance (A-B) in Figure H-2. That was 14.0874' or 14'1⅟16". The distance of the overhang must now be added in. The total run distance to the fascia line then is 16.0874'.

Now we have to find the total rise from line (14) to the ridge line. The distance from line (1) to the ridge line was calculated as 8.2172', or 8'2⅝". Add the HAP of 5⁵⁄16" to get 8.6622'. Now add the fascia line drop of the 4½ in 12 rafter to this distance.

Step 4: The Fascia Line Drop

The fascia line drop relates to the controlling rafter. Figure 10-27 gives the fascia line drop formula as:

Drop = (Tan of controlling rafter) x Run of overhang - HAP

The drop for this 4½ in 12 common rafter will be:

$$\text{Drop} = \frac{4.5}{12} \times 24'' - 5\tfrac{5}{16}''$$

$$\text{Drop} = 3.66'' = 3\tfrac{11}{16}'' = .305'$$

Another method of calculating the fascia drop line is to reason that 2' of run at 4½" is equal to 9". If the HAP for the rafter is 5⁵⁄16", then the fascia line drop equals 9" minus 5⁵⁄16", or 3¹¹⁄16". Both methods yield the same answer.

Step 5: Calculating the Angle

The tangent for the 7 in 12 common rafter is:

$$\text{Tan} = \frac{\text{Math height + HAP + Drop}}{\text{Building line run + overhang run}}$$

$$= \frac{8.2172' + .445' + .305'}{14.0874' + 2'}$$

$$= \frac{8.9672}{16.0874} = .5574 = 29.14^\circ$$

A 7 in 12 common rafter has an angle of 30.26 degrees, but because of the HAP and overhang, the angle changes slightly to 29.14 degrees.

Step 6: The Stair Gauge Setting
The framing square must be set slightly off of 7 and 12 to get the correct angle. Get the right stair gauge setting by using the formula:

$$\text{Setting} = \text{Tan of Rafter x Unit Run}$$

The tangent of the 7 in 12 rafter is 0.5574. Since this is a common rafter, the unit run is 12".

$$\text{Setting} = .5574 \times 12'' = 6.6886'' = 6\tfrac{11}{16}''$$

Laying Out the 7 in 12 Rafter— From the intersection of the 7 in 12 rafter fascia line and line (14) in Figure H-6, construct a 6¹¹⁄₁₆ in 12 pitch line, which represents the top of this rafter. Measure to the right the length of the overhang. Draw the building line. From the corner of the building line and line (1), draw in the two plates. Now add the thickness of the rafter.

The plans call for a 5½" rafter. This would put the rafter 3½" above the rafter plate. If the two-thirds rule is followed, the HAP for the 7 in 12 rafter would be:

$$\text{HAP} = \frac{\text{2/3 thickness of rafter}}{\text{Cos 7:12 rafter}}$$

$$\text{HAP} = \frac{3.66''}{.8735} = 4.1978'' = 4\tfrac{3}{16}''$$

This would require a raised plate of 5½". Here it would be more practical to use studs 97¾" long (rather than the standard 92¼") and add a ledger to the inside of the wall with joist hangers for the ceiling joists.

The 16 in 12 Rafters— Look at Figure H-7. The 16 in 12 common rafter must be drawn from the intersection of the fascia and line (14). If a certain ridge height was called for in this blind valley construction, then you would have to calculate the actual angle for this rafter as you did the 7:12 rafter. In this case, the ridge can be moved upward. Draw the proper slope and then measure to the right the distance of the overhang. Draw the building line.

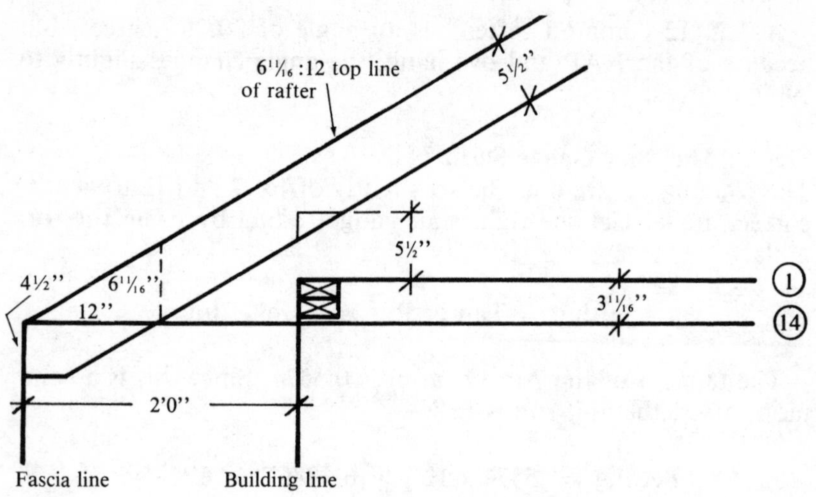

Beginning the second rafter
Figure H-6

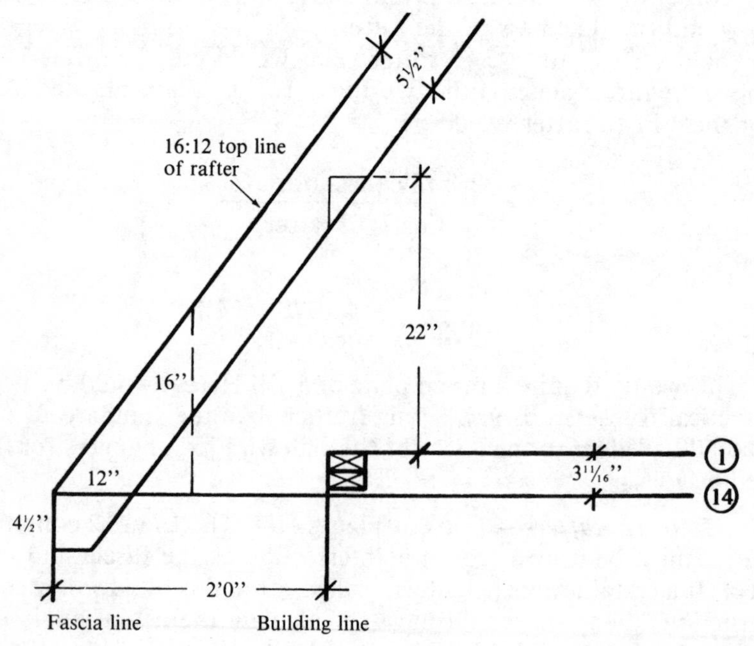

The 6 in 12 rafter
Figure H-7

Add the thickness of the rafter. From this drawing you will see the need for a 22''-high pony wall to fit the birdsmouth. That's indicated with a dotted line in Figure H-7.

It's obvious to you, the roof framer, that the fascia and overhang can't remain constant unless the wall height is adjusted as needed.

For this roof, build a soffit in the 16 in 12 overhang at the 5½'' level of the 7 in 12 bearing wall. This will equalize the outside wall height on this side of the building. Be alert to how exterior trim is applied to the wall as it goes around the building.

There is another way to frame this roof. Since the main roof of this plan was drawn as a gable, the 4½ in 12 side doesn't have a fascia at the same level. Only a 2' overhang was specified. Then the two front pitches could be worked together with the 7 in 12 pitch as the controlling rafter. If the main roof were a hip roof, you would need to make up a section view of the hip.

Step 7: The Main Ridge Framing

Figure 7-23 shows the loss of ridge height. The general formula for the loss of ridge height is:

Loss of height = (Tan common rafter angle) (½ thickness of the ridge)

The roof in question will have a ridge 1½'' thick. The 6¹¹⁄₁₆'' in 12 rafter angle is 29.14 degrees.

Loss of height = (.5574) (.75'') = .4180'' = 7/16''

From Figure 8-9 in Chapter 8, and the text that follows, you know that the 4½ in 12 rafter must be held above the ridge by 1/8''.

A Table of Steps

The following is a synopsis of the steps followed in this appendix.

Step 1:

$$AB = \frac{\text{Tan C (Span)}}{\text{Tan A + Tan B}} \qquad BC = \frac{\text{Tan A (Span)}}{\text{Tan A + Tan B}}$$

Total rise = AB x Unit Rise or: BC x Unit Rise

Step 2:

$$\text{The HAP} = \frac{\text{2/3 Thickness of Rafters}}{\text{Cos Common Rafter Roof Angle}}$$

Step 3:

If: B = Angular thickness of sheathing $= \dfrac{\text{Sheathing thickness}}{\text{Cos Common Rafter}}$

And if: C = Loss of fascia width = (Tan Common Rafter angle) (Thickness of Fascia)

Then for:

1st Method: Plancher Cut = Width of Fascia - B - 1"

2nd Method: Plancher Cut = Width of Fascia + C - B - 1"

Step 4:

$$\text{Fascia line drop} = \left(\begin{matrix}\text{Tan controlling}\\ \text{rafter}\end{matrix}\right)\left(\begin{matrix}\text{Run of}\\ \text{overhang}\end{matrix}\right) - \left(\begin{matrix}\text{HAP controlling}\\ \text{rafter}\end{matrix}\right)$$

or:

Fascia line drop = Total rise of overhang - HAP

Step 5:

$$\text{The rafter angle calculated} = \frac{\text{Total rise + HAP + fascia line drop}}{\text{Total run of fascia}} \; \boxed{\text{INV}} \; \boxed{\text{tan}}$$

Step 6:

Stair Gauge Setting = (Tan Rafter) (Unit Run)

Step 7:

Main ridge framing of lower pitched rafters.

$$\text{Height above Ridge} = \left(\begin{matrix}\text{Tan Low Pitched}\\ \text{Rafter}\end{matrix}\right)\left(\begin{matrix}\text{½ Thickness of}\\ \text{Ridge}\end{matrix}\right) \text{ minus } \left(\begin{matrix}\text{Tan High Pitched}\\ \text{Rafter}\end{matrix}\right)\left(\begin{matrix}\text{½ Thickness of}\\ \text{Ridge}\end{matrix}\right)$$

Also:

Hip backing or dropping = (Tan rafter) (Setback)

$$\text{Angle of the backing cut} = \boxed{\text{INV}} \; \boxed{\text{tan}} \text{ of } \frac{\text{Backing}}{\text{Setback}}$$

Answers

Chapter 1

Problem 1

a) The total run is half the width, or span. The total run is 11'.

b) Total rise is found by multiplying the total run (in feet) by the inches of unit rise. Eleven times 8 equals 88", or 7'4".

c) The unit rise is given as 8", since unit rise is the amount of rise for each 12" of unit run.

d) A gable roof has only common rafters. The one standard for the common rafter unit run is 12".

Problem 2

a) 17' divided by 2 equals 8'6"

b) 12"

c) 4''

d) 8.5 times 4'' equals 34'', or 2'10''

Problem 3

The total run is 30' divided by 2, or 15'. The second column of slope is used for this 4 in 12 roof. No. 2 (usually called ''No. 2 and better'') 2 x 8 at 16'' on center will span 16'3''. That's enough for our 15' of span. If you want to use 24'' spacing and the same grade, use 2 x 10 material.

Problem 4

The total run is given as 10'6''. For an 8 in 12 pitch, the second slope column is used. No. 2 and better 2 x 6 material at 16'' on center can be used.

This table is only an example. Check your local building code.

Chapter 2

Problem 1

To find the number of sixteenths in any number, multiply that number by 16. For example, 0.383 times 16 equals 6.128 sixteenths. Round this off to 6/16, or 3/8 of an inch.

Problem 2

To find the number of degrees when the unit rise and unit run are given, divide the unit rise by the unit run. For example, 10 divided by 12 equals 0.8333. The answer is called the tangent of the angle. With 0.8333 in the display of your calculator, punch [INV] [tan] and 39.80 degrees appears. This is the roof angle.

If you searched a trigonometry table under the column *tan* for the decimal 0.8333, you would find the angle of 39 degrees 48 minutes.

Problem 3

To convert from feet to inches, just multiply the number by 12, since there are 12'' to every foot: 0.3796' times 12 equals 4.5552''. Following Problem 1 then: 0.5552 times 16 equals 8.8832 sixteenths. Round the answer to 4 9/16''.

Problem 4

Follow the mathematics in Appendix B under "The secant function." On your calculator, divide 8 by 12 and then punch [INV] and [tan]. This will give you the angle of the roof. Then push [cos] [1/x] and 1.2018 appears. This is the secant for an 8 in 12 roof.

Problem 5

a) The total run is half the span: 18' divided by 2 equals 9' of total run to the building line. Since the unit rise is 7'', the total rise is 9' times 7'', or 63''. See Figure Answers-1.

Both numbers must be in the same unit, either feet or inches, for the square root method: 63'' divided by 12'' equals 5.25'.

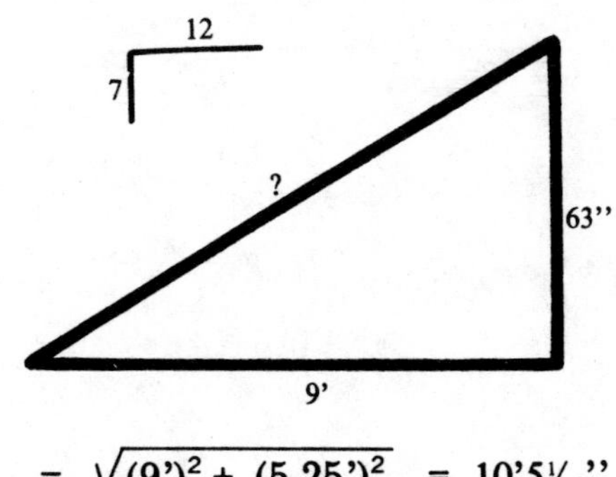

$$? = \sqrt{(9')^2 + (5.25')^2} = 10'5\tfrac{1}{16}''$$

Mathematical length to the building line
Figure Answers-1

?	=	$\sqrt{81 + 27.5625}$	Squaring both numbers
?	=	$\sqrt{108.5625}$	Adding the numbers
?	=	10.4193'	Taking the square root
?	=	10'5 1/16''	

b) The fascia line is 2' farther away than the building line: 9' plus 2' equals 11'. The total rise is 11' times 7'', or 77''. The calculation and answer are given in Figure Answers-2.

$$? = \sqrt{(11')^2 + (6.4166')^2} = 12.7347' = 12'8\tfrac{13}{16}''$$

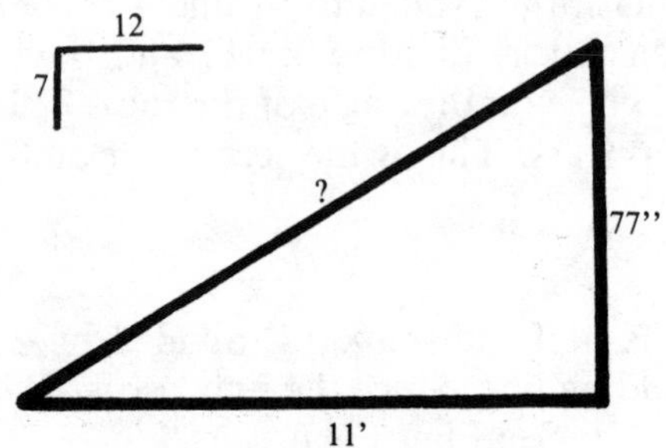

Mathematical length to the fascia line
Figure Answer-2

Problem 6

a) 7'10⅞''

b) 9'5$\tfrac{13}{16}$''

Problem 7

a) 10'0''

b) $\dfrac{\text{total rise}}{\text{total run}} = \text{unit rise} \qquad \dfrac{72''}{8'} = 9''$

The fascia line length is 12'6''.

Problem 8

The secant for a 7 in 12 roof is 1.1577. The total run is 18' divided by 2, or 9'. The rafter length is the secant times the total run: 1.1557 times 9' equals 10'5$\tfrac{1}{16}$''.

Chapter 3

Plan 1, Figure 3-20

Start by drawing lines 1, 2, 3, 4, and 5, based on Rule 2.

Now move to the next section of the drawing, and under the same rule, draw 6, 7, 8, 9, and 10. The broken hips meet in the center of the 10' span. Draw ridge 11, then draw the last two hips.

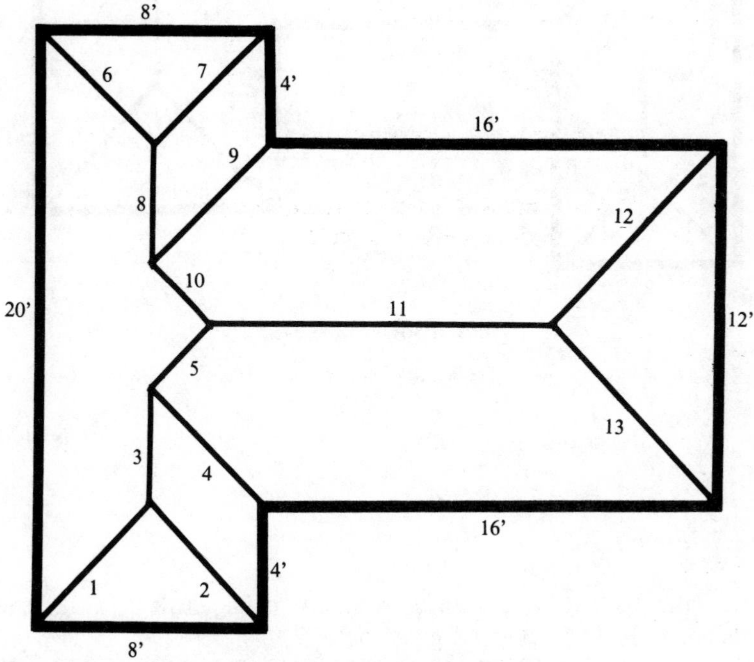

Plan 1 from Figure 3-20

Plan 2, Figure 3-21

Begin with either short span. Using Rule 2, draw lines 1, 2, 3, 4, and 5.

Using the same rule, draw 6, 7, 8, 9, and 10.

Adding hips 11 and 12 gives you the ridge framing points. Draw 13, which is at mid-span.

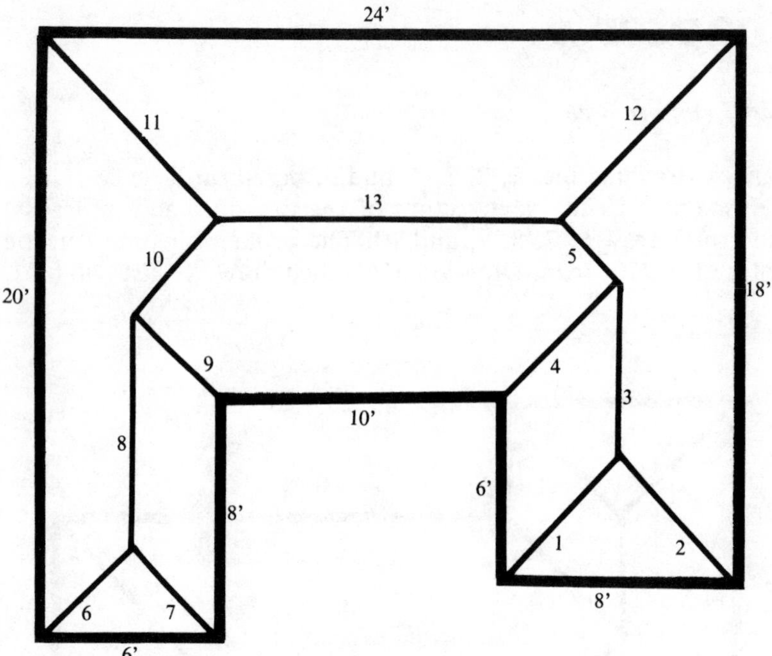

Plan 2 from Figure 3-21

Plan 3, Figure 3-22

Using Rule 3, draw 1, 2, 3, 4, and 5. The ridge can't be along the dotted line. Rule 5 reversed gives ridge 7.

Rule 3 gives 8, 9, 10, 11, 12, and 13. Rule 5 again gives ridge 7. Ridge 7 is centered between the 9' wall and the 10' wall.

Plan 4, Figure 3-23

Following Rule 3, begin with the 16' span and draw 1, 2, 3, 4, 5, 6, and 7. Hip 8 connects to ridge 7, which is between the 22' wall and the 2' wall.

From Rule 2 draw 9, 10, 11 and 12. Rule 5 gives broken hip 13 and closes the roof.

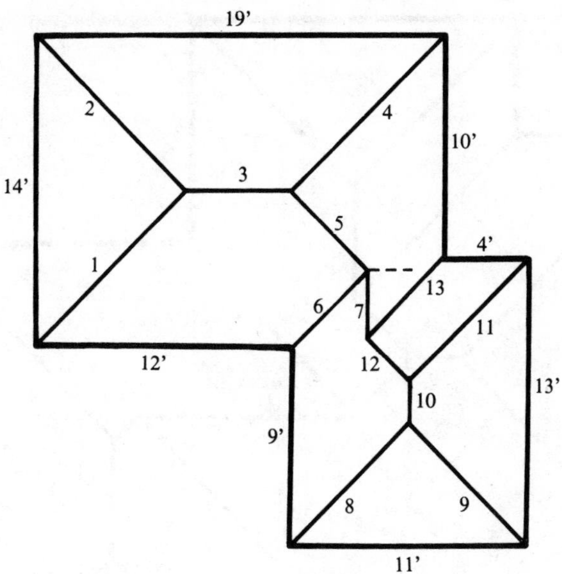

Plan 3 from Figure 3-22

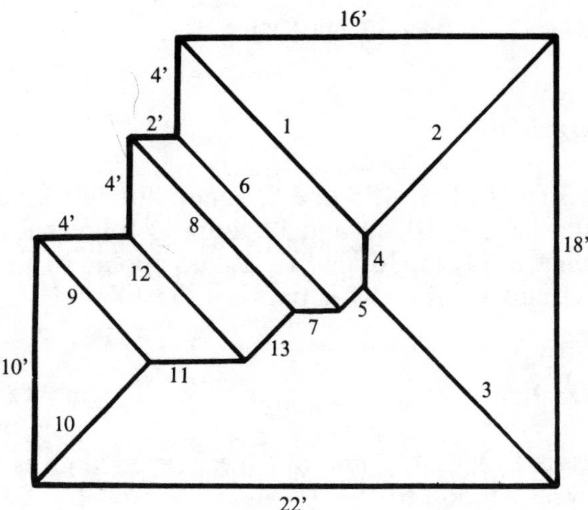

Plan 4 from Figure 3-23

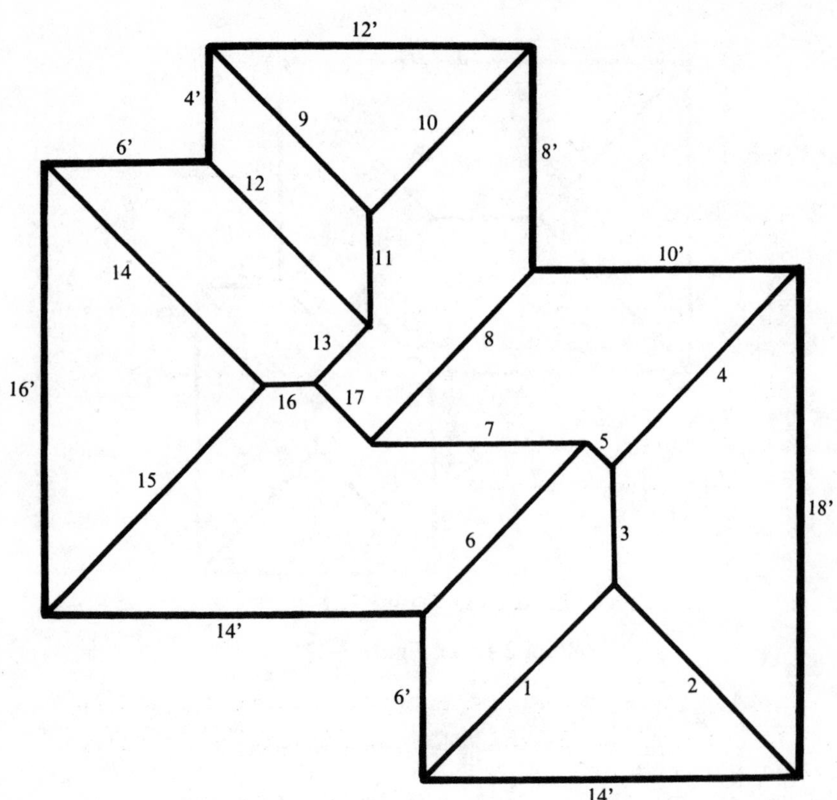

Plan 5 from Figure 3-24

Plan 5, Figure 3-24

Follow Rule 3 for 1, 2, 3, 4, 5 and 6. Line 7 follows Rule 5.

Follow Rule 2 for 9, 10, 11 and 12. Line 13 follows Rule 5.

Rule 2 applies to 14, 15, 16, and 13. Rule 5 applies to 7 and 8, as well as broken hip 17. Ridge 7 is between the 10' and 14' walls.

Plan 6, Figure 3-28

Rule 3a gives us 1, 2, 3, 4, 5 and 6. Rule 5 reversed gives ridge 7, which is between (c) and (d).

With Rule 3a again, draw 8, 9, 10, 11, 12, and 13. Rule 5 reversed gives ridge 14, which is between (f) and (g). Line 15 joins 14 under General Layout Rule (2) and Rule 3b. Line 16 follows Rule

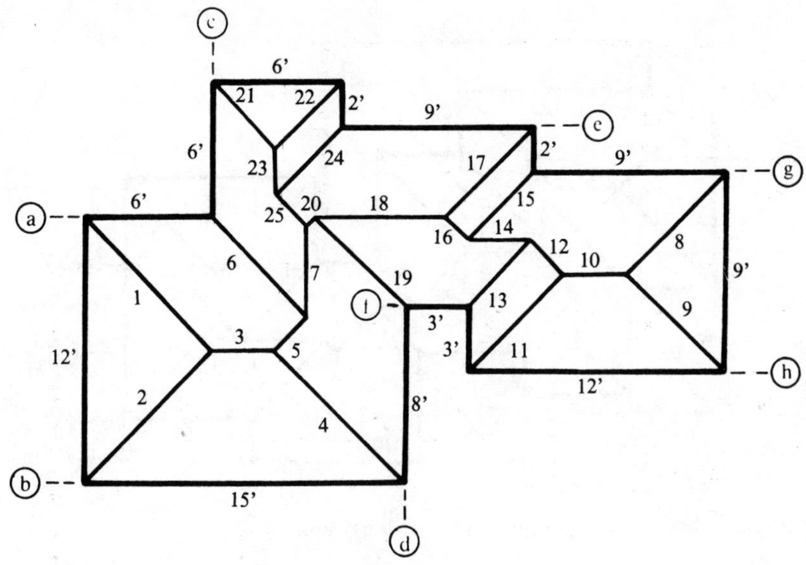

Plan 6 from Figure 3-28

5. Drawing 17 gives the framing point for ridge 18, which is between (e) and (f). Line 19 must terminate ridge 18. Line 20 follows Rule 5 and intersects 7.

Rule 2a gives us 21, 22, 23, and 24. Rule 5 closes the roof with 25. Valley 6 is a part of Rule 2b.

Plan 7, Figure 3-29

Rule 3 yields 1, 2, 3, 4, 5, 6. Rule 5 reversed gives 7, which is centered between (a) and (b). Rule 5 covers 8 and 9. For 10, 11, 12, valley 8 is the near valley and would seem to follow Rule 2. But if 8 terminated ridge 12, any broken hip would lead off-center and Rule 1 would be violated. Draw 10, 11 and 12. Line 9 gives the true termination of ridge 12. If 13 is extended, it meets 14 at the center of (a) and (c). Lines 13, 14 and 15 belong to Rule 5, as well as 15, 16 and 17. Leave 17 pointing toward the center and move to the outside.

Lines 18, 19, 20 and 21 follow Rule 2. Line 22 follows Rule 5. Leave 22 pointing toward the center and move to the next span.

Lines 23, 24, 25 and 26 follow Rule 2 again and 27 follows Rule 5. Lines 27 and 28 meet at the center of (d) and (e). Draw 29. Line

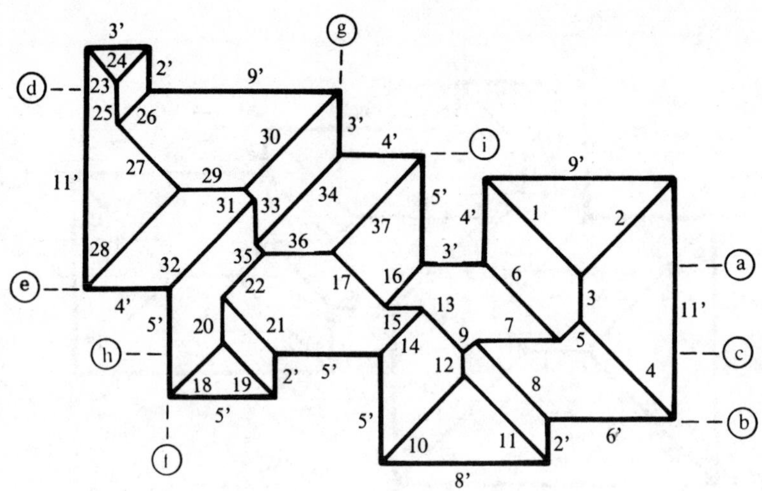

Plan 7 from Figure 3-29

30 intersects 29. Line 31 points toward the center and is intersected by 32. Generate 31, 32 and 33 from Rule 5. Line 34 intersects the far side of 33, which is now centered between (f) and (g). Line 35 follows Rule 5, which intersects 22 midway between (h) and (i). Draw ridge 36. It meets 37 and 17 to close the roof.

Plan 8, Figure 3-30 (The Invisible)

Rule 3 gives 1, 2, 3, 4, 5, 6 and 8. Rule 5 covers 5, 6, 7, 8 and 9. Ridge 7 is between (a) and (b).

With Rule 3, move to 10, 11, 12, 13, 14, 15 and 17. Rule 5 governs 15, 16, 17 and 18. Ridge 16 is between (d) and (e).

Following Rule 2, move to 19, 20, 21 and 22. Line 23 is Rule 5. Since 24 makes a hip to broken hip intersection with 23, the intersection can't follow Rule 4. Rule 4 requires a valley. Line 25 intersects 9 at the center of (a) and (c). Draw ridge 26. Rule 5 gives 26, 27 and 28. Think of point (f) as a ridge of zero length which is centered between (c) and (d).

Plan 9, Figure 3-31 (The Bushwhacker)

Follow Rule 2 for 1, 2, 3 and 4. Line 5 creates a Rule 4b situation, so draw 6, which is centered between (a) and (b). Line 7 terminates 6. Under Rule 5, you begin 8 as shown by the dotted line.

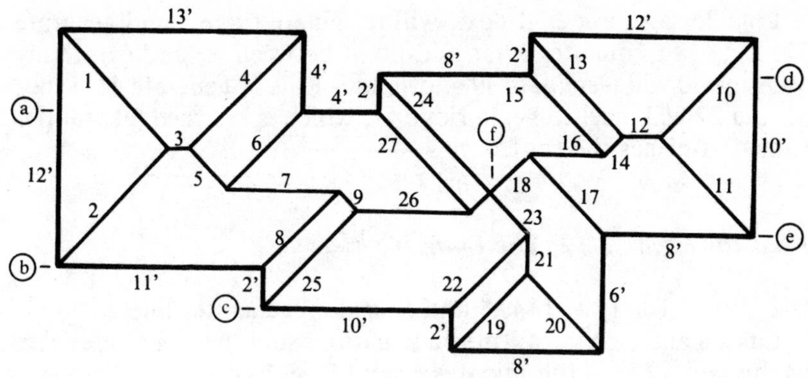

"The Invisible" plan 8 from Figure 3-30

Using Rule 2, draw 9, 10, 11 and 12. With Rule 5, draw 13. It extends to line 7, throwing us into Rule 4b. Draw ridge 8 correctly.

From Rule 2, draw 14, 15, 16 and 17. Use Rule 5 for line 18. Lines 18 and 19 establish the center point for ridge 20. Add 21 and 22 with Rule 5.

Draw 23, 24 and 25. Lines 26 and 27 will make a blind valley. (See Chapter 7.) From Rule 2, draw 28, 29, 30 and 31. Line 32 causes the Rule 4b ridge 33. Terminate it with 34. Under Rule 5, draw 35.

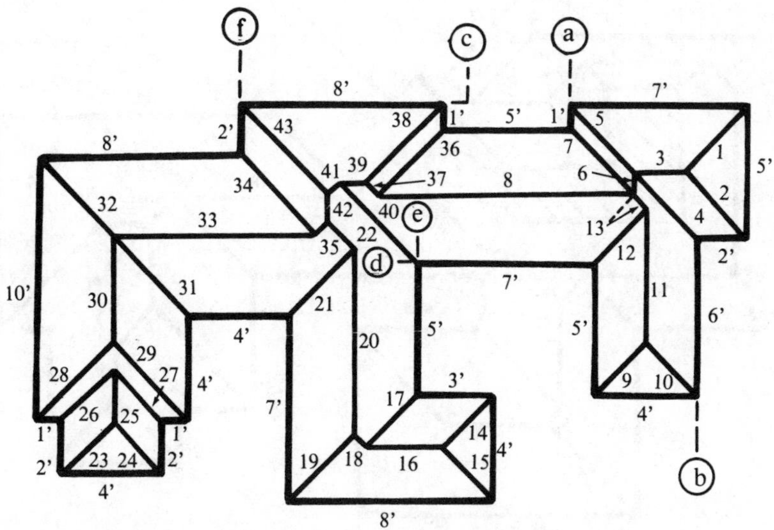

"The Bushwhacker" plan 9 from Figure 3-31

Line 36, also under Rule 5, will terminate ridge 8 and generate 37. Line 38 terminates 37 at mid-point between (c) and (d). Draw 39. Line 40 will terminate 39 and, under Rule 5, generate 41. Lines 35 and 22 will meet to begin ridge 42, which is between (e) and (f). Line 43 finishes the roof.

Plan 10, Figure 3-32 (The Long Nighter)

Use Rule 3 for 1, 2, 3, 4, 5 and 6. Rule 5 generates line 7.

Lines 8 and 9 create a Rule 4b situation, so draw 10 in the center of this wing. Line 11 terminates 10 and Rule 5 gives 12, which connects to 7. Extending 13 and 14 yields a framing point at mid-span. Line 15 is under Rule 5 reversed. Lines 16 and 17 follow Rule 5.

Finish Rule 4b with 18. Add 19. Draw 20 to mid-span. Add 21 and 22 under Rule 5. Lines 23 and 24 also follow Rule 5.

Rule 2 generates 25, 26, 27 and 28. Use Rule 5 for 29. Bring up 30. Draw 31 and 32 under Rule 2b. Ridge 31 is centered between the two 5' walls. From Rule 2, draw 33, 34 and 35. Ridge 35 is centered between the 2' wall and the 11' wall. Lines 35 and 31 form a Rule 4b situation and generate line 36. Rule 5 applies to 36, 37, 38, 39 and 40.

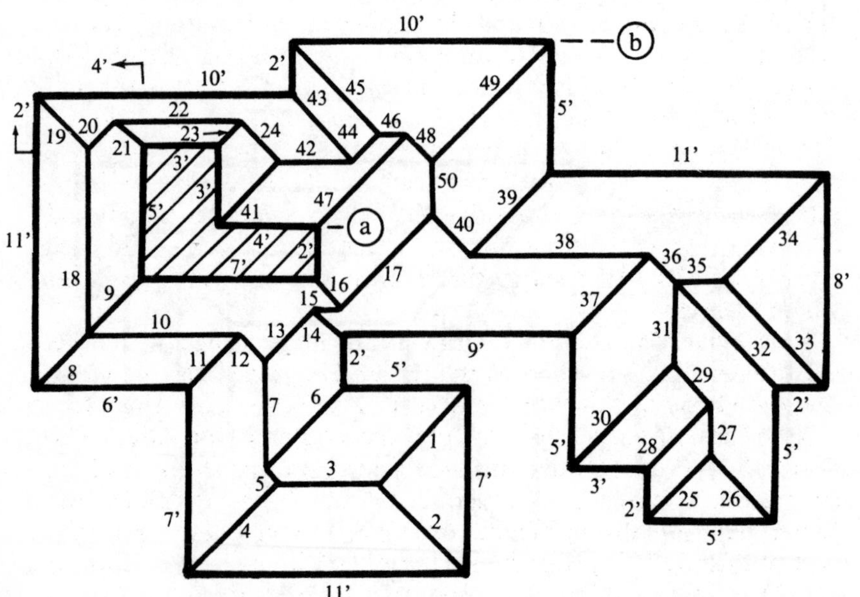

"The Long Nighter" plan 10 from Figure 3-32

Extend 41. It meets 24 at mid-span between the 10' wall and the 4' atrium wall. Draw 42. Rule 5 covers 42, 43 and 44. Line 45 makes a framing point for 46, which is between the upper wall (b) and the atrium wall (a). Lines 46, 47 and 48 follow Rule 5. Extend 17 and add 49. Close the roof with line 50.

Plan 11, Figure 3-33 (The Stumbler)

Rule 2 gives 1, 2, 3 and 4. Rule 5 gives 5. From here, the most obvious rule is Rule 1. Therefore, draw ridge 6. Line 7 intersects at mid-span. So does 8. Lines 9 and 5 cause 10 to be centered. Rule 5 includes 10, 11, 12; 12, 13, 14, and 14, 15, 16.

Rule 5 also includes 6, 7, 17; 17, 18, 19; 19, 20, 21; 21, 22, 23, and 23, 24, 25.

Rule 2 gives 26, 27, 28 and 29. Line 30 is under Rule 5. Extend 31. It may be a Rule 2b ridge, or it may not. We don't know yet. Also extend 32. At this point, the center of the building must be filled in more.

Move to 33, 34, 35 and 36 under Rule 2, and 37 under Rule 5. Rule 1 gives 39. Connect 38.

Move to 40, 41, 42 and 43 under Rule 2. Line 44 can't be under Rule 5. What's going on here? Following Rule 1, draw ridge 45 and 46. Now we see 43 and 44 framed as either blind valleys (Chapter 7) or as valleys to an equal span addition (Chapter 5), since 45 is a Rule 4a ridge. Rule 5 covers 46, 47, 48; and 48, 49, and 50 (which is also Rule 1). Rule 5 continues with 50, 51, and 52, which terminate ridge 39, generating 53 to 54 to properly establish ridge 55 in the center. Rule 5 covers 55, 56, 57. Valley 58 helps establish the mid-point for ridge 59. Rule 5 generates 59, 60 and 61. Leave line 61 pointing toward the center, just as you did with lines 16 and 30. We need to fill in more roof before we discover where these go.

Rule 5 gives 45, 62 and 63. Draw line 64 under Rule 5 with 63 as the broken hip. But where is the ridge? The ridge can't be along the dotted line as this violates Rule 1. Therefore, draw ridge 65.

Rule 2 covers 66, 67 and 68. But what terminates this ridge? The near valley 72 can't, since this would put a ridge off-center. Line 69 must terminate 65 to generate a broken hip 70 toward the center. This hip also terminates 68. Lines 71 and 72 are a reversed part of Rule 5. This properly places ridge 73 in the center. Both 74 and 75 terminate this ridge, creating a Rule 4b ridge 76. Rule 5 gives 76, 77 and 78. Broken hips 78 and 61 point toward each other

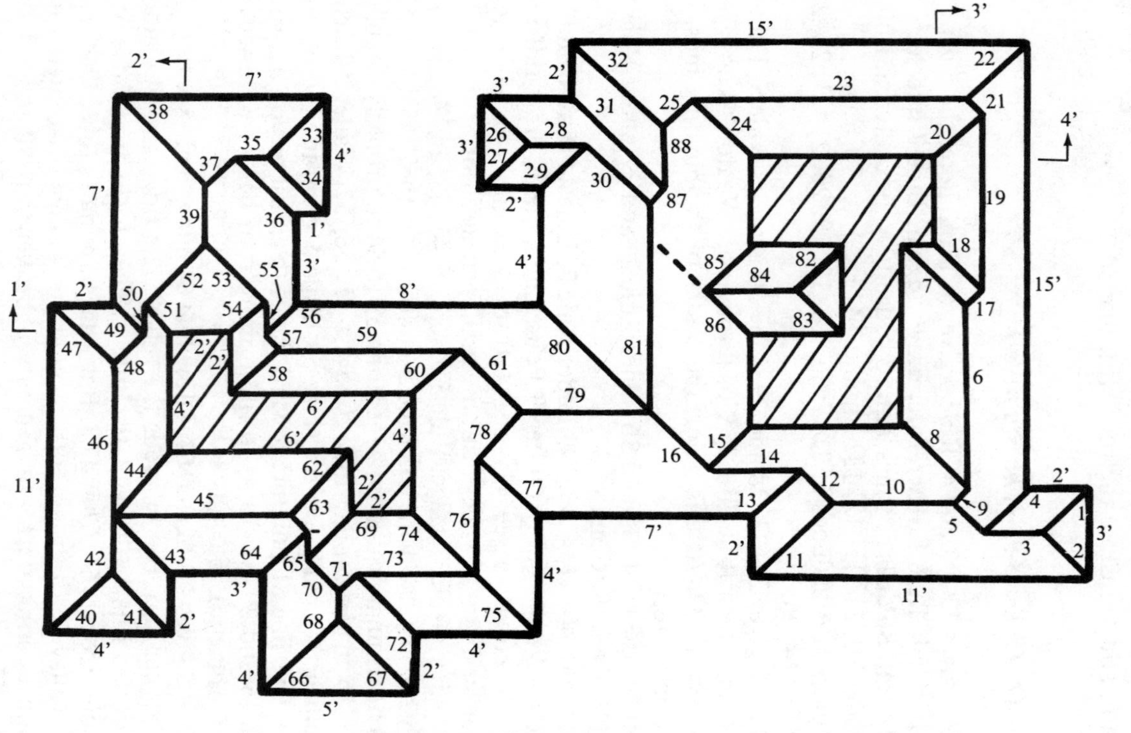

"The Stumbler" plan 11 from Figure 3-33

and the center. Will they join in the center of the span? Yes, generating ridge 79.

Line 16 terminates the other side and also continues to valley 80, thus forming a Rule 4b ridge 81. Line 81 is centered between the atrium wall and the 4' wall. Lines 82, 83, 84, 85 and 86 are from Rule 2. Lines 85 and 86 form a blind valley (Chapter 7), or 86 can be framed along the dotted line as a supporting valley rafter with 85 as the shortened valley rafter (Chapter 5). See General Layout Rule (2), first exception.

Broken hip 30 terminates 81. Line 31 must then terminate 87, forming a point of generation for ridge 88, which will be centered between the 3' atrium wall and the 2' outside return wall. Line 25 terminates 88 and 32 closes the roof.

Mazeltov! You may never have a roof that difficult. But if you do, you're ready for it.

Answers for Table 3-38

Rafters 1 through 8 have a 12" unit run, while rafters 9 through 12 have a unit run of 16.97".

The apparent run is given. If you are using the secant method, multiply the apparent run by the square root of 2 to find the actual run.

Both ends of Ridge 13 are double cheek cut hips. Therefore, each end will grow by half the thickness of the common rafter. 10 feet plus 3/4" plus 3/4" equals 10'1½".

Ridge 14 will have a single cheek cut at one end. This will cause a growth of 3/4" plus 1¹⁄₁₆", or 1¹³⁄₁₆". The other end of the ridge can be framed according to Figure 3-39 A or B.

In Figure 3-39 A, the ridge goes past the broken hip framing point a distance of 1'. See Figure 3-37. From this 12" extension, the ridge must be shortened 3/4", for a total growth of 11¼". Framing by method A gives a ridge length of:

Type		Math Length		Single Cheek Cut Growth				Type A Growth
A	=	12'0"	+	1¹³⁄₁₆"	+	11¼"	=	13'1¹⁄₁₆"

In Figure 3-39 B, the broken hip framing point is the calculated mathematical length of the ridge. Here the ridge is shown to frame against the single cheek cut broken hip, giving a single cheek cut growth at each end:

Type		Math Length		Single Cheek Cut Growth				Type B Shortening
B	=	12'0"	+	1 13/16"	+	1 13/16"	=	12'3 5/8"

No.	Kind	Quantity	Apparent Run	Math Length @ Bldg. Line	Math Length @ Overhang
1	Commons	12	6'0"	7'2½"	8'4 15/16"
2	Commons	10	5'0"	6'0⅛"	7'2½"
3	Hip Jack	10	2'0"	2'4⅞"	3'7¼"
4	Hip Jack	10	4'0"	4'9 11/16"	6'0⅛"
5	Valley Jack	1	3'0"	3'7¼"	4'9 11/16"
6	Valley Jack	1	4'0"	4'9 11/16"	6'0⅛"
7	Valley Jack	1	2'0"	2'4⅞"	3'7¼"
8	Valley Jack	1	1'0"	1'2 7/16"	2'4⅞"
9	Hips	2	5'0"	7'9 13/16"	9'4 9/16"
10	Hips	3	6'0"	9'4 9/16"	10'11 5/16"
11	Broken Hip	1	1'0"	1'6¾"	3'1½"
12	Valley	1	5'0"	7'9 13/16"	9'4 9/16"
13	Ridge	1	10'0"	2- DCC ends	10'1½"
14	Ridge	1	12'0"	*A = 13'1 1/16"	*B = 12'3 5/8"

* A and B are from Figure 3-39

Listing the framing members from Table 3-38
Table 3-38 Answer

Chapter 4

Problem 1

a) The common rafter shortens by half the thickness of the ridge. Since it's stated that all members are 1½" thick, then: 1½" divided by 2 equals 3/4". This is the common rafter shortening.

b) Since the roof is to be a hip roof, we can figure the mathematical length of the ridge by subtracting the span from the width. From 36' subtract 24' to get 12'. (For more help, see Figure 3-1 and the information that follows.)

On a hip roof, find the actual length from the mathematical length by adding the amount of ridge growth to each end. Here we know that there will be double cheek cut hips at each end. Table 4-3 or Figure 4-7 shows that the ridge shortening for a double cheek cut hip configuration is half the thickness of the common rafter. Therefore, add 3/4'' to the mathematical length at each end: 12 feet plus 3/4'' plus 3/4'' equals 12'1½''. This is the actual length of the ridge for this building.

c) Find the hip shortening and setback on Table 4-7, or by following the reasoning in Figures 4-2 and 4-3. This hip shortens by half the 45-degree thickness of the common rafter only ($1\frac{1}{16}$''), not by the thickness of the hip or the ridge.

d) The setback is always half the thickness of the material itself, 3/4''.

Problem 2

a) 6'' divided by 2 equals 3''

b) The mathematical length is 64' minus 44', or 20'. Ridge growth at one end is half the thickness of the ridge plus half the 45-degree thickness of the hip. The ridge is 6''. 6'' divided by 2 equals 3''. The 45-degree thickness of a rough 4'' timber is 4 times the square root of 2, or 5.6568, divided by 2 for half this thickness. That's 2.8284, or $2\frac{13}{16}$''.

Figure Answers-3 shows the proof. On 8½'' x 11'' graph paper, lay out a 4'' square and draw the diagonal. With a ruler, mark a center line across the opposite corners as indicated. The line segment will measure $2\frac{13}{16}$''.

The mathematical length plus the growth equals the actual length of the ridge: 20' plus 3'' plus $2\frac{13}{16}$'' plus 3'' plus $2\frac{13}{16}$'' equals 2'$11\frac{11}{16}$''.

c) 6'' x$\sqrt{2}$ = 8.4852 ÷ 2 = 4.2426 = 4¼''

d) 2''

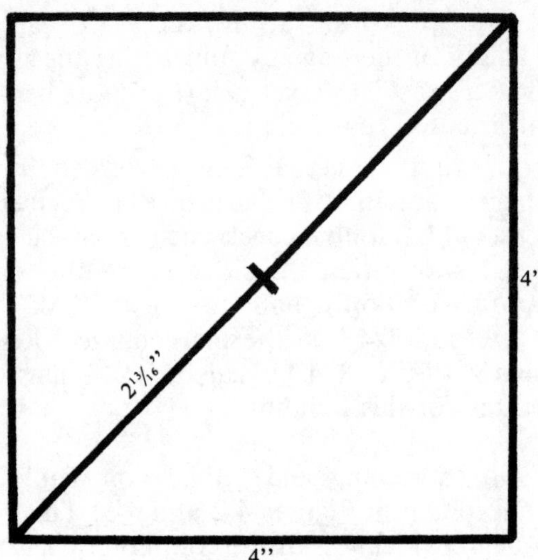

Half the 45° thickness
Figure Answers-3

Chapter 5

Problem 1

See Figure Answers-4. The mathematical length is 2' plus 10' plus 9', or 21'.

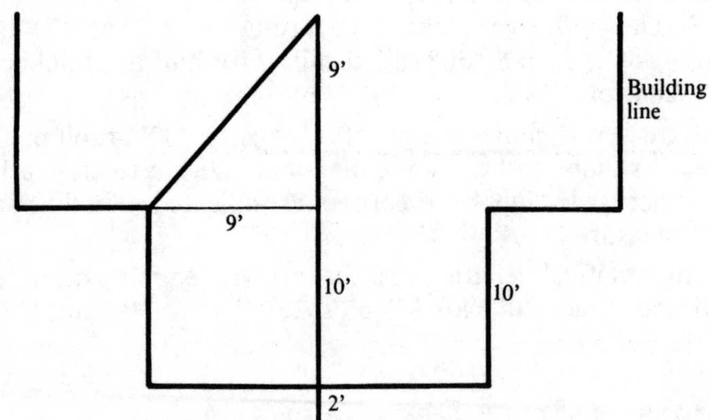

The narrow addition span
Figure Answers-4

Problem 2

2' + 5' + 18' = 25'

Problem 3

7' x$\sqrt{2}$ = 9.8995' = 9'10$\frac{13}{16}$''

Problem 4

11' x$\sqrt{2}$ = 15.5563' = 15'6$\frac{11}{16}$''

Chapter 6

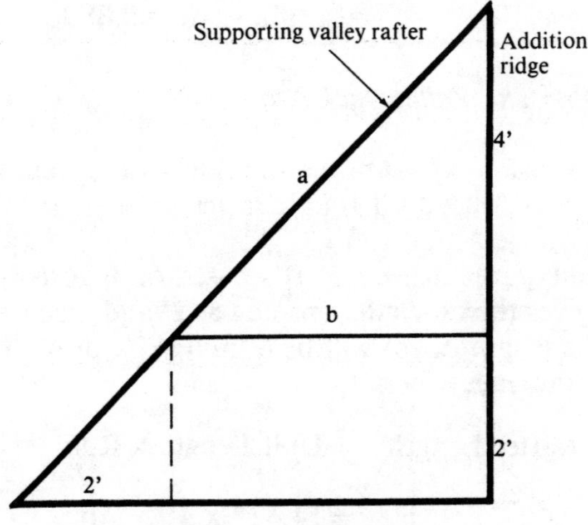

Finding the apparent run
Figure Answers-5

Jack Rafter (a): The Shortened Valley Rafter

Framing Square Method— On the second line of the framing square rafter tables under 8'', you will find a unit length of 18.76''. Multiply this by one-half the addition span, 6' in this case. 18.76 times 6 equals 112.5699''. Divide by 12 to get 9.3808', or 9'4$\frac{9}{16}$''.

Secant Method— To use the secant method, you need to know the actual run: Mathematical Length equals the Secant times the Actual Run.

This shortened valley rafter lies within a 6' span at 45 degrees. Therefore, to find its actual run, multiply 6' by the square root of 2. That's 8.4853'.

The roof is given as 8 in 12, but the secant of the hip is based on 8 and 16.97. Find the secant on your calculator by punching [8] [÷] [1] [6] [.] [9] [7] [=] and 0.4714 appears. Punch [INV] [tan] and you have the angle, 25.24 degrees. Punch [cos] [1/x] and the secant, 1.1055, appears. The secant is also found in Figure 4-65.

Mathematical length = Secant x Actual run

= 1.1055 x 8.4853 = 9.3808'

or

Mathematical length = Secant x $\sqrt{2}$ x Apparent run

= 1.5635 x 6' = 9.3808'

Jack Rafter (b): The Valley Jack Rafter

Framing Square Method— On the first line of the framing square rafter table under 8", you'll find a rafter unit length of 14.42". Multiply this by the actual run.

Look at the Figure Answers-5. It's based on Figure 6-23. Since the problem gives the on-center spacing as 2' and member (a) is at 45 degrees, the length of (b) will be 6' minus 2', or 4'. This is the actual run of the valley jack.

Rafter Length = Unit Length x Run

= 14.42" x 4'

= 57.6888 ÷ 12

= 4.8074'

= 4'9 11/16"

Secant Method— The roof pitch is 8 in 12. To find the secant on your calculator, punch: [8] [÷] [1] [2] [=] and 0.6667 appears. Punch

[INV] [tan] and 33.69 degrees appears. This is the roof angle. Then punch [cos] [1/x] and 1.2018 appears. This is the secant.

Mathematical Length = Secant x Actual Run

= 1.2018 x 4'

= 4.8074'

= 4'9$\frac{11}{16}$''

Jack Rafter (c): The Supporting Valley Rafter

Do this the same way you did Jack Rafter (a), but use the run for this rafter.

Jack Rafter (d): The Hip-Valley Cripple Jack

The run of this jack is the same as the length of the offset wall.

Jack Rafter (e): The Hip Jack

Do this the same way you did Jack Rafter (b), but use the run for this rafter.

Jack Rafter (f): The Valley Cripple Jack

The valley cripple jack has a run that is twice the run of the jack rafter it meets.

	Rafter Name	**Apparent Run**	**Actual Run**	**Mathematical Length**
a	Shortened valley	6'	6' x $\sqrt{2}$	9'4$\frac{9}{16}$''
b	Valley jack	4'	4'	4'9$\frac{11}{16}$''
c	Supporting valley	10'	10' x $\sqrt{2}$	15'7$\frac{5}{8}$''
d	Hip-valley cripple jack	4'	4'	4'9$\frac{11}{16}$''
e	Hip jack	2'	2'	2'4$\frac{7}{8}$''
f	Valley cripple jack	4'	4'	4'9$\frac{11}{16}$''

Answers from Figure 6-23

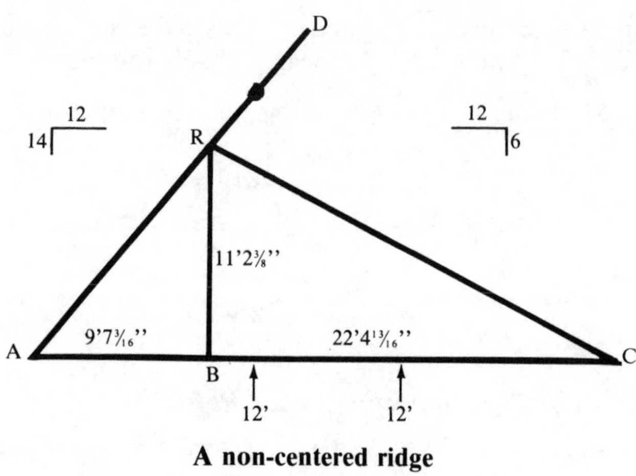

A non-centered ridge
Figure Answers-6

Chapter 8

Problem 1

Layout Method— Figure Answers-6 shows the solution using the layout method. On 8½'' x 11'' graph paper with 1/4'' squares, mark off 32 squares to represent 32'. Draw an arrow 12' to the right of (A) and draw another arrow 12' to the left of (C). Find the point 14' above the left arrow and 6' above the right arrow. Make a dot at each of these points. Draw line (ARD) and line (CR). Then draw (RB). Measure the three segments.

Trig/Algebra Method—

$$AB = \frac{\text{Tan C (span)}}{\text{Tan A + Tan C}} \qquad \text{Tan A} = \frac{14}{12} = 1.1667$$

$$\text{Tan C} = \frac{6}{12} = .5$$

$$AB = \frac{(.5)\ (32')}{1.1667 + .5} = \frac{16'}{1.6667} = 9.6' = 9'7\tfrac{3}{16}''$$

$$BC = \frac{\text{Tan A (span)}}{\text{Tan A + Tan C}} = \frac{(1.1667)\ (32')}{1.6667}$$

$$BC = \frac{37.3333'}{1.6667} = 22.4' = 22'4\tfrac{13}{16}''$$

Proof:

$$
\begin{array}{rlcl}
 & AB & = & 9'7\tfrac{3}{16}'' \\
+ & \underline{BC} & = & \underline{+22'\,4\tfrac{13}{16}''} \\
 & AC & = & 32'0''
\end{array}
$$

The Total Rise:

Total Rise	= Total Run x Unit Rise			
	= 9.6	x	14" = 134.4"	
	= 22.4	x	6" = 134.4"	
134.4" ÷ 12	= 11.2"	=	11'2⅜"	

Height Above Ridge:

$$\left(\frac{14}{12}\right)(2) = 2\tfrac{5}{16}''$$

$$\left(\frac{6}{12}\right)(2) = 1''$$

6:12 rafters held up $2\tfrac{5}{16}'' - 1'' = 1\tfrac{5}{16}''$

Problem 2

Tan A = 1.3333

Tan C = .6667

AB = 10'8"

BC = 21'4"

RB = 14'2$\tfrac{11}{16}$"

8:12 rafter held up ½"

No.	Kind	Q	Degrees	Secant	Bldg. Line Actual Run	Math Length Bldg. Line	Fascia Line Actual Run	Math Length Fascia Line
1	Common	7	33.69°	1.2018	18''	21⅝''	24''	28⅞''
2	Reg. hip jack	4	33.69°	1.2018	6''	7 3/16''	12''	14 7/16''
3	Reg. hip	2	25.24°	1.1055	18'' x $\sqrt{2}$	28⅛''	24'' x $\sqrt{2}$	37½''
4	Regular hip jack	2	33.69°	1.2018	10''	12''	16''	19¼''
5	Irr. side hip jack	2	41.63°	1.3380	3''	4''	9''	12 1/16''
6	Irr. common	1	41.63°	1.3380	12''	16 1/16''	18''	24 1/16''
7	Irr. hip	1	28.07°	1.1333	20''	22 11/16''	30''	34''

Mathematical length cutting list from Figure 10-47

No.	Ridge Cut			HAP	Plancher	Backing		Stair Gauges	
	Shortening	Setback	Cut Degrees			Amount	Degrees	Body	Tongue
1	3/4''	--	0°	2 3/4''	2''	--	--	12	8
2	1 1/16''	3/4''	45°	2 3/4''	2''	--	--	12	8
3	1 1/16''	3/4''	45°	2 3/4''	2''	5.5/16''	25.23°	16.97	8
4	1 1/4''	1''	53.13°	2 3/4''	2''	--	--	12	8
5	15/16''	9/16''	36.87°	2 5/8''	2''	--	--	12	10 11/16
6	1 1/2''	--	0°	2 5/8''	2''	--	--	12	10 11/16
* 7	15/16''	9/16''	36.87°	2 3/4''	2''	5/16''/9/16''	22.62°/36.87°	16.97	9 1/16

* Tail setback and angle for each cut 9/16'' at 36.87° and 1'' at 53.13°

The shortening and setback cutting list from Figure 10-48

Rafter Layout: 6'', 18'', 30'', 42'', 54'', 6''

Ridge Length	=	**Mathematical Length** 30''	+	**Growth** 1/4''	+	**Growth** 1 1/2''	= 32 1/4''

Raised Plate = 34 1/2'' long point to long point

Index

T

U

V

W

Cost Records for Construction Estimating

Shows how quick and easy it is to collect, organize, and use your actual job costs to create the most accurate estimates possible for future projects. Explains how to track costs for sitework, footings, foundations, framing, interior finish, siding and trim, masonry, subcontract expense and more. Includes sample customized forms to demonstrate the recordkeeping methods, and provides blank copies for your use. **208 pages, 8½ x 11, $15.75**

Construction Estimating Reference Data

Collected in this single volume are the building estimator's 300 most useful estimating reference tables. Labor requirements for nearly every type of construction are included: site work, concrete work, masonry, steel, carpentry, thermal & moisture protection, doors and windows, finishes, mechanical and electrical. Each section explains in detail the work being estimated and gives the appropriate crew size and equipment needed. Many pages of illustrations, estimating pointers and explanations of the work being estimated are also included. This is an essential reference for every professional construction estimator. **368 pages, 11 x 8½, $18.00**

Contractor's Year-Round Tax Guide

How to set up and run your construction business to minimize taxes: corporate tax strategy and how to use it to your advantage, why you should consider incorporating to save tax dollars, and what you should be aware of in contracts with others. (Includes sample contracts). Covers tax shelters for builders, write-offs and investments that will reduce your taxes, accounting methods that are best for contractors, what forms of compensation are deductible, and what the I.R.S. allows and what it often questions. Explains how to keep records and protect your company from tax traps that many contractors fall into. **192 pages, 8½ x 11, $16.50**

Masonry & Concrete Construction

Every aspect of masonry construction is covered, from laying out the building with a transit to constructing chimneys and fireplaces. Explains footing construction, building foundations, laying out a block wall, reinforcing masonry, pouring slabs and sidewalks, coloring concrete, selecting and maintaining forms, using the Jahn Forming System and steel ply forms, and much more. Everything is clearly explained with dozens of photos, illustrations, charts and tables. **224 pages, 8½ x 11, $13.50**

Building Cost Manual

Square foot costs for residential, commercial, industrial, and farm buildings. In a few minutes you work up a reliable budget estimate based on the actual materials and design features, area, shape, wall height, number of floors and support requirements. Most important, you include all the important variables that can make any building unique from a cost standpoint. **240 pages, 8½ x 11, $10.00. Revised annually**

Builder's Guide to Accounting

Explains how to set up and operate the record systems best for your business: simplified payroll and tax record keeping plus quick ways to make forecasts, spot trends, prepare estimates, record sales, receivables, checks and costs, and control losses. Loaded with charts, diagrams, blank forms and examples to help you create the strong financial base your business needs. **304 pages, 8½ x 11, $12.50**

Spec Builder's Guide

Explains how to find the right lot in the right area and at the right price, get the financing you need, plan, build, and then sell the house at a price that earns a decent return on the time and money you've invested. Includes professional tips to ensure success as a spec builder: how to cut costs at every opportunity without sacrificing quality, avoiding losses, anticipating buyer preferences, and taking advantage of construction cycles. Every chapter includes checklists, diagrams, charts, figures, and estimating tables that make this an invaluable reference for every speculative or custom home builder. **448 pages. 8½ x 11, $24.00**

Builder's Office Manual

This manual will show every builder with from 3 to 25 employees the best ways to: organize the office space needed, establish an accurate record-keeping system, create procedures and forms that streamline work, control costs, hire and retain a productive staff, minimize overhead, shop for computer systems, and much more. Explains how to create routine ways of doing all the things that must be done in every construction office in a minimum of time, at lowest cost and with the least supervision possible. **208 pages, 8½ x 11, $13.25**

Craftsman Book Co.

6058 Corte Del Cedro
P.O. Box 6500
Carlsbad, CA 92008

Phone Orders

In a hurry?
We accept phone orders charged to your MasterCard or Visa.
Call (619) 438-7828

Mail Orders

We pay shipping when your check covers your order in full.

10 Day Money Back GUARANTEE

- ☐ 10.00 Building Cost Manual
- ☐ 12.50 Builder's Guide to Accounting
- ☐ 13.25 Builders Office Manual
- ☐ 11.75 Building Layout
- ☐ 6.95 Carpentry
- ☐ 18.00 Construction Estimating Ref. Data
- ☐ 16.25 Contractor's Guide To The Building Code
- ☐ 16.50 Contractor's Year-Round Tax Guide
- ☐ 15.75 Cost Records For Construction Estimating
- ☐ 14.00 Estimating Home Building Costs
- ☐ 10.50 Finish Carpentry
- ☐ 13.50 Masonry & Concrete Construction
- ☐ 16.00 National Construction Estimator
- ☐ 16.75 National Repair & Remod. Estimator
- ☐ 10.75 Rafter Length Manual
- ☐ 10.25 Reducing Home Building Costs
- ☐ 9.25 Roofers Handbook
- ☐ 14.50 Rough Carpentry
- ☐ 24.00 Spec Builders Guide
- ☐ 12.75 Stair Builders Handbook
- ☐ 9.75 Wood-Frame House Construction
- ☐ 19.50 Roof Framing

Name ______________________

Company ______________________

Address ______________________

City ______________ State ______________ Zip ______________

Send check or money order
Total Enclosed ________ (In California add 6% tax)
If you prefer, use your ☐ Visa or ☐ MasterCard

Card no. ______________________

Expiration date ______________ **Initials** ______________